Movements in Time

Movements in Time:
Revolution, Social Justice and Times of Change

Edited by

Cecile Lawrence and Natalie Churn

CAMBRIDGE
SCHOLARS
PUBLISHING

Movements in Time:
Revolution, Social Justice and Times of Change,
Edited by Cecile Lawrence and Natalie Churn

This book first published 2012

Cambridge Scholars Publishing

12 Back Chapman Street, Newcastle upon Tyne, NE6 2XX, UK

British Library Cataloguing in Publication Data
A catalogue record for this book is available from the British Library

ISBN (10): 1-4438-4016-5, ISBN (13): 978-1-4438-4016-3

TABLE OF CONTENTS

Part V: Creating/Making Time

Part VI: Revolutionary Time

ACKNOWLEDGEMENTS

This book would not likely have come into existence were it not for the Philosophy, Interpretation and Culture Graduate Program at Binghamton University (S.U.N.Y.). In the Spring of 2011, Cecile Lawrence was the main convener for the annual conference of the program which was its 21st and final conference. We thank all those who played a role in making the conference happen, including James Stanescu, who originally devised the theme and call for papers for the conference, and Donna Young who smoothed the way for all the details that needed to be addressed in order for anything to happen. We thank all the presenters who came to the conference for one or two days.

With regard to the book itself, we particularly thank Christian Garland, one of the conference presenters, for his early work in helping draft the new Call for Papers for the book, and in sending out the CFP to his contacts. We thank all those who responded to the Call and regret that we could not include them all.

INTRODUCTION

NATALIE CHURN AND CECILE LAWRENCE

The year 2011 marked a notable increase in the global turn towards acts of and ideas about revolution, not the least with the amazing and unexpected participation by people in the U.S. in the form of the Occupy Wall Street movement. The dominance of Western culture and media was reflected in the categorisation of the uprisings in Egypt, Libya, Yemen, Syria and other nations as "the Arab Spring." Perceptions about revolution can have a wide influence not only in cases of national self-determination but also in terms of how we conceptualise time and our place within history. Yet revolution does not take part only on the national stage: radical social change is constantly being called for globally on the levels of gender, race and class, reflecting a future-oriented view of time that aims to change the thrust of history.

Merely looking into the future is itself a limited way of evaluating approaches through which we can create a more just society. The concept of revolution recalls alternative notions of time, such as the cyclical movements of celestial bodies, suggesting that inherent to the notion of revolution is the idea of repetition of what has already been, and the question as to whether humanity can learn from history, or even perceive plural histories and multiple possibilities for the future. This collection of essays brings together not only discussions on revolutions as social protests and movements that aim to bring about radical political, social and economic transformation; but also essays on industrial revolutions; the need for revolutions in ways of thinking, stereotyping and perceiving; revolutions in technology; in human labour time; and even in the ways in which theory itself can be revolutionary.

Crucial to all of these conceptions of change is how we come to understand time in relation to revolution: as history, as cycle, as arrow, as a series of moments to be seized, as a social construction that we can manipulate, or as something maintained by a political system to control us. Time in revolution is at once opportunity and constraint, yet it is in time that changes occur, and thus time becomes a tool, a vessel through which social change and a more just society can be thought and acted upon.

Philosophers have long critiqued the patriarchal, linear notion of time reflected in national narratives and teleological worldviews, which often function only to reinforce the status quo. Marx himself calls for an end to temporal limitations, while Negri considers the possibilities of *kairos* time, and Deleuze and Guattari focus on the importance of becoming. The concept of becoming is crucial when considering revolutions after the year 2000, with the millennial anxieties around the turning of the century, leading up to more intense apprehension about the Mayan calendar ending in 2012 contorted into U.S. fundamentalist Christian ideas about The End of Times complete with The Rapture. The notions of messianic time featured in Agamben and Benjamin presage these anxieties and anticipations.

Reorienting such thinkers' ideas for current situations, the essays in this volume seek to rupture Western modernity's standardized, patriarchal conception of time, radically intervening in dominant ideas of teleology and linearity. They consider instead marginalised and creative understandings of time, such as the alternative notions of temporality that have been put forward by non-Western belief systems. Indigenous cosmologies, for instance, portray time as cyclical, while Buddhism separates time into tiny moments or even offers possibilities of transcending time.

Time is thus not simply socially constructed notions of clock time and linear conceptions of history, but rather is an encounter that differs according to human experience. Julia Kristeva's work on women's time, for example, outlines the cyclical temporalities and subjective experiences unique to women, while Robert Levine suggests that climate can have an effect on the pace of life in various countries, although postcolonial writers have critiqued this perspective as problematic. Literary, postcolonial and media studies conceive time as something that can be reversed or stopped altogether, portraying history as plural and emphasising the subversive and oppressive facets of time ideologies. Nations are held together by popular conceptions of shared times which often function to exclude minorities and repress their actual histories, while class antagonisms are partly characterised through ideas of productive time, exchange value and leisure time.

This book explores the importance of alternative philosophical and theoretical understandings of time for radical politics in the 21st century. Essays within various disciplines, as well as essays that make connections between two or more disciplines, rework several philosophical concepts of the time of revolution and apply those concepts to the concerns of social change. The goal is to determine which ideas are still relevant and which must be adapted in order to address the ever-changing political conditions and situations of social inequality. The growing interconnections between academic disciplines and between theory and practice have to a large extent been neglected by modern scholarship. Many of the essays in the book cut across several disciplines, presenting an accessible introduction to the work of key thinkers covered in specific pieces, whilst also breaking down barriers between disciplines that can be somewhat forbidding to the non-specialist.

Movements in Time: Revolution, Social Justice and Times of Change explores case studies that consider ways in which ideologies of time and alternative temporalities can be (or are already) useful for solving conflicts and exposing and overcoming stereotypes and boundaries created around questions of gender, race, ethnicity and socio-economic inequality. Time-perception is often used as a tool for marginalisation, but the alternative temporalities of the subaltern may also provide a way out of current restrictive policies around the world. With a focus on time as an element of radical activism, this book asks: how can visions of the future and the past, becoming time, untimely time, protest time and political time be implemented both theoretically and practically in order to change the way in which time functions as a vital element of social, political and cultural revolution?

Part I examines the relationship between time, revolution and protest, bringing together essays that consider the power of the protester to bring about social change, but also the extent to which protest itself is governed by concepts of time. Christian Garland's opening chapter explores the historical memory of the May 1968 rebellion in France and the eventful year of 1968 itself through the lens of Walter Benjamin's conception of history. Placing the May 1968 fortieth anniversary events within the context of more recent social protests worldwide, he argues that the memory of May 1968 is an animating force that constantly returns to invigorate present protests with the spirit of past social struggles, such that the past, rather than remaining a mere nostalgic memory, becomes a source of action and hope.

Examining activists themselves from a more empirical point of view, Kamilla Pietrzyk evaluates the results of a study on Canadian activists' understandings of and relationships with time. These results, when explored in the context of what the activists are fighting against—globalisation, capitalism and environmental damage—suggest that protesters themselves are not immune from the controlling time concepts of 21st century society. She highlights the importance of time management skills for activists, as well as the effect that past- or future-oriented mentalities could have on the success of activism.

John Streamas explores the invention of "colored people's time" (CPT) as a means through which people of colour could survive oppression and even use time as a force for creativity and resistance. Relating CPT to revolution, he suggests that these concepts of time may be useful for theorising movements such as Occupy in that they present alternate visions of history and the future. CPT releases people of colour from the oppressive linearity of history, enabling grassroots versions of history and giving communities, rather than individuals, the tools with which to create futures. In this way, Streamas argues, people of colour have an important role to play in the Occupy movement.

Concluding the discussion of protest time in Part I, Maria Prodromou considers the relevance of Lyotard's politics of resistance for the radical politics of the Occupy movement. She suggests that the logic of exchange in capitalism must be broken in order to free the present from its dependence on the future, interrupting the oppressive temporality of capitalism. This can be seen in the examples of the People's Library and the provision of free food in occupied Zuccotti Park, where people can meet and protest together in moments freed from the future and thus moments where future events are yet to be determined.

Part II presents three essays on the importance of Deleuze's theory of becoming, from the Eternal Return, to the privileged position of women in regard to transformation, to the technique of resisting the present, concluding with an exploration of nostalgia and retrospective views in relation to the future. Concepts of the past, the present and the future frame our understanding of change and revolution, and it is from the idea that nothing is fixed but rather constantly "becoming" that social transformation can be perceived and enacted. Applying Deleuze's concept of "becoming" to history itself, Allison Merrick considers the importance for Deleuze of Nietzsche's Eternal Return, arguing that this refers not to the cyclical repetition of the past but rather a return of the Dissimilar, giving the past a redemptive potential. She argues that it is this virtual excess of that which has not yet happened which provides the conditions for revolution.

Anupa Batra considers Deleuze's concept of "becoming-woman" as revolutionary as it aims to bring about a transformation in the way our whole existence is structured. Following Deleuze, she argues for a move away from the concept of gender itself, which shapes so much of Western thought, suggesting that subject-forms are repressive and that it is in the process of becoming, rather than subjectivity, that creativity in life can be found. Suggesting that women are in a privileged position in regard to bringing about transformations in structures of thought, she explores becoming-woman as the process in which this revolutionary exit from gender, and other oppressive categories of Western thought, could occur.

Also following Deleuze, Bryan Nelson explores the revolutionary possibility of resisting the present in order to bring about a new earth and new people. He claims that philosophy, with its task of creating new forms of thought and a future different from the past, has a role in this resistance of the now by rendering democracy's becomings thinkable. Becoming-democratic opens a space for minorities in its function of becoming-minoritarian, challenging the abstract, majoritarian logic of democracy. Becoming-democratic should not be conceived of in relation to an impossible future, but rather only in our resistance to the conditions of the present, in the hope of creating an entirely other, possibly more democratic future.

In Rochelle Green's contribution, nostalgia, as a kind of homesickness for the past, can be influenced by judgements of present conditions and thus used to support an active turning towards an unrealistically glorified past, as is often seen in extreme politics. In her analysis of the nature of nostalgia as a full emotion in relation to concepts of hope in the work of Ernst Bloch, she asserts the importance of hope as revolutionary time. She argues that hope, despite its ideological similarity to nostalgia, involves a different understanding of time that is not afraid to look ahead; a radical concept of time that asserts the potential of the future in the present, thus providing a revolutionary mentality with which we can work to overcome exploitation and alienation.

Part III explores Untimely Time, namely times which may be deemed by those in power as a form of threat or subversion, as embodied in the concept of revolution. Natalie Churn considers the Australian example of contradictory notions of "bogan" time. She suggests that the hypocritical stereotyping of classes based on how they use time functions to reinforce existing social hierarchies. Leisure time is scorned when enjoyed by the working classes, yet at the same time leisure forms an integral part of the Australian stereotype. While carnivalesque times in the Australian calendar allow for these hierarchies to be briefly exposed, social movements should make overcoming these working class time stereotypes a major part of their agenda by making public the real questions behind the stereotypes.

Time stereotypes are also a major concern for Dalbir Sehmby, who discusses the stereotyping of Sikhs following the 9/11 World Trade Centre bombings. The Sikhs' memory of the 12th of September as the anniversary of Sikh bravery in the Battle of Saragarhi has been corrupted by stereotypes and violence based on the turbaned image of Osama bin Laden, reinforcing the Western view of the Sikh male as associated with temporal concepts of primitivism and dirtiness rather than progress and civilisation. While Sikh beliefs emphasise moving with the flow of time, Western ideologies of terrorism and fear emphasise cycles of hate and the control of time. Sehmby thus calls for more inquiry into plural visions of time in order to counter the momentum of hate and discourage future cycles of violence and stereotyping.

Amanda Conroy suggests that the Western right-wing movements responsible for much of this hate and stereotyping attempt to create an alternative historical trajectory based on nostalgic nationalism, essentially reversing time and characterizing themselves as the saviours of history. She identifies the white male as the majority member of these movements. Her essay reinforces the need for change, just as 21st century revolutions have been cognizant of the need to confront and engage with the effects of centuries of patriarchy.

Alison Hulme describes China as embodying two contrasting visions of time: Western perceptions of a fast-growing economy and a huge-scale industrial revolution conflict with the "slower" traditions of Confucianism and communism. Yet China's fast revolutionary time can only be enjoyed as long as China does not exert too much power on the world stage. As far as the West is concerned, the path towards capitalist democracy has already been written, reflecting a logic that believes all aspects of change follow Western neo-liberal logic. On the other hand, Hulme suggests we should view this interweaving of fast development and slower foundations as forging a unique path to social change that is being negotiated from both above and below.

Returning to the Occupy movement, Enqi Weng considers that Singaporean attempts at Occupy Raffles Place did not succeed partly because, despite many Singaporeans' discontent with the government, the protestors had no long-term plan for change to back up their protest. She explores the popularity of Wikipolitics, suggesting that the making of important political decisions cannot necessarily keep up with the increasing speed of technology and social network activism. She thus warns that revolutionary protest, especially when encouraged through social media, should also take into account the long-term implications of bringing about fast change in a world where decision making is still governed by the human need to ponder wisely.

Part IV encompasses essays that consider the different ways in which politics and history can be explored from various philosophical or political perspectives. The first two essays focus on capitalist labour time as a source of struggle and as something to be overcome in order to create a more just world, while the next three contributions consider history, progress and ideas of "perfecting" time from a range of perspectives.

Considering Antonio Negri's argument that capitalism conceives of time in terms of productivity, Bradley Kaye theorises unproductive time as not only a form of resistance but also of empowerment for marginal groups. He suggests that subjects labelled disabled or mad are excluded from work and other aspects of society as a result of capitalism's ideologies about work. Using the example of the active refusal of work in Hermann Melville's novella *Bartleby the Scrivener: A Story of Wall Street*, he connects this unproductive time with Negri's conception of Love. With the possibilities of advancing technology positing a revolutionary shift in the mode of production, a future metrics of time may be possible where unproductive time does not mean the exclusion of marginal groups.

Revolutions and protests around the world have highlighted the precarious situations of many people and peoples in the early 21st century. Alexander Brown explores precarity as a temporal phenomenon where not only the time of labour but also the other, social times of life are exploited for capital. Using Negri's concept of liberated time as a departure point, he suggests that it is in the present moment that revolution must liberate time from capital. Drawing on examples from the 2011 revolutionary movements, he argues that revolution is not an event situated in the future but rather a way through which the potentiality of the present can be realised. This liberated temporality is being realized daily in struggles on the streets, despite the ruling classes' attempts to deny the people their own future.

Brad D. Hume takes an agnostic position, arguing that messianic time necessitates a kind of weak teleology that foresees the possibility of humans creating a more just world in the face of current conditions. He sees the abundance and technological innovations of today as not necessarily leading to more justice and equality. Hume suggests that late modernity attempts to prevent revolution by promoting the myth of free-market economics as non-political, whereas in reality change is not given a chance as it threatens those in power. History may well provide a crisis which will bring about revolution, but by then it will be too late. Thus we must critically examine the material production of the world today if we are to find a way to make the world a more just place.

M.F.N. Giglioli explores the relevance for the early 21st century of the writings of two nineteenth-century thinkers, Georges Sorel and Vilfredo Pareto. Despite their many differences, both promoted political theories that downgraded the importance of political change and the concept of modern society, positing progress as an ideological myth. Giglioli considers the relevance of their writings for thinkers at the turn of the century who announced the end of history. Both thinkers represent two differing yet similar attempts to problematize the perception of revolution as constituting a fundamental break with history. Sorel questions the value of duration as a mark of success and sees revolutionary politics in the glorious instants of collective action, while Pareto maintains that a necessary condition of an intellectually defensible investigation in social matters is a radical separation between theory and practice.

Taking a theophilosophical standpoint, Christos Retoulas argues that understanding revolutions and their relation to time necessitates an inter-encapsulating approach. He describes progressivism as a mere palliative of the modernist economic system in capitalism and argues that as a result of enlightenment, the right and left remain in a state of deep fusion. Exploring the concepts of *aeon* and *chronos*, he suggests ecumenical Romanity as an example of an incommensurable revolutionary path. He argues that revolution is not just about structures and forms but also functions on a deeply personal level.

The essays in Part V present various revolutionary possibilities for creating or expanding time, as well as new ways of viewing or understanding time in order to bring about social change. Łukasz Kamieński identifies the increasing speed of technology as a phenomenon with which the human body cannot keep up, a deficit that is being countered by the growing use of drugs that enhance mental performance and remove the need to sleep for several days. Yet our obsession with finding extra time to be productive signals a problem for human freedom in a society that is increasingly coming to reflect the drugged worlds of dystopian novels. This chapter examines whether such ideas of physically cheating time are being (or will be) forced on workers, what the implications of such a future could be, and whether organized resistance is required to prevent this.

Rhiannon Firth and Andrew Robinson consider the possibility of utopia as a temporal phenomenon with implications for social change. Through an investigation of the positions of Deleuze, Benjamin, Negri, Agamben and Bonanno, who all locate the experience of revolution in the present, they argue that revolution must have a temporal utopian dimension if it is to overcome the entrapment of modern societies within the homogeneous empty time of capitalism. Revolutionary theory must restore the subjective aspect of time and advance multiple temporal zones over linear teleologies if it is to overcome the homogenizing tendencies of time in capitalism.

Identifying some of the problems of the tendency towards present-mindedness in today's world, Linda Levitt discusses the Clock of the Long Now as a symbolic representation designed to remind viewers of the importance of thinking about the future in the present, thus creating a "long now." She suggests that environmental issues and problems of inequality must be conceived within a deep time perspective if we are to save humanity from its own destruction. She calls for a revolution in thinking about time, and while the Clock of the Long Now may be merely representative, its makers emphasise the importance of a hopeful, future-oriented mentality in order to bring about social change.

Martin Savransky explores the relationship between modern, Western scientific knowledge as based on dualities of subject-object and nature-culture, and the scientific, historically linear conception of temporality which informs it. He argues that the present of this modern temporality is in fact destabilised and betrayed by the many hybrid temporalities which inhabit it, by an ecology of temporalities. He suggests we can envision a non-modern notion of revolution as a constructive, creative event that does not prescribe a radical break with the past but rather a rearrangement of the relations between entities, knowledge-practices and their multiple temporalities into new, unanticipated configurations.

In Part VI, the essays look at particular examples or scenarios when considering the notions of time necessary to understand and enact revolution. Considering the 2011 revolutions in some Arabic-speaking countries through the lens of Benjaminian thought, Antoine Chollet suggests that revolutions must abandon the homogeneous, empty, future-oriented time of progress of the oppressors in order to bring about change in the present. He suggests that the revolutions in Tunisia and Egypt were not planned with an outlook to the future in terms of defining a future form of government but only aimed to overthrow the current regime, focusing on a "now-time." Acknowledging that abandoning ideas of the future and progress can result in new regimes of terror (as in the case of the Paris Commune), Chollet argues that many revolutions may not have happened at all if people had deliberated on the future consequences of radical shifts in power.

Moving away from two dominant concepts of time in revolution, namely the Hegelian idea of time as future-oriented or Benjamin's messianic time as suspended or frozen in an eternal present, Tania Rispoli proposes a third possibility that describes the specific temporality of 21st century struggles. She views time as related to the flexible, contingent nature of things—each event has its own time. In this conception, the time of revolution is governed by the logic of encounter, but also results from the intersection of different times. Rispoli describes the idea of a federation of independent times which she proposes could change the mode of production and create new times of life.

Thomas Pieragastini continues the book's exploration of the relationship between time and revolution by posing some important questions about the place of theory when talking about real revolution. As many historical revolutions have resulted in police states that were as controlling and terrifying as those that they sought to overthrow, he asks what it is about the concept of revolution that comprises its inherent tendency towards autodestruction. He identifies this failure as a temporal issue in that the material amplification of conflict occurs in material moments with real effects, whereas the non-particular conceptualization of revolution itself is an atemporal notion, and it is in the conflation of these two manifestations that the autodestruction of the revolutionary project presents itself. The attempt to materialize atemporal concepts in material moments must be abandoned: revolution must be separated from teleology, in order to avoid setting up revolution again for failure.

Using the example of the Mau Mau revolution, Mickie Mwanzia Koster draws a comparison between African time, which is said to emphasise the past and present yet has the aim of transforming the future, and revolutionary time, which also aspires to a new vision of the future. She suggests that the oath rituals of Mau Mau warriors opened portals to the past through which warriors could be connected to their ancestors, but also emphasised the urgency of addressing the present aims of the revolution. This subaltern, African treatment of time and revolution is significant when considered in the greater context of the "Arab Spring" events of 2011 and 2012, especially for interpretations of power and protest not covered by the Western media.

Placing Carl Schmitt, Theodor Adorno, and Jacques Derrida in conversation with each other, Lewis Levenberg considers the importance of the intersections of their thought as well as their silences for addressing political and temporal problems. Identifying three streams of focus in this dialectic relationship—alterity, aesthetics, and futurity—he argues that these delineate the conditions required to consider revolution in its historical context. Levenberg suggests that it is only at the limits of this discourse that revolutionary shifts in historical perspective can be conceived, as the exceptional ruptures in modernity, or revolutions, transform our conceptions of temporality, modernity and reason into the very rules of politics.

Concluding these discussions on time and revolution, Cecile Lawrence explores the nature and genesis of the Occupy Wall Street movement in the U.S. She concludes that after progressives in that country had come close to giving up on there ever being any movement against the increasing amount of economic, social and political assaults being perpetrated against the country's residents, people rose up in a manner that ironically reflected the very strategies of the indoctrinations to which they began to object. The people were sent an "order," so to speak, to "occupy" a specific place at a specific time and they did, so as to reject Capital's decades long orders to the people as to where to be, what to do and at what time.

The chapters summarised above conceive of time and revolution in multiple, complementary and contrasting ways. Yet it is precisely this multiplicity of views and treatments that form the strength of this collection. As an undercurrent that flows through and connects human life on so many levels, time is at once both hidden and controlling, reminding us that the moment of change must be seized before time itself, our creation, escapes us, or that to enact change we must escape or recreate time, or do something totally new with time. There has never been a better time to consider how ancient and modern, philosophical, indigenous and marginal conceptions of time and temporality might be employed in a quest to reconcile alternative histories, and to bring about radical social change.

PART I

PROTEST TIME

A Secret Heliotropism of May '68: Historical Postponement, Mimesis, and Nostalgia

Christian Garland

> The class struggle, which is always present to a historian influenced by Marx, is a fight for the crude and material things without which no refined or spiritual things could exist. Nevertheless, it is not in the form of the spoils which fall to the victor that the latter make their presence felt. . . . They manifest themselves in this struggle as courage, humour, cunning, and fortitude. They have a retroactive force and will constantly call into question every victory, past and present, of the rulers. As flowers turn toward the sun, by dint of a secret heliotropism, the past strives to turn toward the sun which is rising in the sky of history.
> —Walter Benjamin

Graffiti from the 2006 revolt in France, which defeated the proposed CPE (*Contrat Première Embauche*) allowing employers to fire younger employees more easily and without reason given, put it best: "Groping for the lost thread, humanity will rediscover itself" (Bureau of Public Secrets 2006). This chapter will seek to offer a critical and historical indictment of the fortieth anniversary recollections of the events of May 1968 in France, recalled as one such process of nostalgic commemoration or recreation.[1] Indeed, as Walter Benjamin noted, "Reflection shows us that our image of happiness is thoroughly coloured by the time to which the course of our existence has assigned us" ([1940] 1999, 245).

The events themselves are of key historical and political significance, remaining to this day an image of the future that brings to mind Marx's "spectre that is haunting Europe" ([1848] 2004, 3). The events were the first major rebellion "against modern consumer and technical society," as

[1] For the best first-hand participant accounts of the events of May 1968, see Brinton (1968) 2005; Gregoire and Perlman (1970) 2005; Quatrocchi and Nairn 1998; Vienet 1992. Good critical analyses and reflections include Feenberg, Freedman and Kellner 2001; and Willener 1970, which also include documents of the time, unavailable elsewhere.

President de Gaulle noted with some trepidation at the time (quoted in Vienet 1992, ch. 8). May ’68 also offers what might be called “a secret heliotropism,” to use Benjamin’s words ([1940] 1999, 246), mapping a topography of moments between past and present and, it will be argued in the course of this paper, the appearance of a weak messianism—the recurring return of the repressed. In the course of this paper it will be further contended that nostalgia, while it can be reassuring, is indeed always illusory: in its false sense of security as much as in its retrogressive narrative in which is found not so much a *temps perdu*, but a circumnavigation of the past, a substitute for remaking the present and creating the future: a future that breaks with that version of history assigned for us, which as Marx observed “weighs like a nightmare on the brains of the living”(Marx [1852] 2010, ch. 1).

With each successive decade that is added to “remembering” the events of May 1968, there seems to be a renewed effort to banish the awareness of their real historical meaning, but even as this sets out to explain the past, it denies it, seeking instead to offer an assurance that cannot be guaranteed. The affirmation of the present, the existent, aims to recognise a vision of the past, but only one in its own image imputed with its own conditions and demands—as can only be expected. Reviewing the anniversary commemorations taking place from the safety of 2008, one could be forgiven for believing the events under discussion to not in fact be “the concern of history” but a clichéd retrospective of distant events apparently unconnected with the present, the meaning of which cannot be grasped or understood. An attempt to counter this version of events might take as its standpoint Benjamin’s maxim that nothing that has ever happened can be regarded as lost for history, discerning in the smug commentaries and chiding op-ed pieces, the obligatory testimonies of wondrous left-to-right journeys and conversions, a definite discomfort that despite their own gloating reassurances, what happened across France in May 1968 could happen again, that however much they would seek to sanctify and eternalize the present, history demands otherwise.

2010 and 2011: The Global Return of the Repressed and the Unsettled Score of History

2010 and 2011 marked a radical shift in perspective for the social subject. This could be seen as a generational realisation and revolt, though those who remember the similarly diffuse and radical searching of the 1968 revolts for a non-doctrinaire practice have seen more than a few similarities.

From 2010 and especially 2011 onwards, social rebellion has appeared and ignited across the world, and a direct historical line can be traced to that of May 1968. From the UK student protests of November and December 2010 to the momentous "Arab Spring" that opened 2011 by toppling at least two entrenched dictatorships, to the global Occupy movement (Kingsley and Gabbatt 2011) which began with Occupy Wall Street and spread to over 900 cities and saw thousands of protests worldwide closing the year, the spurious notion that rebellion and contestation of the conditions of existence is "from another time," has been conclusively debunked across the globe. In the UK, 2010 closed with mass student protests destroying both the image of contentment and passivity of a generation and literally Conservative Party HQ. In England in August 2011, riots erupted in Tottenham, a poor area of north London, after the death of yet another black man at the hands of armed police, before spreading like wildfire to more than forty areas of the UK capital, and then virtually every major city and a number of smaller places over the nights of 6th-10th August.[2] In the intervening six months since and following major studies into the disturbances, the overriding reasons for involvement from participants have been anger at their experience of social deprivation, poverty, and police repression.[3] At the time, these rebellions were dismissed as nihilistic looting and "not political," but the accounts of many of those involved give a different perspective on what happened. These voices, of the ignored and the invisible, also warn of "more to come" this year and in the years ahead.[4]

Moving to a global context, Occupy has succeeded in throwing into glaring contrast the widespread discontent with a global system which in the movement's famously direct words, cares only about "the 1%" of billionaires—the capitalist class—and which sees "the 99%" merely as dispensable and disposable labour to be exploited for profit, so far as their labour is needed at all. Occupy is especially significant for spreading *globally*: the commonalities of different people unifying and linking their particular experiences in particular locations—superficially at great distances from each other, but compressed across space and time and virtually overcome by the unified linkage of struggles. The Occupy Movement remains unbowed and undefeated as an ongoing developmental

[2] See Garland 2011 and Harper 2011 for more detailed analyses of the events.

[3] For analyses on the causes of the riots, see Lewis et al. 2011b; the video by Afshar et al. 2011; Jackson and Jeffrey 2011; and Prasad 2011.

[4] Such voices sometimes find an audience in the news media. See for instance the articles in *The Guardian* by Lewis et al. 2011a and Dodd 2011 regarding both riot participants' and a police chief's warnings regarding the future.

process of 2012, while the harshness of winter— just like the harshness of state responses—means a relative but only temporary hiatus to re-group for the spring and summer ahead.

Occupy may have come seemingly "from out of nowhere"—as social rebellions frequently appear to—but as the subterranean material forces which manifested in this and other faces of social rebellion are themselves the return of the repressed in history: the class struggle, struggles around gender, ethnicity and sexuality and other fights against oppressive social relations of all kinds, *never went away*, there has been no end of history, and there never will be however much the apologists and defenders of this present "state-of-things" may wish in vain that there were.

Elsewhere, the rebellion spoken of here has manifested in the outright refusal of and resistance to the "necessity" for austerity measures to shift the burden for the ongoing crisis of capitalism back onto society in order to try to maintain profitability following the first paroxysms of 2007-08, illuminating the gloom of trying to survive under such "painful but inevitable" conditions of precarious material survival and submission to such an order-of-things. As Mike Davis has noted,

> As the fates of previous *journées révolutionnaires* warn us, spring is the shortest of seasons, especially when the communards fight in the name of a 'different world' for which they have no real blueprint or even idealised image. . . . It's a frighteningly long road just to reach the starting points of earlier attempts to build a new world. But a new generation has at least bravely initiated the journey. (2011)

The rebellions of 2011, in some of its specific manifestations winningly referred to by Davis as the "'hot' Iberian and Hellenic summers," and "the 'occupied' fall in the United States" (2011) have been conspicuous in their militancy but also the absence of doctrinaire party lines and ready-made ideologies. The many media and academic critics of these specific faces of the rebellion have dismissed them as lacking any clear programme or defined alternative, but have also paradoxically tried to dismiss them for their non-ideological "incoherence," as if previous revolts, such as those of 1968, were defined by Communist Parties or Trotskyist sects. The somewhat inchoate nature of the rebellions in Europe, and indeed the Occupy Movement, can be seen as a mark of their authenticity, however, which Rosa Luxemburg understood as the way in which such social rebellion develops and takes its own form through spontaneity, noting "the spontaneous element has played such a predominant role not because the . . . proletariat is 'unschooled,' but because the revolution can't be school-mastered" (quoted in Frölich [1939] 2010).

It is no small coincidence that this worldwide revolt should appear at the same time as life is becoming that much harsher for so many, with populations told by their governments that they can only hope things might get better by first enduring yet more prolonged social suffering. This is the reality behind such apparently unfortunate facts as unemployment, cuts, and indeed, all "austerity" measures: in order that a minority with much to lose can keep all of it and accumulate yet more, the majority with nothing to lose must be made to suffer further still. To be sure, many, in fact most of the 1% also want to change existing society, though certainly not in the way hoped for by the 99%—the overwhelming majority. What is meant by this is that even as this system enters further crisis, the 1% want interventions by the state and others to better adapt to the imperatives of capital and sustain profitability at all costs. In practice this means all remaining social threads unpicked, all collective social goods eliminated, and for the 99% to suffer even more austerity for even longer, leaving an even more fragmented, atomised and "flexible" army of precarious, hireable and fireable labour so far as it is required at all; the state maintaining this order through force at all times, ideally with as many as possible believing this to be natural, inevitable and most of all a "painful but necessary" adjustment to the "reality" of 2012. The rebellious social body fragmented and disorientated by more than 30 years of neoliberal hegemony, and the re-composition of class relations, demands otherwise however: "For this is the sound of the quick that is heard in the ears of the dead" (Swinburne 1860, line 2). None of these individual manifestations of social rebellion are by themselves enough to bring about the social revolution, however authentic they are, and however much they may strive to, but they are the recurring and jarring reminder to the (temporary) and imagined victors of history that their own version of events remains in question, that their days of plenty are numbered, and that their world will not last.

May 1968 to the Twenty-First Century "Time-of-the-Now" (*Jetztzeit*)

The rebellions of the beginning of the second decade of the twenty-first century marked the re-appearance of a rupture of the "homogeneous, empty time" (Benjamin [1940] 1999, 252-253) playing out the hollow continuum of the would-be Masters of the Universe, and their supposedly untouchable world. In such rebellions can be discerned a definite chronicity back to May '68 and its as-yet unfulfilled promises. The significance of May '68 should not be ascribed solely to the events themselves but in how this connects to the intervening four decades and their chronicity with the present, if we are to develop our understanding of it and once more find that historical moment of "rupture" in the continuum of what passes for history, which as Marx famously observed "weighs like a nightmare on the brains of the living" ([1852] 2010, ch. 1). Indeed it seems a grasp and sense of history is vital if we are to make sense of our present situation, which is, of course, the starting point for contestation over the form of life that we all live, and, to a great degree, are forced to live in our own era.

"May 1968" continues to provide a unique example for those wishing to understand contemporary history in the context of a revolutionary rupture, which called into question every aspect of life under late capitalism. The movement of occupations, strikes and "action committees" —at its high point encompassing more than 10 million people—no less than the sense of joyful festivity lucidly expressed in poetic graffiti, and passionate revolt expressed in riots amid the inverted architecture of barricaded streets, remains the largest-scale and furthest-reaching social revolt of its kind in recent memory. The events are often misleadingly referred to simply as "student protests" as if the rest of French society merely watched itself be consumed by revolution from the safety of their TV screens. The "explosion" of May 1968—to use President De Gaulle's own description—may have found its initial spark in the student revolt, but only in so far as this was symptomatic of the underlying tensions in wider French society. May 1968 also comprehensively debunked the myth that in the affluent society of purchase power and technocratic efficiency, the concept of class struggle was all but dead. Indeed, it may be argued that prior to May 1968, consumer capitalism had few real critics outside the group of thinkers loosely referred to as the "Frankfurt School," or in the equally unsparing theory of the Situationists, whose trenchant critical diagnoses were largely overlooked until they found direct expression on the streets during the month of May. Of course, a number of those associated with the "'68 Generation," came to refer to the events as a turning point in the sense that they lost hope in the possibility of revolutionary social change, and the development and subsequent dominance of postmodern and post-structuralist thought in France during the 1970s can be seen as indicative of this. The May 1968 rebellion illustrated the ability of a significant number of the population to take direct control of their own lives through collective action. It also demonstrated the spontaneous power of mass social revolt unencumbered by the proclamations of parties and political game playing.

One of the most intriguing articles reflecting on the legacy of 1968, but unfortunately displaying a numbingly a-historical fatalism, was written two years before the inevitable slew of 40th anniversary commemorative articles, in April 2006, amid the anti-CPE (*Contrat Première Embauche*) revolt that shook France. Writing in *The Guardian*, David Drake argued that May 1968 "aspired to make people's dreams a reality" (Drake 2006), believing the movement of 2006 which defeated the CPE, a legislative programme aimed at creating a compliant, flexible youth labour market, was of course entirely unrelated. Drake noted perhaps without realising the full implications of his words, "It is tempting to see this movement as a re-run of the student-worker revolts of 1968 but this would be a mistake" (2006). It would indeed be a mistake to see the movement of 2006 in those terms, despite the preferences of individual commentators. Drake, unsurprisingly perhaps, makes no mention and has no apparent awareness of similar movements in France in 1986-87, 1994, 1995, 1998, and 2003.[5] The first was sparked by proposed reforms to French universities and found further catalysts in unfolding railway strikes, while the movement of March 1994 was a wide scale youth revolt against workfare measures. Those of December 1995 and 2003 involved mass nationwide strikes, and the jobless revolt in 1998, like that of 1994 and 2006, was initiated by those locked out of the social factory, exemplifying their rebellion against the non-existence of being capital's unwanted social and economic surplus, that is being unemployed.

These revolts quickly overwhelmed the country and rapidly combusted into full-scale social rebellions, providing many examples of the poetic revolutionary graffiti which appears in France, subverting both representation and mediation apparently from out of nowhere, being as much immediate textual and visual interventions into everyday life as commentaries on it. Similarly, all six revolts had a keen eye for history, with graffiti observed in 1986/7 noting "Beneath the tarmac, the paving stones" (quoted in Blob and Combustion 2009).[6] The 2006 movement, like

[5] For more on the 1986-1987 workers struggles, see Blob and Combustion 2009. For an interesting reading of the 1994 youth revolt, see "Nous Sommes Tous des Casseurs" 2006. For the 1995 movement, see "The Class Struggles in France" 1996. For the 1998 movements, see "Tentative Impressionistic Notes" 2008.

[6] This is a knowing nod to 1968, when one of the most famous signature pieces of graffiti was "Beneath the paving stones, the beach." Indeed, in keeping with the long tradition of the French ruling class's very real fear of revolution and attempts to assuage it which date back literally centuries, planners were on hand to cover the paving stones of central Paris with tarmac by 1986, in order to make

so many later struggles against capitalism, could be seen at least in straightforwardly political terms as being defensive, or at least the issues at stake that catalysed it were. However the successful anti-CPE movement was much more than a mere protest at measures intended to "modernise" the French labour market. In the movement of 2006, there was the recognition that life as it is lived day to day is only bearable at all because we are aware that it could be different, that this present state of things is not a "given" and is not invulnerable. "In May 1968 the future looked rosy and full of promise; today it looks dark and foreboding," Drake (2006) added, emphasizing but missing the point just mentioned. Such a de-historicized perspective is extraordinary for its obliviousness to the historical moment found for a couple of months in 2006 as much as for its reductive view of May 1968.

In 1968 consumer capitalism was in its relative infancy, as were spurious ideological confections preceding it, proclaiming classless social peace such as "The End of Ideology" prefiguring of course "The End of History" thirty years later. The point here is to highlight the fact of historical continuity between 1968 and 2008: the revolts of 1968, particularly at the high point of the French *evenements*, were not, for the most part, precipitated by material poverty or unemployment, much less the effects of capital's restructuring at the expense of the majority, the issue in 2011 on, complete with one of its full-blown economic crises. Instead, the *evenements* were a collective challenge to the fledgling stages of what has since been periodised as "late capitalism:" the society of consumption which demands we work in order to consume and consume if all or most of us are to work. But it would be mistaken to see the society thus described as inherently different from what is understood as "capitalism." The gradual overcoming of material necessity but arrested and contained by the continuum of history defined by capitalism is not in question here; rather what is at stake is how '68 and '06 ('08 and '11) and the time in between form themselves into a constellation of lived moments reawakening the "time of the now" (*Jetztzeit*), to "blast open" that same continuum. It is these moments which display the *weak* Messianic power of which Benjamin speaks, "a power to which the past has a claim" ([1940] 1999, 246).

To be sure, in the events of May '68 it was possible to glimpse, in the interruption of the quotidian banality of life shaped by consumer capitalism, "the marvellous"—so beloved by the Surrealists—but this

---

readymade weapons less available during moments of social rebellion, doing little, however, to stop these actual moments themselves.

moment was also glimpsed in the revolt against neoliberal austerity measures fought in 2006 and everywhere in between. Beneath the paving stones the beach. What is striking about almost all the commentaries revisiting May '68 is the absence of any real understanding of history: it is as if the events themselves "happened" (somehow, for some reason) but bear no relation—and have no bearing to—anything before or since. This seems strongly indicative of this epoch of capitalism frequently defined as "postmodern," which found many of its ideological explanations in the disillusion and resignation of certain thinkers in the years after 1968, but remains crucial for those of us who see in May '68 and its legacy an unfinished and ongoing project to make sense of where we are now in order to go beyond this miserable world of the present.

Writing in *The Times* in May 2008, Adam Sage noted,

> Then, the country exploded in search of a better future. Now, instead of revolution, demonstrators want pragmatism and stability—to stop things from getting worse. Today's young demonstrators look to the state to find ways of delivering them from a precarious and uncertain future. . . . In May 1968, the students rallied behind the call 'Let's ask for the impossible.' In 2006, backing the demand for the withdrawal of the CPE, it had become 'Everything we are asking for is possible.' (Sage 2008)

Steven Erlanger, writing in the *International Herald Tribune*, went one better, arguing "Today, students worried about finding jobs and losing state benefits are marching through the streets demanding that nothing change at all" (2008).

Commentators often misunderstand or misinterpret events, especially those seeking to challenge a society and hence challenge specialist roles within it, but such an assertion that the participants of the movement of 2006 sought for "the state to deliver them from a precarious and uncertain future," instead of challenging it, is particularly extraordinary for its naïve incomprehension. The anti-CPE movement which defeated a piece of legislation to which the French ruling class had already publicly committed itself, aimed at "freeing up rigidities" in the economy, was, if nothing, else an electric burst of class struggle illuminating its own visibility with lightning intensity across Europe.

The imposition of political measures such as the CPE toward offering capitalism a more "flexible," efficient model of precarious and disposable wage slaves underlines the structural realignment and metamorphoses of the global capitalist economy in the years post-68, which in a very real sense was in response to the seismic shocks generated by the worker and student revolts of the 1960's and 70's. Class struggle, class conflict, after all, works both ways.

The "defensive" nature of the movement of 2006, rather like that of March 1994, which also defeated a similar piece of legislation imposing neoliberal austerity measures on young people, cannot be dismissed as a "retreat" or an attempt to "hang onto" something that has passed. The movement imposed its own demands from its own needs and won: such is the nature of class struggle. To quote from a participant account in the 2006 movement:

> Anyone who knows anything knows that if they get away with this, it'll be extended to all workers, as has happened in Germany. At the same time, there are many workers who are on short-term contracts far worse than the CPE. The CPE's just a little straw—but, for a lot of people, it's the last one. And, without wishing to make a simplistic equivalent, there have been many uprisings that have begun from fairly small beginnings (May '68, for example, partly began against university dormitories being segregated according to gender). ("All Quiet on the French Front" 2006)

To return to the discussion of 1968, one of the greatest victories of that movement was, as Ben Trott rightly argued, the social battles waged and won by myriad different social forces from below. The emancipatory impulses of the late 60's, that Marcusian "eros effect" which overwhelmed the old society's boundaries of "order" and inherited so-called morality and decency cannot be ignored, but they should not be overestimated. Whilst the appetite of May '68 was very much a libidinous, Dionysian one, the sad recuperation of such impulses in institutionalised identity politics and corporate equal opportunities policies is everywhere in evidence. It remains a task of all future movements to try and reclaim these emancipatory impulses and reawaken their subversive quality: the normalising or domesticating of sexual freedom, feminism, black power, and gay liberation is as much a defeat as it is a victory.[7]

[7] On the mixed results of the momentous struggles of 1968, and May '68 in particular twenty years on, see Caute 1988.

Conclusion

This essay began with an attempt to define May 1968 in historical terms against the general view of the events as nostalgic mimesis, capable of being understood (if at all) as a phenomenon reflecting a world that has passed. Despite the incomprehension and the exercises in generational distancing and learning the lessons of the past, May 1968 was a moment that can be mapped into a constellation of such moments before and since, as a piece of graffiti commented at the time of the 2006 movement (tellingly in English): "We are winning" (quoted in "Humour of Students" 2006). The words may seem overly optimistic, but become less so, and more like a fledgling dawning of *a sense of the possible*, when recognised as direct appropriation of those written on the walls of Seattle in November 1999 during the momentous anti-WTO protests, in which the alter-globalisation and anti-capitalist mobilisations faced down the forces of state repression, and succeeded in the disruption and early closure of this supra-national "bourgeois talking shop." Indeed, May 1968 did not "fail" because its animating force was not and cannot be confined to a particular space-time, something made glaringly clear in 2011. It exists wherever the exploited and oppressed resist their condition and envision, however falteringly, a different world; this same spirit must once more return in the present to settle the score of history. Indeed as graffiti from 2006 puts it, "L'avenir nous appartient" (the future belongs to us, quoted in Bureau of Public Secrets 2006).

Works Cited

Afshar, Sara, Christian Bennett, Paul Lewis, Luisa Miller, Sarah O'Connell, Cameron Robertson and Jason Parkinson. 2011. "It Was a War, and We Had the Police Scared: Video." *The Guardian*, Dec. 5. http://www.guardian.co.uk/uk/video/2011/dec/05/reading-riots-video.

"All Quiet on the French Front." 2006. Libcom.org. Accessed Feb. 25, 2012. http://libcom.org/library/all-quiet-french-front.

Benjamin, Walter. (1940) 1999. "Theses on the Philosophy of History." Translated by Harry Zohn. In *Illuminations*, edited by Hannah Arendt, 245-255. London: Pimlico.

Blob, B.M. and B.M. Combustion. "1986-1987: France Goes Off the Rails." 2009. Libcom.org. Accessed Feb. 22, 2012. http://libcom.org/library/france-goes-rails.

Brinton, Maurice (1968) 2005. "Paris: May 1968." Libcom.org. Accessed Feb. 25, 2012. http://libcom.org/library/May-68Solidarity.

Bureau of Public Secrets. 2006. Documents from the Anti-CPE Uprising in France. Accessed Feb. 26, 2012. http://www.bopsecrets.org/recent/france2006.documents.htm.

Caute, David. 1988. *'68: The Year of the Barricades*. London: Paladin.

Davis, Mike. 2011. "Spring Confronts Winter." Editorial, *New Left Review* 72 (Nov-Dec). http://newleftreview.org/?page=article&view=2923.

Dodd, Vikram. 2011. "December Police Face Years of Public Disorder, Former Met Chief Warns." *The Guardian*, Dec. 6. http://www.guardian.co.uk/uk/2011/dec/06/police-years-public-disorder-warns.

Drake, David. 2006. "France 2006 is not May 1968." *The Guardian*, Apr. 6. http://www.guardian.co.uk/commentisfree/2006/apr/06/france2006isnotmay1968.

Erlanger, Steven. 2008. "May 1968: A Watershed in French Life." In *IHT*, Apr. 29. http://www.iht.com/articles/2008/04/29/europe/france.php.

Feenberg, Andrew and Jim Freedman. 2001. *When Poetry Ruled the Streets: The French May Events of 1968.* Stony Brook: SUNY Press.

Frölich, Paul. (1939) 2010. *Rosa Luxemburg: Ideas in Action.* Chicago: Haymarket.

Garland, Christian. 2011. "Simulating Events as They Happen: Media Spectacle, Ideology, and Readymade Bogeymen." Text of a Presentation given at UCL London, 16 August 2011, Theme 3: Language and the Media. Accessed Jan. 2012. http://fuberlin.academia.edu/ChristianGarland/Papers/1143145/Simulating_events_as_they_happen_media_spectacleideology_and_readymade_boogeymen.

Gregoire, Roger and Fredy Perlman. (1970) 2005. "Worker-Student Action Committees, France May '68." Libcom.org. http://libcom.org/library/worker-student-action-committees-france-1968-perlman-gregoire.

Harper, Stephen. 2011. "Riots in the UK: Echoes of the Past and a Glimpse of the Future." *Relative Autonomy*, Aug. 10. Accessed Feb. 20, 2012. http://www.relativeautonomy.com/2/archives/08-2011/1.html.

"Humour of Students Evokes Spirit of 1968." 2006. *The Telegraph*, Mar. 20. Accessed Apr. 24, 2012. http://www.telegraph.co.uk/news/worldnews/europe/france/1513520/Humour-of-students-evokes-spirit-of-1968.htm.

Jackson, William and Bob Jeffery. 2011. "Pendleton: A Political Sociology." Paper for the Conference "Urban Unrest, Social Resentment and Justice," Centre for Urban Research, University of York, Sept. 22-23. http://uclan.academia.edu/WilliamJackson/Talks/61050/Pendleton_A_Political_Sociology.

Kingsley, Patrick and Adam Gabbatt. 2011. "Occupy: We Are the World." *The Guardian*, Nov. 30. http://www.guardian.co.uk/world/2011/nov/30/occupy-movement-we-are-world

Lewis, Paul, Tim Newburn, Matthew Taylor and James Ball. 2011a. "English Rioters Warn of More to Come." *The Guardian*, Dec. 5. http://www.guardian.co.uk/uk/2011/dec/05/more-english-riots-to-come.

—. 2011b. "Rioters Say Anger With Police Fuelled Summer Unrest." *The Guardian*. Guardian-LSE Study, Dec. 5. http://www.guardian.co.uk/uk/2011/dec/05/anger-police-fuelled-riots-study.

Marx, Karl and Friedrich Engels. (1848) 2004. *Manifesto of the Communist Party*. Harmondsworth: Penguin.

Marx, Karl. (1852) 2010. *The Eighteenth Brumaire of Louis Bonaparte*. Chapter One. http://www.marxists.org/archive/marx/works/1852/18th-brumaire/index.htm

"Nous Sommes Tous des Casseurs." 2006. Libcom.org. Translated by Saul. Accessed Feb. 26, 2012. http://libcom.org/library/1994-nous-sommes-tous-des-casseurs.

Prasad, Raekha. 2011. "English Riots Were 'A Sort of Revenge' Against the Police." *The Guardian*, Dec. 5. http://www.guardian.co.uk/uk/2011/dec/05/riots-revenge-against-police.

Quatrocchi, Angelo and Tom Nairn. 1998. *The Beginning of the End: France May 1968, What Happened, Why it Happened*. London and New York: Verso.

Sage, Adam. 2008. "Spirit of 1968 is Invoked on Streets of Sarkozy's France to Demand Stability." *The Times*, May 3. www.timesonline.co.uk/tol/news/world/europe/article3863864.ece

Swinburne, Algernon. (1860) 2012. "A Song in Time of Revolution." The Swinburne Project. Accessed Mar. 3, 2012. www.letrs.indiana.edu/cgi-bin/acs-idx.pl?type=section&rgn=level1&byte=218409.

"Tentative Impressionistic Notes on the Movements in France, June 2003." 2008. Accessed Feb. 25, 2012. www.revoltagainstplenty.com/index.php/archive/16-archive-global/98-french-rebellion-2003.

"The Class Struggles in France." 1996. *Aufheben* 5 (Autumn). Libcom.org. http://libcom.org/library/france-aufheben-5.

Vienet, R. 1992. *Enrages and Situationists in the Occupations Movement, France May '68*. Translated by Loren Goldner and Paul Sieveking, Chapter 8. New York: Autonomedia. Accessed Feb. 20, 2012. http://www.cddc.vt.edu/sionline/si/enrages.html.

Willener, Alfred. 1970. *The Action Image of Society: On Cultural Politicization and Revolt*. London: Tavistock.

SURVEYING ACTIVISTS' ATTITUDES TOWARD TIME: A CANADIAN CASE STUDY

KAMILLA PIETRZYK

Recent decades have seen an increase in scholarly as well as popular interest in the question of time and temporality, a welcome recognition of the dialectic between time and space that had been forgotten, if temporarily, in the midst of the globalization debate. Thanks to Manuel Castells' theorization of "timeless time" (2001), David Harvey's "time-space compression" (1989), Anthony Giddens' "time-space distanciation" (1991) and Robert Hassan's "network time" (2009), as well as popular works with titles like *Faster: The Acceleration of Just About Everything* (Gleick 2000) and *In Praise of Slowness* (Honoré 2004), a general framework for understanding the nexus between capitalism, technology, and the structuring of time-space relations has been constructed. However, as Judy Wajcman has observed, though temporality "has become a central issue in the social sciences and much has been written about the increasing lack of control over time experienced in postmodern societies," there remains "a dearth of empirical research informing these debates" (2008, 73). For instance, while researchers have found that the conception of time differs across cultures, few have studied the temporalities, including shared "cognitive temporal structures," of the subcultures within them (Brodowsky, Granitz and Anderson, 2008, 233). To fill this lacuna, this study draws upon the results of an exploratory empirical survey carried out by the author and modelled after a design used by several past researchers in relation to various social groups. Here, the aim of the survey was to examine the temporal orientations and preferences shared by members of a loosely bound yet distinct subculture in Canadian society, specifically the radical activist Left, drawn in this case from the ranks of the so-called "anti-globalization" movement.

As Barbara Adam (1995) and other scholars have noted, contrary to Newton's conceptualization, time is a social construction, rather than a God-given reality; therefore, not only individuals but whole societies, cultures and subcultures conceptualize and organize time in their own ways (see Hall 1983; Gurvitch 1964; Levine 1997). This applies also to social movements as distinct social formations, since shared temporal norms and values constitute a crucial, yet largely neglected component in the socially constructed reality that forms the basis of collective action. According to Tarkowska (1992), a fundamental rationale and property of social movement actors are their characteristic temporal preferences and orientations—the different relative emphasis put on the past, present and future—separating them from the rest of society.

For some, a temporal orientation that extends beyond the present and into the past and the future is a central characteristic of left-wing activism. The environmental movement, for instance, can be defined in terms of its mission to displace the hegemonic short-term outlook promoted by consumer capitalism with a commitment to sustainability and a long-term vision of the future. The ideological commitment of anarchism is to the ethos of "pre-figurative" politics, which emphasizes seeking change on the micro level of interpersonal relations in the here and now, "building a new world in the shell of the old" rather than awaiting the fruits of a far-off revolution. It is often contrasted with the *telos* attached to the socialist tradition, which upholds the notion of progress and emphasizes that the present and future struggles are building upon a historical legacy of revolutionary resistance. Echoing the latter theme, Canadian historian Ian McKay argues that:

> Leftists have traditionally been—and, I am tempted to say, by definition always are—haunted by spectres from the past. They are "non-contemporaneous" with themselves, insofar as they have internalized a sense of being involved in a process of history that extends long before, and will endure long after, their individual life-times. They are not "all there," in the present, because they are in their minds living as much in the past and in the future. Even in the most day-to-day activities, they carry with them the idea of the more general historical pattern into which these activities fit. This intense historicism means that each left formation comes to live its history in an intense, partisan, and inescapable way. (2005, 120)

Likewise, in an article entitled "Resisting the Acceleration Society," Carmen Leccardi describes the "anti-globalization" movement as "capable of escaping from the shrinking of the global present (the time of cosmopolitan capitalism) in order to look equally at the past, by shedding light on the roots of contemporary social inequalities, and at the future, by underlying the responsibilities to unborn generations" (2007, 33).

This optimistic assessment contrasts with the extant critiques of the effects of time-space compression on the capacities and potentials of contemporary social movements (Pietrzyk 2010). In his famous essay "A Plea for Time," Canadian communication theorist Harold Innis (1951) argued that the dark side of cultures dominated by time-annihilating, "space-biased" media is a dangerous neglect of time and duration: what he saw as a suicidal "obsession with present-mindedness" characteristic of modern Western civilization. While the imperative toward social acceleration can be traced back to the advent of modern society (see Rosa 2009; Leccardi 2007; Rifkin 1987), many scholars concerned with time and temporality argue that advanced capitalist societies have in recent decades experienced a profound shift in the structuring of time-space relations underwritten by neoliberal globalization and the ICT revolution. This shift involves a transition from the linear and relatively predictable "clock time" of the modern machine age to what Robert Hassan has termed "network time"—a "new and powerful temporality" predicated upon the unprecedented penetration of daily life by networkable devices (2003, 134). Through the proliferation of mobile phones, personal digital assistants, laptops and other portable electronic devices, the daily lives of more people in more parts of the world are becoming more closely attuned to the exigencies of the post-Fordist regime of flexible accumulation, as "[f]lexibility becomes possible against the background of a previously unattained degree of constant temporal availability" (Nowotny 1994, 100). In light of the pressures of living in high-speed capitalist society, "both the past and the future as mental categories are threatened by the tyranny of the moment" (Eriksen 2003, 3).

Informed by this critical perspective, the hypothesis animating this study is that as individuals living in a technologically advanced, capitalist society wherein the temporal horizon has shrunk to an everlasting present, and where speed and efficiency are predominant values, social movement activists in Canada are bound to manifest similar temporal preferences and orientations in their own lives. Do the survey results support the hypothesis? Or do they bear out the assessment offered by McKay and Leccardi? How do social movement actors in high-speed society valorize different areas of time? To begin to answer these questions, this paper discusses the results of a temporal attitude survey carried out by the author in the fall of 2011 with a group of Canadian "anti-globalization" activists. Since the survey constitutes the first stage of a larger research project that will seek additional information from follow-up interviews, the insights offered below should be seen as necessarily tentative and preliminary.

Research Design

A total of forty-four social movement activists were asked to fill out a two-page survey instrument, a slightly altered version of the scale that has been used consistently by several past researchers (see Brodowsky, Granitz and Anderson 2008, 243). Temporal orientations were measured using 35 questions or "items," divided into eleven categories or "factors" of time. The first factor assessed the values and practices associated with time management (scheduling and planning), and consisted of ten items; between one and four questions were asked in each of the subsequent time "factors:" Time Pressures (e.g. I would like the pace of my life to be slower; Never enough time to do things I enjoy), Time Flies (When I enjoy myself, I lose track of time; I like to spend time with friends and family), Time Scarcity (Time lost can never be regained; I treat time as a scarce resource), Infinity (Time heals all wounds; What I can't do today I can do tomorrow), Time Direction (I think of time as a straight line; I think of time as a circle), Punctuality (I am very punctual; No excuse for being late), Future (The future goes on indefinitely), Focus (I like to stay on task till it is done), and Wasting Time (People should not waste time).

Each of the 35 items was measured on a five-point Likert scale anchored by strongly agree (5) and strongly disagree (1), with 3 indicating a neutral position. To enrich the discussion and ensure that shared temporal norms were fully captured with the 35-item scale, the survey also included a separate section that measured temporal preferences, using a nine-item, seven-point Likert scale (distant future-future-near future-present-near past-past-distant past). The questions in this section asked which area of time the respondents "most frequently talk about" and "almost never think about," which represents the best time of their lives, and which they would prefer to live in. The section also probed the respondents' temporal preferences concerning movies and stories.

Respondents were instructed to pick only one number or answer in response to items or questions in both sections of the survey. For the sake of brevity, and unless specified otherwise, the responses registered on "distant future," "future," and "near future" and the equivalent three values concerning the past have been aggregated into three overarching areas of time: the past, present, and future.

Although the size of the sample precludes a more in-depth demographic analysis, it is helpful to get an overall sense of the group's basic demographic attributes. Among the forty-four respondents, all are currently active organizers, based in several major Canadian cities, though mostly centered in Toronto. Twenty-four among the respondents are men and twenty are women, most of them white and under the age of forty. Almost half hold advanced degrees. Politically, the biggest group represented are perhaps the anarchists, based on my background knowledge of the respondents' primary political identity and affiliation—but there are also environmentalists, feminists, labour and anti-war activists, in addition to a number of professional as well as "unaffiliated" activists.

Though the politics across the sample are diverse, in keeping with the heterogenous quality of the "movement of movements," the activists surveyed here can be considered to belong to the political left activist subculture, endowed with its own temporal values and social norms, as suggested above by McKay and Leccardi. Despite often brutal sectarianism, left-wing activists in cities everywhere tend to share the same social norms and values, and belong to the same, locally-based counter-cultural social scene or network, which, as many of them would surely attest, can sometimes be so tightly knit as to become insular. It is upon this conceptualization of the Canadian radical Left as a counter-cultural group with its own socio-temporal norms that the following analysis is built.

Analysis

The survey's very first question got right to the point in asking whether "things will/are/were better" in the past, present, or future. On this question, a majority of the respondents proved optimistic, with sixty percent indicating that things will be better in the future. Optimism is certainly a vital motivating force driving activists—if there is no hope that things can get better, why bother trying to change them? In light of this, the percentage of optimists should perhaps be higher still. Instead, 12 percent of respondents indicated that things are better in the present while 28 percent chose the past.

Interestingly, only one person indicated that things will be better in the *near* future. Since the respondents live in Canada, and many of them are residents of Toronto, it is possible that some of their pessimism vis-a-vis the near future comes from the newest phase of the neoliberal "age of austerity," currently being implemented in this country by the conservative government of Stephen Harper, and locally in Toronto, by the equally conservative mayor Rob Ford. Since the responses were collected in the fall of 2011, less than a year since the beginning of Arab Spring and with a full third of the responses gathered after the world first learned of Occupy Wall Street, it would be interesting to find out to what extent these events influenced the activists' levels of optimism—a task to be pursued during follow-up interviews.

Elsewhere, other pieces of data suggest that activists do think about the future. Notably, in the survey's second strongest show of shared sentiment, 85 percent of respondents disagreed that they spend little time thinking about the future, with 30 percent feeling very strongy about this. Also, 82 percent of respondents indicated that the future is an object of their daydreams. This potentially problematizes the hypothesis regarding present-mindedness posed at the outset of the research—but confirms a couple of previous studies which proposed that Western nations, in contrast to Asian ones, tend to be future-oriented (Hall 1976, Yau 1988 cited in Brodowsky, Granitz and Anderson 2008, 238).

However, even if over half of the respondents claimed to see the future brightly, only nine percent of them chose the future as their preferred time to live in. In the strongest area of agreement on this section of the survey, a solid majority at 73 percent indicated that they prefer to live in the present. Similarly, 59 percent of the respondents assert that the best time of their lives is right now. The sense of satisfaction with the present situation demonstrated on these two items is consistent with a previous study that compared temporal orientations across Mexicans, Americans, and New Zealanders, and found that *all* cultures thought that the best time in their lives was the present (Anderson and Venkatesan cited in Brodowsky, Granitz and Anderson 2008, 238).

Another area of temporal preference tested on the survey pertained to the activists' tastes in stories and movies. Interestingly, optimism for the future declared by the majority does not translate here: only 12 percent of respondents prefer stories about the future, while only 13 percent prefer movies about it. When it comes to tastes in such things, nostalgia rules the day: a decided majority, at sixty-one percent, indicated that they prefer stories about the past, while 47 percent opted for movies about the past. A discrepancy seems at play here: if activists daydream mainly about the future, as they claim to do, why is it that they do not like stories or movies about it? Two plausible if partial explanations come to mind. First, much of the cultural industry, especially mainstream cinema, tends to wax apocalyptic when it comes to the future, arguably both manufacturing as well as giving expression to more deeply rooted cultural fears embedded in the collective imaginary. With titles like *The Day after Tomorrow*, *Deep Impact*, and now *2012*, the visions of the future produced by Hollywood are nightmarish and dystopian, negating the hopes and dreams of a better tomorrow preferred by most of the respondents. The second explanation points to the strong nostalgic tendency permeating postmodern culture, which Huyssen (2000, 30) connects to the successful mass marketing of memory in an age of uncertainty. In his words, "the past is selling better than the future." It is no wonder then that movies about the past, including documentaries about the political turmoil in the 1960s and 1970s like *The Weather Underground* and *A Grin Without A Cat,* or dramatic adaptations of the Spanish Civil War like *Land and Freedom,* continue to be popular within Canadian activist circles.

To elaborate on the respondents' temporal *orientations*, in the most decisive point of agreement on this part of the survey, an overwhelming majority, at ninety-one percent, indicated that they strongly or very strongly enjoy spending time with their family and friends. This result can almost be taken for granted; yet, however much a person may enjoy this activity may be increasingly irrelevant in the face of a widespread sense of time scarcity connected to the acceleration of everyday life. As technological, economic and socio-cultural processes continue to quicken, work time increasingly impinges on available leisure time, resulting in a widespread sense of extreme busyness; this in turn reduces the amount of time individuals feel they can devote to family and friends—and also, to civic engagement (Scheuerman 2005).

To assess the extent to which activists experience time pressures, the survey posed the following three questions: "I feel a great amount of time pressure," "I would like the pace of my life to be slower," and "Never have enough time to do things I enjoy." In confirmation of one of the study's guiding hypotheses, all three questions generated a majority affirmative response: 77, 55, and 50 percent of respondents, respectively, agreed or strongly agreed. Furthermore, an even half of the respondents believe that people used to have more time than we do now while 77 percent of them disagreed (of these, 30 percent did so strongly) that they only rarely glance at their watch or a clock. As expected, this confirms that Canadian activists are not immune from the hegemonic temporal structures, imperatives and pressures of living in a high-speed society. On the other hand, asked whether they feel they have little control over time, more people disagreed than agreed (43 compared to 32 percent, with the rest remaining neutral). This discrepancy can perhaps be attributed to the high scores registered in the "time management" sub-section of the survey.

To reiterate, the time management part of the survey consisted of ten questions about planning and scheduling. In the strongest response registered here, 70 percent of the respondents agree or strongly agree that "managing time is the key to success." A majority also concurred that they always like to know how long a task will take, that they try to group activities together to save time, and that they plan each and every day. The remaining questions registered a slightly lesser but still strong affirmation of the value placed on time management. Given that the majority of activists within the sample—and within social movements more broadly—are unpaid volunteers, it seems plausible to assume that time management skills are particularly highly regarded because they play a key role in allowing movement actors to negotiate increasing demands on their time in a way that still allows for political engagement.

Conclusion

Though the results of the survey are by no means conclusive, they do offer new insights, and confirm a number of guiding hypotheses posed at the outset of this ongoing study while problematizing and challenging others. Notably, the posited preoccupation with present-mindedness failed to manifest with anticipated force, as the respondents indicated a strong preference for the past or the future in a number of ways that warrant closer attention. The results do, however, bear out the hypothesis concerning the time pressures and the sense of extreme busyness associated with everyday life in fast-paced, technologically advanced society like Canada, both of which are bound to influence the ways individuals go about "doing" political activism. The resultant "addiction to urgency," and the tendency to perpetually reinvent the proverbial wheel, arguably diminish social movement's ability to wage an effective battle against capitalist globalization. If activists are to develop the kind of intellectual and creative capacities necessary to correctly identify and analyze the deeply rooted systemic problems of capitalism as well as their own biases that give rise to and support these tendencies, reflexive thought must be at the forefront of contemporary activist praxis.

By way of conclusion, it is precisely because the hegemonic temporal norms are rarely contemplated or acknowledged that survey data alone cannot suffice in providing a full picture of the temporality commonly shared among members of the Canadian activist Left. Since the nature of research into time and social temporalities is exploratory, the correct approach will yield "rich" data that can be interpreted and critically investigated, but not easily measured and statistically processed. It is therefore necessary to supplement insights discussed above with qualitative analysis based on in-depth, semi-structured interviews with the sample of activists surveyed here. By using a combined methods approach, it is possible to advance a more complete and nuanced understanding of how time and temporality enter into activists' attitudes, and by extension, into *collective* temporal orientations and practices characteristic of contemporary Canadian activism. It is to this empirical task that this researcher's own optimistic future gaze is directed.

Works Cited

Adam, Barbara.1995. *Timewatch: The Social Analysis of Time.* Cambridge: Polity Press.

Brodowsky, G., Neil Granitz and B. Anderson. 2008. "The Best of Times is Now: A Study of the Gay Subculture's Attitudes Toward Time." *Time & Society* 17 (2/3): 233-260.

Castells, Manuell. 2001. *The Internet Galaxy: Reflections on the Internet, Business and Society.* Oxford: University Press.

Hall, Edward T. 1983. *Dance of Life: The Other Dimension of Time.* Garden City: Anchor/Double day.

Harvey, David. 1989. *The Condition of Postmodernity: An Enquiry Into the Origins of Cultural Change.* Oxford: Blackwell.

Hassan, Robert. 2009. *Empires of Speed: Time and the Acceleration of Politics and Society*. Leiden: Brill.

—. 2003. *The Chronoscopic Society: Globalization, Time and Knowledge in the Network Economy.* New York: Peter Lang.

Honoré, Carl. 2004. *In Praise of Slow: Challenging the Cult of Speed.* New York: HarperCollins.

Huyssen, Andreas. 2000. "Present Pasts: Media, Politics, Amnesia." *Public Culture* 12 (1): 21-38.

Innis, Harold A. 1951. *The Bias of Communication.* Toronto: University of Toronto Press:

Giddens, Anthony. 1991. *The Consequences of Modernity.* Stanford: Stanford University Press.

Gleick, James. 2000. *Faster: the Acceleration of Just About Everything.* New York: Vintage Books.

Gurvitch, Georges. 1964. *The Spectrum of Social Time.* Dordrecht: Reidel.

Leccardi, Carmen. 2007. "New Temporal Perspectives in the 'High-Speed Society.'" In *24/7: Time and Temporality in the Network Society,* edited by Robert Hassan and Ronald E. Purser, 25-36. Stanford: Stanford University Press.

—. 2003. "Resisting the Acceleration Society." *Constellations* 10 (1): 34-41.

Levine, Robert. 1997. *A Geography of Time: the Temporal Misadventures of a Social Psychologist; Or How Every Culture Keeps Time Just a Little Bit Differently.* New York: Basic Books.

McKay, Ian. 2005. *Rebels, Reds, Radicals: Rethinking Canada's Left History.* Toronto: Between the Lines Press.
Nowotny, Helga. 1994. *Time: The Modern and Postmodern Experience.* Cambridge: Polity Press.
Pietrzyk, Kamilla. 2010. "Activism in the Fast Lane: Social Movements and the Annihilation of Time." *Fast Capitalism* 7 (1). http://www.uta.edu/huma/agger/fastcapitalism/7_1/pietrzyk7_1.html.
Rifkin, Jeremy. 1987. *Time Wars: The Primary Conflict in Human History.* New York: H. Holt.
Rosa, Harmut. 2009. "Social Acceleration: Ethical and Political Consequences of a Desynchronized High-Speed Society." In *High-Speed Society: Social Acceleration, Power, and Modernity,* edited by Harmut Rosa and William E. Scheuerman, 77-111. University Park: The Pennsylvania University Press.
Scheuerman, William E. 2005. "Busyness and Citizenship." *Social Research* 72 (2): 447-470.
Tarkowska, Elżbieta. 1992. *Czas w Zyciu Polakow: Wyniki Badan, Hipotezy, Impresje* [Time in Polish Life. Results of Research, Hypotheses, Impressions]. Warsaw: Polska Akademia Nauk.
Wajcman, Judy. 2008. "Life in the Fast Lane? Toward a Sociology of Technology and Time." *British Journal of Sociology* 59 (1): 59-77.

The Temporality of Double Consciousness

John Streamas

In Julian Barnes's 2011 Man Booker Prize-winning novel *The Sense of an Ending*, the narrator and his school friends, coming of age in the sixties and seventies, wear their watches facing the insides of their wrists, effecting the "personal time" that, the narrator says late in the book, "is the true time . . . measured in your relationship to memory" (133). Yet, as he ages, the narrator becomes increasingly conscious of the inadequacies of this personal time, conceding that, "when we are young, we invent different futures for ourselves; when we are old, we invent different pasts for others" (Barnes 2011, 88). Barnes the author seems to say what Tony Webster the narrator cannot say: that somehow the sum total of everyone's personal time does not add up to what is called history.

At four moments in the first half of the short novel, Webster comes close to realizing that history is differential and variable. I wish briefly to examine these passages for their insights into the default temporality of history, a temporality that is racialized white, against which I will discuss alternatives. Early in the book, Webster's schoolmate Adrian Finn, claiming to quote a Frenchman named Patrick Lagrange, says, "History is that certainty produced at the point where the imperfections of memory meet the inadequacies of documentation" (Barnes 2011, 18). Later, considering the suicide of Finn, Webster thinks, "History isn't the lies of the victors. . . . It's more the memories of the survivors, most of whom are neither victorious nor defeated" (Barnes 2011, 61). A few pages and several decades later, Webster comes close to understanding:

> We live in time, it bounds us and defines us, and time is supposed to measure history, isn't it? But if we can't understand time, can't grasp its mysteries of pace and progress, what chance do we have with history—even our own small, personal, largely undocumented piece of it? (Barnes 2011, 66)

His deepest insight comes in his brief remark that, in England, "shadings of class resist time longer than differentials in age" (Barnes 2011, 80). This claim is made at almost the very midpoint of the novel. In the book's second half, while Webster unravels the sordid mysteries of his schoolfriend's suicide and of his and Finn's relationships with a woman and her "damaged" child, Barnes's story retreats into "personal time." Readers may be excused for suspecting that, after the book's startling ending, a chastened yet cynical Webster will forever after wear his watch on the inside of his wrist. But in the first three of those early passages he struggles against a time that is not only subjective and personal but even atomized and isolated, and that isolation is reinforced and deepened by memory's vagaries. Only in the fourth passage, in his flickering notice of class, does he move toward a sense that it may not be individual humans at all but rather clusters of humans—classes, in perhaps every sense—whose collective times add up to history. It is time in this collective sense that I wish to consider here, to argue that, in societies in which race determines profit and power, racially marginalized peoples create their own time-sense both for surviving racism and for envisioning a revolutionary future free of racism.

Senses of time in hard sciences such as biology and physics usually mean time as it is experienced by whole species and as it functions within natural laws. Humans may not be alone in experiencing time self-consciously, yet an implication of Marx's theories is that, though time as temporary release from wage-labor might be experienced in personal terms, true liberation can be realized only in communal and class terms. Here I regard communities of color in the United States, victims of a racism the origin of which is mutually imbricated with the origin of a rigid and vertical class structure. Critics and activists do not speak of a "black class" or a "brown class" only because the structure is, by design, porous enough to allow, even to encourage, the occasional person of color to ascend the class hierarchy. In fact, it is this calculated porousness, this limit of upward mobility, that prompts conservative and libertarian whites—and even the occasional "self-made" black or brown man such as Herman Cain—to blame the poor for their poverty and the dark-skinned for their complaining. Meanwhile, young black men, though no likelier to commit felonies than young white men, are seven times likelier to go to prison. And after their release they lose their voting rights and face, for the rest of their lives, legalized discrimination in employment, housing, education, and social services (Alexander 2010, 2).

Communities of color, dreaming of liberation while still in thrall to "master" races, invented "colored people's time"—CPT—both to survive through oppression and to envision a better world. Historian Mark Smith credits CPT to slaves, as a sort of silent, implicit language of resistance (1997, 130). But variations exist in all communities of color, so that, for example, "Indian time" and "island time" bind indigenous peoples. In certain contexts, especially in interactions with whites, CPT and its variants are a joke on late arrivals and languid performance; but, within the community, it is the very sort of public but implicit time-sense that Barnes's narrator fumbles to desire but never recognizes, the time-sense that constructs a history. It is also, I would argue, a parallel time, an alternative time that is not necessarily either linear or cyclical, that may be discontinuous, even contingent and strategically atomistic and particularized.

I have argued elsewhere that CPT is the time-sense of what W. E. B. Du Bois called "double consciousness," that it constructs an alternative temporality in narratives by writers of color, and that this is most evident in narratives of apocalypse and end-times (Streamas 2010, 230-32). My aim here is to apply CPT to visions of social justice in a revolutionary future—that is, to visions that may be called proleptic narratives. Double consciousness, as Du Bois first and most explicitly defined it, refers to a black person's "sense of always looking at one's self through the eyes of others, of measuring one's soul by the tape of a world that looks on in amused contempt and pity" ([1903] 1973, 3). This seems less gloomy when it is understood as a means of bearing the constant and ubiquitous burden of racism, as a survival strategy and a "know your enemy" injunction. Narratives constructed by writers and witnesses who practice double consciousness might understandably be expressed in the second, layered temporality of CPT. In that layer, that temporality, imagination may roam freely, even—especially—toward a desire for a world liberated from the condition that made double consciousness a necessary survival skill. This is not to suggest that, if that world were ever realized, CPT would necessarily vanish into superfluity. One hope of revolution is, after all, the construction of new conditions in which to work and thrive; and CPT might well be retained as a new condition of the post-revolutionary world's narratives, a rejoinder to cynics who charge that art would cease to exist in an ideal world.

At the very least, CPT might jog the post-revolutionary world's memory, reminding revolutionaries of the old bad conditions under which they suffered. For the first time, the time-sense that had been the sufferers' secret for centuries might now surface, a heuristic for memory and history. Left-leaning historians such as Howard Zinn have sought for decades to write history from the bottom up, and the fields of ethnic studies and gender studies have centered the perspectives of victims of racism and sexism. But the temporal scale and pacing of such histories have remained those of dominant histories. An easy example is historical periodization. Historians, social scientists, and cultural critics alike—but also people of color and revolutionaries—parse time in centuries and decades, eras and epochs. "The long 1960s" might have begun in 1964 and ended in 1975, but even this altered decade remains familiar as the sixties. Christian scholars may abandon the calendar when they speak of prelapsarian and postlapsarian ages. Literary critics argue over whether Joyce or Dujardin or Hardy was the first modern novelist, but they agree that modernism suffuses the spirit of a certain literary age. Periodization is enabled by linearity. Postmodernism follows modernism, the postbellum follows the antebellum. Scholars trying to examine, for example, European exploration of the Americas from the perspective of the indigenous may reject the language of discovery but cannot as easily discard the language of its periodization. Labor historians may easily discard celebrations of the Industrial Revolution, but their examinations of strikes and unionization seem inescapably tethered to the periodization of industrialism.

The dangers of linear periodization are the sense, first, that chronology becomes destiny and, second, that chronology assumes progress. But arguments for cyclical time—whether from anthropologists' studies of ancient cultures or from chronobiologists' observations of organic life-patterns or from theologians' interpretations of myths of eternal returns—are even likelier to find patterns than are arguments for linearity. Circularity asserts predictability. The discovery of recurring patterns of constellations in the night sky gave astronomy such precise and dependable predictability that, for centuries, nonspecialists have been reading the calendar in the stars.

CPT is not so dependably predictive. A few years ago I invited a Blackfeet doctoral student to speak to a graduate class about "Indian time." At first she spoke of her family, and she drew arcs on the board, and as she brought together various relatives' stories the arcs converged into a circle. But then she spoke of other people in her community, people related less by blood than by culture and history, and the new lines she drew could be interpreted as giving depth to the circle, expanding it so that it became three-dimensional, a sphere. On a sphere a person, and that person's time, can be linear and circular at once. On a sphere the person may even zig and zag, may more easily tailor movement to conditions. A person may experience concurrent events in parallel temporalities. She may even construct memories and experiences that move in opposite directions on the sphere. Moreover, a sphere "gives," that is, it can absorb, reflect, and respond to materials and contingencies in the spaces around it.

Oppressors and oppressions change; CPT anticipates and adjusts to those changes.[1] Time may thus be responsive, contingent. This sphere is, I argue, part of what becomes the revolutionary space of CPT.

To demonstrate, I wish briefly to examine temporality in both Octavia Butler's time-travel novel *Kindred* ([1979] 2003) and the Occupy movement. In Butler's book a young black woman, Dana, very much a contemporary woman of the bicentennial summer of 1976, finds herself thrust several times into the antebellum South of the 1810s to the 1830s. She comes to realize that the young man Rufus, whose life she saves on her trips back in time, is her ancestor, though he is also the son of a brutal slaveowner. Saving Rufus becomes her mission, her only way of ensuring a life of her own. And, as Butler establishes in two early passages, Rufus saves her too, if only by existing. This is different from the stereotypical "diversity" theme that promotes unity based on a notion that "we are all in this world together, brothers and sisters under the same skin," for after all Rufus still casually calls slaves "niggers" and assumes their inferiority. Yet it is also different from an anti-presentism that would impose a tolerance for ancestors who "knew no better," as Butler's graphic depictions of slaveowners' violence argue that such practices are intolerable everywhere and in all times.

In one of these passages, Rufus asks why and how Dana appears at moments of crisis, and when she says, "We're from a California that doesn't exist yet," he replies incredulously, "But it's eighteen nineteen. It's eighteen nineteen everywhere," to which she falls back on a helpless explanation: "We come from a future time and place. I don't know how we get here. We don't want to come. We don't belong here. But when you're in trouble, somehow you reach me, call me, and I come" (Butler 2003, 62). Rufus fails to understand, and Dana thinks, "I wondered why I had bothered to try to convince him. After all, how accepting would I be if I met a man who claimed to be from eighteen nineteen" (Butler 2003, 63). The irony is that, of course, Dana *has* met a man from 1819, and that saving his life is all that enables her own existence. In the other, earlier passage, just as Dana is being swept back to the nineteenth century, Rufus hears a bit of her conversation with her partner Kevin, a conversation from 1976: "How had Rufus heard Kevin and me across time and space?" Dana wonders (Butler 2003, 31). She does not know, and Butler refuses to explain. But this suggests a possibility that the two time periods are more

[1] For the point I make in these last two sentences, I thank Kelvin Monroe, Siskanna Naynaha, and Sky Wilson, who, after reading this paper's middle paragraphs, separately and independently recommended that I stress CPT's capacity to conform to the spaces around it and even to anticipate and respond to change.

than mutually and equally dependent. It suggests that, just as Dana and, later, Kevin, travel back to the nineteenth century, maybe Rufus could, under only slightly different circumstances, travel forward to the twentieth. Butler constructs a 1976 America without slavery, but not without subjugations that relegate Dana to unfulfilling labor: "I was working out of a casual labor agency—we regulars called it a slave market. Actually, it was just the opposite of slavery. The people who ran it couldn't have cared less whether or not you showed up to do the work they offered. They always had more job hunters than jobs anyway" (Butler 2003, 52). The opposite of slavery is thus as unfulfilling as slavery, just differently so. That Rufus catches glimpses of this post-slavery time suggests that it may be not as distant as Dana thinks it is, that it may be a matter of stepping out of one room into another, though the time crossed is measurable in centuries. This is why the book's first sentence—"I lost an arm on my last trip home" (Butler 2003, 9)—is remarkable for the act of mutilation rather than for its setting of "home." For home is, by the end of the book, just as much the 1830s as it is 1976. And that act of mutilation is a literal crossing from one room to another:

> I was back at home—in my own house, in my own time. But I was still caught somehow, joined to the wall as though my arm were growing out of it—or growing into it. From the elbow to the ends of the fingers, my left arm had become a part of the wall. I looked at the spot where flesh joined with plaster, stared at it uncomprehending. It was the exact spot Rufus's fingers had grasped. (Butler 2003, 261)

Narratives of time travel assume time's linearity, so that a traveler, as if standing at a station waiting for a ride, may take a train moving in one direction into the future or in the opposite direction into the past. But, like the Native student drawing circles and then spheres on the board, Butler's traveler experiences time as if it moves variously, perhaps even in parallel, so that 1976 and 1819 may occupy parallel concurrent spaces. *Kindred* may thus be ascribed a temporal dialectic. Within her additional temporality, Dana experiences a doubling of her double consciousness. When time travel becomes so mobile, variable, flexible, and contingent, it becomes dangerous to those who would enforce linearity.

This is why Slavoj Žižek, addressing a crowd of Occupy movement protestors in Zuccotti Park in New York on 9 October 2011, charged that only a half year earlier China's government had banned "all stories that contain alternate reality or time travel. . . . These people still dream about alternatives, so you have to prohibit this dreaming" (2011, 67). Žižek told his American audience that they had been so thoroughly oppressed that they had lost the capacity to dream, and now they could only imagine an end of the world by asteroids but not by capitalism (2011, 67). As if to underscore this point, another participant-witness of the Occupy movement, Nikil Saval, left a gathering in Philadelphia to see the film *Melancholia* and to register its nihilism:

> People at a country estate watch as a rogue planet, hidden behind the sun until now, speeds along on a collision course with the earth—all to the tune of the Prelude to *Tristan und Isolde*. In the final minutes, Charlotte Gainsbourg suggests that they all have a glass of wine on the verandah, as a sort of final act. Kirsten Dunst does not want to, however. "You know what I think of your plan?" Kirsten Dunst says. "I think it's a piece of shit." (Saval 2011, 161)

Whether earnest witness or satire, this view of the world surrenders itself to forces of destruction, unable to imagine, much less to inhabit, alternate times and spaces of resistance. Of course Butler's time traveler Dana, while no Marxist, understands slavery as an expression of capitalism; and she understands that time travel itself, this window into double consciousness and CPT, has taught her this.

But, ostensibly at least, to occupy is to fill a space, not a time. Angela Davis acknowledges the dissatisfaction expressed by some indigenous activists over the word "occupy," which, after all, names the very strategy of colonization; but she also urges a transformation of the word, into something that names acts of community and hope (2011, 133). As the Occupy movement spread across the United States and the world, it did so unevenly—perhaps unavoidably, given its apparent suddenness—and so not everyone who should have benefited experienced its exhilarations. Audrea Lim knows that Chinatown is, by bicycle, a short ride to Zuccotti Park, and yet its garment workers—sweatshop laborers who live in rundown and overcrowded tenement buildings—feel left out of the movement (2011, 101, 102). Because uneven development results in various oppressions, making resistance difficult to organize among variously suffering communities, the joys of resistance—and, eventually, the fruits of revolution—may be experienced differently by different peoples, and at first some may experience none of those joys. A few—for example, the upwardly mobile in communities of color, the black and brown middle classes—may even discourage revolutionaries. In Butler's novel, Rufus comes closer to traveling forward in time than any of his slaves ever come to traveling anywhere in time. And yet Dana, against the dictates of time-travel conventions, violates protocols by teaching some of Rufus's slaves. Their antebellum literacy is sure to alter the postbellum world, including Dana's 1976 California. This is only to suggest that time travel, as an expression of CPT and double consciousness, can work for revolution. Rules in the service of injustice demand to be broken in the service of the suffering.

How, then, can resistance occupy time as well as space? Time travel happens in fiction, after all. I argue that CPT, as an alternative temporality, releases people of color from the single-track, linear temporality that is all that oppressors know and enforce. While it may not deliver anyone to particular points in time, as time travel does for Dana, it constructs alternative temporal perspectives that enable memories of past injustices—that enable, in other words, conceptual versions of history from the bottom up—and visions of future justice. As the Occupy movement evolves, then, people of color should contribute considerably, if unevenly, to its various visions. Julian Barnes's narrator mistakenly claims that, in "personal time," it is individuals who "invent different futures for ourselves" (2011, 88). Rather, it should be whole communities who invent these futures.

Works Cited

Alexander, Michelle. 2010. *The New Jim Crow: Mass Incarceration in the Age of Colorblindness*. New York: New Press.

Barnes, Julian. 2011. *The Sense of an Ending*. New York: Knopf.

Butler, Octavia. (1979) 2003. *Kindred.* Boston: Beacon.

Davis, Angela. 2011. "(Un)occupy." In *Occupy! Scenes from Occupied America*, edited byAstra Taylor, Keith Gessen, and editors from *n + 1*, *Dissent*, *Triple Canopy*, and *The New Inquiry*, 132-33. London: Verso.

Du Bois, W. E. B. (1903) 1973. *The Souls of Black Folk.* Millwood, NY: Kraus-Thomson.

Lim, Audrea. 2011. "Chinatown Is Nowhere." In *Occupy! Scenes from Occupied America*, edited by Astra Taylor, Keith Gessen, and editors from *n + 1*, *Dissent*, *Triple Canopy*, and *The New Inquiry*, 99-104. London: Verso.

Saval, Nikil. 2011. "Scenes from Occupied Philadelphia." In *Occupy! Scenes from Occupied America*, edited by Astra Taylor, Keith Gessen, and editors from *n + 1*, *Dissent*, *Triple Canopy*, and *The New Inquiry*, 110-15. London: Verso.

Smith, Mark M. 1997. *Mastered by the Clock: Time, Slavery, and Freedom in the American South.* Chapel Hill, NC: University of North Carolina.

Streamas, John. 2010. "Closure and 'Colored People's Time.'" In *Time: Limits and Constraints*, edited by Jo Alyson Parker, Paul A. Harris, and Christian Steineck, 219-35. Leiden: Brill.

Žižek, Slavoj. 2011. "Don't Fall in Love with Yourselves." In *Occupy! Scenes from Occupied America*, edited by Astra Taylor, Keith Gessen, and editors from *n + 1*, *Dissent*, *Triple Canopy*, and *The New Inquiry*, 66-69. London: Verso.

MARX "AFTER" LYOTARD: TIME, CAPITAL AND REVOLUTION

MARIA PRODROMOU

The *différend* with Marxism

Lyotard had a long love affair with Marx. As with most love affairs, it was no stranger to the experience of disappointment. As a teacher in Algeria from 1950 to 1952 he wrote a series of articles for the French revolutionary Marxist group *Socialisme ou Barbarie* hoping to create the conditions for a spontaneous revolution against the injustices the Algerian people suffered under French colonialism. He continued writing mostly political analyses on colonialism and class struggle in Algeria for the group's journal until the schism of the group in 1964 occasioned by Castoriades' attempt to steer the group towards thinking of revolution in terms of critical subjectivity rather than an objective force of historical materialism. Although sympathetic to Castoriades' position, he sided with his militant friend Souyri and joined the splinter group "Pouvoir Ouvrier" which continued working towards a proletarian revolution to overthrow capitalism. Lyotard resigned from the group in 1966, for reasons too complex to be settled with a simple "intellectual differences with Marxism and revolutionary action." The differences were profound: not only personal—"Nothing else, with the exception of love, seemed to us worth a moment's attention during those years" he wrote in 1982 in "A Memorial to Marxism" ("*Pierre Souyri: Le Marxisme qui n'a pas fini*" [1982] 2006, 232)—or conceptual—was Marxism ("and with which Marxism?" as he points out in that essay) capable of understanding and transforming "the new direction taken by the world after the end of the Second World War" (233). Rather, the differences marked a *différend* with Marxism, a *différend* that was never settled once and for all in Lyotard's work but remained an unfinished "business."[1]

Simplifying to the extreme, this *différend* had to do with Lyotard's suspicion of Marxism as a universal language with a claim to absolute universality and a value above all others. A *différend* is not a matter of disagreement between two parties that could present their case in front of a judge, their merits debated, and a verdict reached by appealing to a common idiom. In the absence of a common idiom, a *différend* signals the occasion of radical differences and incommensurable positions, an intolerable suffering where something that cannot be phrased (i.e. presented, articulated) asks to be put into words (Lyotard 1988, 13). The suspicion of Marxism did not have so much to do with the "failures" of the revolutionary movement but rather with a deeper suspicion of the grand narratives of modernity, where the latter is understood as a mode of thought which organises events (history) around the Idea (in the Kantian sense) of a universal subject progressing towards the realisation of an end: freedom. This Idea is told in different ways by different grand narratives, yet the promise of emancipation has failed in every form it took to realise itself:

> Everything real is rational, everything rational is real: 'Auschwitz' refutes speculative doctrine. This crime at least which was real, was not rational. Everything proletarian is communist, everything communist is proletarian: 'Berlin 1953, Budapest 1956, Czechoslovakia 1968, Poland 1980' . . . refute the doctrine of historical materialism: workers rise up against the Party. Everything democratic is by and for the people, and vice versa: 'May 1968' refutes the doctrine of parliamentary liberalism. The social in its everydayness puts representative institutions in check. Everything that is the free play of supply and demand is favorable for the general enrichment and vice versa: the 'crises of 1911 and 1929' refute the doctrine of economic liberalism. And the 'crisis of 1974-1979' refutes the post-Keynesian revision of that doctrine. (Lyotard 1988, 257)

It is only proper to add to this list: a free market economy produces more efficient government, liberty, personal responsibility and wealth that "trickles down." The "global financial crisis of 2008–?" refutes the doctrine of neo-liberal capitalism.

---

[1] Lyotard's fascinating and provocative reading of Marxism and capital in *Libidinal Economy* in terms of desire and within the framework of libidinal materialism, is beyond the scope of this paper which opts for a more critical (i.e. Kantian) approach, and as such focuses instead on Lyotard's reading of Marx and capital in *The Differend.*

Raising questions on the defaillancy of Marxism was deeply frightening for Lyotard, not only because of the "formidable theoretical tasks they promised" but because "they seemed to condemn anyone who gave himself over to them to the abandonment of any militant practice for an indeterminate time" (Lyotard 2006, 234). As I hope to show here, these questions do not lead to quietism or a withdrawal from politics and the political, but rather a radical rethinking of the political as a persistent attention to incommensurable difference and, in the context of Marxism, a radical critique of the *différends* that capitalism creates and covers through the hegemony of the genre of exchange.

Bearing witness to the *différend* through reflective judgment (judgment in search of its rules) becomes the political task par excellence for Lyotard. Bearing witness entails the creation of new idioms and finding imaginative ways to articulate the inarticulable:

> To give the *differend* its due is to institute new addressees, new addressors, new significations, and new referents in order for the wrong to find an expression and for the plaintiff to cease being a victim. This requires new rules for the formation and linking of phrases. (Lyotard 1998, 21)

Giving the *différend* its due is an interminable task, which means that the business of politics—understood as the demand and search for justice—is without end. This has practical consequences for any revolutionary praxis: we will never be done with justice and this requires rethinking what revolution means both practically and theoretically. Let us turn to this task.

Modernity and Revolution

Lyotard examines modernity as a mode of temporality based on diachrony and the principle of revolution. Its conception of time is linear and forward moving, in other words, it involves a spatial representation of time as a line composed of a series of moments spread along the axis of past, present and future. The temporal organization of modernity finds expression in the grand narratives whose legitimacy is grounded in a future to be realized (in an Idea which is legitimating because it is supposed to be universal) which will return the subject to a pre-lapsarian state, reconcile it with itself and overcome its separation. The eschatological principle, as Lyotard points out, induces a political rather than a ritual attitude (Lyotard 1991, 68).

The modern political attitude entails a mode of temporal synthesis which Lyotard examines under the aegis of Freud's notion of remembering.

Like Oedipus' attempt to avert the oracle of parricide and incest and change his fate by enquiring into the origin of the calamities of the present, the patient's attempt to remember the past does not bring an end to the compulsive repetition of destiny but rather repeats it anew. For Lyotard, this is one way in which re-writing modernity is understood: "in the sense of remembering, as though the point were to identify crimes, sins, calamities engendered by the modern set-up—and in the end to reveal the destiny that an oracle at the beginning of modernity would have prepared and fulfilled in our history" (1991, 27). Lyotard uses the example of Marxism that locates the source of the plague in the capitalist system of private property whose abolition was thought to lead to the overcoming of the alienation of man and his reconciliation with his species-being. What Lyotard questions here rather rapidly (and as a result blurs the differences between the early and late Marx) is the Manichean political logic of humanist Marxism whose essentialism produces the sort of oppositional thinking and politics of revolution which has its basis in a gesture of wiping the slate clean, setting the clock back to zero and starting anew. Such politics ends up reaffirming and reproducing the power of authority it seeks to overcome. Emancipation, especially for the Marx of the 1844 manuscripts, consists in the return to the origin, to a state before the fall (unalienated, free creative labor) that constitutes the essence of human nature from which man is thought to be alienated from in capitalist society. The trap with such an understanding of re-writing modernity lies in the repetition of what it seeks to overcome as has been bitterly proved from the experience of socialist communism in the past century:

> The localization and diagnosis may change, but the same illness re-emerges in this re-writing. Marxists believed that they worked to disalienate humanity, but the alienation of man has been repeated in scarcely displaced form. (Lyotard 1991, 28)

Demystifying the Necessity of Capitalism: The Temporality of Exchange

For Lyotard, capital is not totalitarian in a political sense, that is, the experience of totalitarianism in contemporary capitalist societies does not take place as a violation or deprivation of human rights, as persecution through political trials with a system of informants and secret police (such as, for example, in the experience of communism). Rather, the experience of totalitarianism occurs much more inconspicuously at the level of civil society by reducing all phrases to the notion of exchangeability. What we are witnessing in our so-called postmodern world (what Lyotard calls the realism of the "anything goes" which is a camouflage for the realism of money; 1992, 17) is a much more surreptitious form of totalitarianism that is achieved through mass consumption and the mass media where terror is not exercised in the name of freedom, but in the name of "'our' satisfaction" (Lyotard 1992, 317). If there is the overwhelming sense that capital is everywhere it is because it is totalitarian in terms of language: "it calls for the complete hegemony of the economic genre of discourse" (Lyotard 1992, 69). Capital needs "civil society to repeat its cycle. Society is for capital the indispensable moment of the destruction (consumption) of singular *thises or thats*" (69).

In terms of legitimation, capital appears to be self-legitimating: based on necessity rather than any kind of finality. Simply think of the discourse of the crisis in 2012: the burning question is "how are we going to save the economy" not, let's say, "how are the terms of the troika destroying workers' rights and undermining the sovereignty of Greece." As the former vice-president of the ECB and unelected prime minister Lucas Papademos said in his TV address to the Greek people on the eve of the parliamentary vote for the fifth austerity package drafted and demanded by the troika (as the ECB, the EC and the IMF are collectively known) as a condition for a second bailout loan to keep Greece from defaulting on its debts and declaring bankruptcy: "Our main and essential goal is the sustainable development of our economy." Unsurprisingly, the speech was a brilliantly executed exercise in Machiavellianism, carefully manipulating an economy of fear by balancing the melodrama of catastrophology with the sombre tone that is called for when addressing a nation that is, in his own words, "a breath away from ground zero."

Now, the appearance of a necessity to capitalism is not accidental but a distinctive feature of the genre of exchange. Under capitalism's demand for development and performativity, the system aims to control time so as to foreclose the possibility of the unexpected happening which would threaten its performance and so the present is subordinated to the future, a future that must be controlled and predicted and as such predetermined. There is no better model for this mode of temporality than that of exchange since in its practice the present is not merely subordinated to the future but *conditioned* by it. This is how Lyotard puts it:

> Someone (X) gives someone (Y) an object a at time t. This giving has as its condition that Y will give X an object b at time t' . . . There are many 'language games'—I prefer to say 'genres of discourse'—in which a later defined occurrence is expected, promised, etc., at the time the first takes place. But in the case of exchange, the 'second' occurrence, the payment,

> is not expected at the time of the first, it is presupposed as the condition of the 'first'. In this manner, the future conditions the present. (1991, 65-66)

Herein lies an important distinction between the temporality of the grand narratives of modernity and capital: modernity projects the future, capitalism programmes it. Capitalism depends upon the countable time of diachrony: the "days, hours, etc." used to measure use-value (in terms of labour time) (Marx [1867] 1990, 129). But on top of this linear-time, congealing it (since in capitalism only exchange-value matters) by twisting it is the *convoluted* temporality of exchange. The distinctive feature of the genre of exchange is that it needs two moments ("phrases" in the language of *The Differend*) to complete a cycle: X gives Y object a at time t on the condition that Y gives X object b at time t'. The two moments are linked by necessity in capitalism.

To understand the significance of this it is necessary to briefly overview some of the basic themes of *The Differend*. Here Lyotard takes the "phrase" as his basic unit of analysis. A phrase occurs and presents the four instances of an addressor, an addressee, a sense and a referent. In the mere presentation of a phrase universe these four instances are undetermined. Their relation is determined in the phrase that links onto the original; hence the latter can only be presented in a representation (a situation, i.e. a relating of the four instances in a phrase universe). The original phrase is open to a myriad of possibilities of linkages or responses. Different genres of discourse, such as science or literature, provide their own set of possible responses (governed by the end to be attained), that is, ways to link onto the initial phrase. So the type of response to the initial phrase is a contingent one: "A linkage may reveal an equivocalness in the previous phrase. *The door is closed* can give rise to *Of course, what do you think doors are for?*, or to *I know, they're trying to lock me in*, or to *All the better, I have to talk to you*, etc." (Lyotard 1988, 81). The meaning of the original phrase is determined in the phrase that links onto it.

What Lyotard calls the *event* and the time of the event is the occurrence of a phrase before its situation in a (re)presentation. The event is the singular occurrence of a phrase's presentation: an *it happens* in contrast to *what happens*. The time of the event is a now that is "absolute;" it is always already too soon or too late to grasp this "now" in any identifiable manner. The event (or receptivity to the event's occurrence) interrupts and suspends linear temporality, the linkage of phrases, and the putting together of time in a diachronic series. The phrase as an event is also what withstands universal doubt: "You cannot doubt that something happens when you doubt: it happens that you doubt" (Lyotard 1988, 104).

In the same manner, silence is also a phrase; in other words, even silence is a link (105). Thus, a phrase is not just articulated language but a gesture or even a cat's raised tail can be a phrase (123).

The contingency of linkage and the absence of a metalanguage that would regulate the linking of phrases entails that different genres of discourse enter into a conflict over *how* to link onto a phrase that happens. The *différend* is thus tied (arises in relation) to the contingency of linkage, not only because only one link can take place at a time (a *différend* between the phrase that actually takes place and the unactualised possibilities) but also because different genres of discourse provide their own set of (incommensurable) responses. The *différend* arises because there is no right way to link even though it is necessary to make a link (even silence is a link): "To link is necessary, but how to link is contingent" (Lyotard 1988, 102).

The appearance of a necessity to capitalism is thus tied to the genre of exchange which takes time as its object: the gap between two phrases is filled in by necessity in capitalism and the contingency of linkage is smothered. In the convoluted temporality of exchange, what comes after (the future) is not anticipated (as in the temporality of the modern project) but expected as a condition for the present. By making the necessity of linkage its internal logic, capitalism appears an inevitability, which explains its spectacular success: with the hegemony of capitalism over all other genres of discourse, we have internalised its logic and all life is reduced to the logic of exchange. Capital does not just capitalise, capital temporalises by capitalising time: money is not just time, time is money.

On the question of value at which we have just arrived, Lyotard rids Marx's answer in terms of social labour time from its Aristotelian metaphysics of energy that permeate the notion of production (Marx [1867] 1990, 125-177). On Lyotard's account "work is not an expenditure of energy, but an expenditure of time" (Lyotard 1988, 244). Production takes time, time which is stocked in the object. This time is not just labour time since the time spent after production with the object remaining in stock up until the time of exchange also raises costs. Production puts delays on exchange and this "lost time" must be "annulled" (recuperated) in exchange. Production involves the concatenation of phrases from the imagination and technical effectuation (transformation of raw materials into an object). Both of these are heterogeneous with the economic genre that subordinates them to the rule of exchange (Lyotard 1988, 245). The speeding up of time experienced by communities that come under capitalism occurs with the imposition of this rule for recuperating time over genres whose stakes are heterogeneous to it.

Money is not the equivalent of commodities but rather money is time. It is the time spent and "lost" during production, which is subtracted from the exchange; credit is stocked-up time and what is ceded by the lender is an advance of time while what is gained by the other party is borrowed time. In credit transactions the counter-cession (repayment) is deferred and the potential cycles of exchange blocked by credit (the time lost by the lender when say, he could have refused credit and put it back into circulation) are discharged in the form of interest (Lyotard 1988, 248). The enjoyment or ownership of objects is rather inessential to the economic genre—their destruction in a cycle of exchange is more pertinent. What David Harvey calls the "capital surplus absorption problem" in the context of the importance of the continuity of the flow of capital is relevant here: "the process cannot be interrupted without incurring losses" which leads to an acceleration of the speed of circulation:

> Those who can move faster through the various phases of capital circulation accrue higher profits than their competitors. Speed-up nearly always pays off in higher profits . . . Any interruption in the process threatens the loss or devaluation of the capital deployed. (Harvey 2010, 41)

Exchange is "a negotiation of time" (Lyotard 1988, 246). Money is the sign of "abstract" time that can be put into circulation (cycle of exchange) at any moment. Wealth is the possession of such "abstract" time: those in possession of wealth can afford to lengthen the interval between the two moments of exchange (though as Harvey shows this is not beneficial from a profit point of view). Those dispossessed of such abstract time have nothing but their "real" time to exchange and so have to speed up the interval between the two moments.

With what is called "wage earning" we have arrived at the *différends* at the heart of capitalism which the hegemony of the genre of exchange covers up and which Marx sought to expose: there is the *différend* that arises from subjecting work (production) to the stakes of the economic genre (gaining time). Here, what for the worker is a transformative activity is turned by the logic of capital into so much quantifiable time to be exchanged. But there is a radical asymmetry between the idiom of the worker that exchanges his "real" time and the idiom of the boss that buys this time with "abstract" time. There is an ineluctable *différend* between the time spent ("real" time) by the wage earner and its exchangeability for money (abstract time) in that "real" time cannot be recuperated (Lyotard 1988, 250; see also Readings 1992, 129-133 and Bennington 1988, 169-171). The time spent "earning a living" is radically incommensurable with the abstract time that is gained, i.e. money. If Marxism has not come to an end for Lyotard it is because:

> There is an insoluble differend between working and gaining time. The feelings (sadness, anger, hatred, alienation, frustration, humiliation) that accompany the said working conditions are born from this differend and signal it. (Lyotard 1988, 250-251)

Class struggle is being rewritten as the *différend* between the time that the boss gains and the time that the worker loses. The genre of exchange smothers the *différend*: you can't complain, you get paid for it!

Politics of Resistance and Temporal Alterity

In "A Memorial to Marxism" Lyotard wrote that "it is not enough to understand" the radical oppression born from the hegemony of capital over incommensurable genres of discourse "and be its philosopher; one must also destroy it" (2006, 251). How do we begin?

As a response to this demand just cited, I would like to conclude this paper by way of some experimental and unfinished thoughts.

. . . A crisis in capitalism occurs when there are significant blocks to the flow of capital. The current crisis is caused by such blocks. In terms of the temporality of capital, this translates to a disruption of the internal necessity of the logic of exchange: the future is unhinged from the present and as a result the acceleration of time characteristic of capitalistic societies slows down. There is perhaps nothing more threatening to capital than an interruption or suspension of its flow; in such moments, as Harvey puts it "we encounter a crisis of capitalism in which daily life can no longer go on in the style to which we have become accustomed" (2010, 6). These moments of crisis can be understood as *liminal* in the sense that they mark an in-between or indeterminate point of transition between the existing political and socio-economic configuration of the regime which is failing and being called into question and the one being prepared to replace it. Such moments precisely because of their indeterminate character are full of possibility and are opportunities for radical changes that range from socialist alternatives to extreme neoliberal ones.

. . . The "threat" of revolution emerges when the flow of capital is interrupted or suspended, and the necessity of the capitalistic logic of exchange is raised to the status of urgency. The rapid introduction of reforms within the system seeks to re-establish the continuity of the flow of capital as soon as possible and avert or manipulate a full-blown or existing crisis of legitimation that might lead to revolution.[2] As Harvey puts it, "the longer the uncertainty and the misery are prolonged, the more the legitimacy of the existing way of doing business will be questioned and the more the demand to build something different will escalate. Radical as opposed to band-aid reforms to patch up the financial system may seem more necessary" (Harvey 2010, 225). It is in this context that we understand the apocalyptic discourse of catastrophology and the Thatcherite "there is no alternative" propagated by politicians and conservative media during crises, which seeks to convince people by manipulating an economy of fear to accept reforms. In the context of the Greek financial crisis, this discourse is being used to implement an aggressive form of neoliberal capitalism and a thanatopolitics which simply renders a large portion of the population useless in relation to the reproduction of capital, and therefore redundant and left to die (see Foucault 2003 and Agamben 1998). As Naomi Klein has documented, crises pave the way for implementing economic shock therapy in the interests of neo-liberal capitalism (Klein 2007).

. . . A resistance to capitalism for Lyotard would have to "introduce," as Readings correctly points out, a "*temporal alterity*, an *unaccountable time*" (1991, 133). The word "introduce" could be misleading though: it carries the connotations of an intentional act of consciousness, a desire to manage a situation or an encounter by *leading* into it (from the latin *ducere*), such as when we make introductions in social settings. In contrast, the event *happens* as an encounter with something unexpected that disrupts pre-existing frameworks of reference and its radical singularity resists representation by the rules of those frameworks.

. . . The condition of possibility of the event lies in the void in between phrases, the gap that capitalism hijacks and fills in with the demand to "save time." A resistance to capital begins in disrupting this necessary linkage, and freeing the present from its hinges to the future. The suspension of the logic of exchange makes possible encounters between people that take on the character of an event. This should count as one of the accomplishments of the Occupy movement. From September 17 until November 15, 2011 when the NYPD violently evicted the Occupy Wall Street movement's occupation of Zuccotti Park, a space was being co-

---

[2] Although Lyotard is clearly suspicious of revolutionary politics – and as I hope to have indicated, not without good reason—it is possible, I think, to perform a reading of Lyotard 'against' Lyotard which considers the event of revolution precisely as an event, in the sense that what is made of revolutions, i.e. the meaning attached to them, comes 'after' the event. This reading is performed in a longer and revised version of this paper.

created right in the heart of capitalism, free from the logic of exchange where 3000 meals were served daily for free and a People's Library with about 4500 catalogued volumes operated on a no-checkout system (Sharlet 2011, 60). The convoluted temporality of capital was interrupted and people could come together *unencumbered* from the logic of exchange. By suspending the logic which reduces relations between human beings to relations of exchange, OWS was creating a space where the unexpected could happen: meetings between strangers were taking place and people were taking time to talk and listen to one another. What I found most remarkable was this: no introductions—if names were exchanged, it was usually at the end. These "moments" cannot be measured by any measure of time. They are the sorts of moments that Malick enacts so singularly in his films: a protracted present that interrupts and disarms consciousness and its drive to control and determine the events that come to it. It is in this context that we should understand Lyotard's remark that "in a world where success means gaining time, thinking has a single but irredeemable fault: it's a waste of time" (1992, 47).

. . . A re-writing of modernity that would escape the blind repetition of what it seeks to re-write would take the form of a perlaboration of what has remained hidden in past events, covered over by diachrony and suppressed by the temporality of capital. This is a work of anamnesis and not of remembering: "a procedure of analysis, anamnesis, anagogy and anamorphosis which elaborates an 'initial forgetting'" (Lyotard 1992, 92). The temporality of the event has the structure of trauma in the sense that what returns is some-*thing* in the event that was suppressed, excluded, and not integrated in a meaningful context. An unconscious timeline is set into relation with a conscious one where an unconscious event returns unexpectedly and erupts in the present in the form of a feeling that cannot be accounted for. In this sense, two events are put together in a relation that links by breaking apart: the temporality of *Nachträglickheit* is structured like a temporal loop where continuity and discontinuity are introduced together.

. . . An anamnesis of the event would consist in a ceaseless attempt at recuperation and reactivation of the event, in opening-up and maintaining a dimension of temporal alterity, irreducible to the capitalist reduction of time to an exchangeable commodity. Bearing witness to the *différend* would entail a multiplicity of links and the creation of new idioms that remain faithful to the indeterminacy of the event. One could see "signs" of this in the Occupy movement which multiplied across the world, and gave rise to local signifiers which responded to intensely local injustices: from "Occupy the Hood" to "Occupy the Buffer Zone" in Nicosia, Cyprus, where activists from both communities on the island (divided by a UN controlled buffer zone since 1974) came together to rearticulate the "Cyprus problem" as the result of antagonisms created by the global capitalist system that serves the interests of local and international forces in the region (Occupy Buffer Zone, 2012).

. . . The Occupy movement still has a long way to go in articulating both theoretically and in practice a politics of resistance to the hegemony of neoliberal capitalism that heralds a new form of totalitarianism which has unashamedly and explicitly rendered political institutions redundant as evidenced in the neologism "corporotocracies." It can only be hoped that the violent suppression by the police which the movement has come up against, both in New York and Nicosia, and which once again confirmed Marx's insight that "the executive of the modern state is but a committee for managing the common affairs of the whole bourgeoisie" (Marx and Engels [1850] 2010, 69), will act as a learning curve marking the movement's end of the beginning rather than the other way around.[3]

Works Cited

Agamben, Giorgio. 1998. *Homo Sacer. Sovereign Power and Bare Life.* Translated by Daniel Heller-Roazen. Stanford: Stanford University Press.

Bennington, Geoffrey. 1988. *Lyotard: Writing the Event.* Manchester: Manchester University Press.

Foucault, Michel. 2003. *"Society Must Be Defended." Lectures at the Collège De France, 1975-76.* Translated by David Macey. New York: Picador.

Harvey, David. 2010. *The Enigma of Capital and the Crisis of Capitalism.* Oxford: Oxford University Press.

Klein, Naomi. 2007. *The Shock Doctrine: The Rise of Disaster Capitalism.* New York: Metropolitan Books.

Lyotard, Jean-François. 1988. *The Differend: Phrases in Dispute.* Translated by Georges Van Den Abbeele. Manchester: Manchester University Press.

—. 1991. *The Inhuman: Reflections on Time.* Translated by Geoffrey Bennington and Rachel Bowlby. Cambridge: Polity Press.

[3] I would like to thank James Williams for his valuable feedback on an earlier version of this paper.

—. 1992. *The Postmodern Explained to Children: Correspondence 1982-1985*. Translated by Julian Pefanis and Morgan Thomas. London: Turnaround.

—. (1982) 2006. "A Memorial to Marxism." Translated by Cecile Lindsay. In *The Lyotard Reader and Guide*, edited by Keith Crome and James Williams, 231-253. Edinburgh: Edinburgh University Press.

Marx, Karl. (1867) 1990. *Capital. Volume 1*. Translated by Ben Fowkes. London: Penguin.

Marx, Karl and Engels, Frederick (1850) 2010. *Manifesto of the Communist Party*. Translated by Samuel Moore. In *Karl Marx, The Revolutions of 1848, Political Writings Volume 1,* 62-98. Edited by David Fernbach. London: Verso.

Occupy Buffer Zone. 2012. "Press Release Occupy Buffer Zone." Last modified April 13, 2012. Accessed Apr. 19, 2012. http://occupythebufferzone.wordpress.com/2012/04/13/press-release-occupy-buffer-zone-obz/.

Readings, Bill. 1992. *Introducing Lyotard: Art and Politics*. London: Routledge.

Sharlet, Jeff. 2011. "Welcome to the Occupation." *Rolling Stone*, Nov. 24, 58-64.

PART II

BECOMING AND FUTURE TIME

TOWARDS A FUTURE PREGNANT WITH BECOMING: OR DELEUZE, NIETZSCHE AND THE ETERNAL RETURN

ALLISON M. MERRICK

> And are not all things bound fast together in such a way that this moment draws after it all future things? Therefore—it draws itself too? For all things that can run must also run once again forward along this long lane. And this slow spider that creeps along in the moonlight, and this moonlight itself, and I and you at this gateway whispering together, whispering of eternal things—must we have not all have been here before?
> —Nietzsche, Thus Spoke Zarathustra

Introduction

As is widely assumed, "Nietzsche provides Deleuze with his chief philosophical resource for thinking creative time" (Hallward 2006, 149). More specifically, perhaps, it could be maintained Deleuze looks to Nietzsche for the intellectual apparatus of the eternal return in order tap the redemptive potential of the past. Or, as Alain Badiou suggests: "one can argue that most of Deleuze's work is devoted to defending, unfolding, and understanding ever more comprehensively the founding intuition of Nietzsche concerning the eternal return" (2000, 67). As such, the central puzzle, which will be the subject of this essay, is best expressed in the form of the following question: what, for Deleuze, is the philosophical import of Nietzsche's eternal return of the same? That is: "Does [the eternal return] stand for factual repetition, for the repetition of the past which should be willed as it was, or a . . . return-reactualization of that which was lost in the past of occurrence, of its virtual excess, of its redemptive potential?" (Žižek 2006, 193). I return to this question in the final section of this essay, whilst in an attempt to properly situate such a discussion I begin, in section one, by examining Nietzsche's thought of the eternal return, before, in section two, turning to look at Deleuze's interpretation of Nietzsche's greatest weight. My overarching ambition in this essay is to lay bare the manner in which this view of the historical may harbor the necessary resources for the possibility of revolutionary actions. In other words, if Deleuze is correct about Nietzsche's fundamental thought, then we may make a break with the teleology of the linear, secular, progressive conceptions of history and open up the space for a new way of conceiving of the past or, as Deleuze puts it, the "conditions of historical action itself" (Deleuze 1994, 91).

Nietzsche: the Greatest Weight

There are points at which Nietzsche makes much of the eternal return. For instance, he heralds himself "teacher of the eternal return" (Nietzsche 1990, 121), argues it represents *Thus Spoke Zarathustra's* "most fundamental conception" (Nietzsche 1989, 294), claims it "the highest formula of affirmation that is at all attainable" (Nietzsche 1989, 294), suggests it may be viewed as a kind of "confirmation and seal" (Nietzsche 1974, 274) and, in a posthumously published note, asserts it may "work as a breeding agent: strengthening the strong, paralyzing and destructive for world-weary" (Nietzsche 1968, 458). Moreover, and perhaps most notably, he treats the subject at length in *The Gay Science* amidst a section entitled "The Greatest Weight" in which presumably one is tested to determine if one can, as it were, bear such a weight:

> *The greatest weight.*—What, if some day or night a demon were to steal after you into your loneliest loneliness and say to you: 'This life as you now live it and have lived it, you will have to live once more and innumerable times more; and there will be nothing new in it, but every pain and every joy and every thought and sigh and everything unutterably small or great in your life will have to return to you, all in the same succession and sequence—even this spider and this moonlight between the trees, and even this moment and I myself. The eternal hourglass of existence is turned upside down again and again, and you with it, speck of dust!'
>
> Would you not throw yourself down and gnash your teeth and curse the demon who spoke thus? Or have you once experienced a tremendous moment when you would have answered him: 'You are a god and never have I heard anything more divine.' If this thought gained possession of you, it would change you as you are or perhaps crush you. The question in each and every thing, 'Do you desire this once more and innumerable times more?' would lie upon your actions as the greatest weight. Or how well disposed would you have to become to yourself and to life *to crave nothing more fervently* than this ultimate confirmation and seal? (Nietzsche 1974, 273-274)

Commentators have generally, and in using rather broad strokes, painted the eternal return as something of a cosmological theory, a psychological test, or a "decision-test" (Risse 2009, 226). Put briefly, interpretations of the first kind seek to take seriously Nietzsche's *Nachlass* claim, which suggests the eternal return is the "most scientific of all hypotheses" (Nietzsche 1968, 33). Claims of the second sort, by contrast, suggest the eternal return is best understood as, for want of a better expression, an "existential imperative" (Magnus 1978, 1). Accordingly, such an imperative is meant to test one's response to questions of the following sort: can you affirm your life, denuded of transcendent value, without presumably, and in rather dramatic fashion, beginning to gnash your teeth? Or, to put the point slightly differently and perhaps more positively, if one affirms one's existence, then one delights at the prospect of the eternal hourglass of existence being "turned upside down again and again" and, hence, labels the demon divine (Nietzsche 1974, 274). Lastly, claims of the third order take the eternal return to be a "decision-test" akin to Kant's Categorical Imperative (Risse 2009, 226). Milan Kundera neatly teases out the nuance of this final interpretive trajectory:

> If every second of our lives recurs an infinite number of times, we are nailed to eternity as Jesus Christ was nailed to the cross. It is a terrifying prospect. In the world of the eternal return the weight of unbearable responsibility lies heavy on every move we make. That is why Nietzsche called the idea of the eternal return the heaviest of burdens . . . (Kundera 1984, 5)

According to this third line of interpretation one can test one's actions against the rule "act always in accordance with that maxim you can at the same time will as eternally returning" and hence evaluate your actions in terms of your conformance with such a rule (Owen 1994, 73).

Irrespective of the interpretive trajectory one initially chooses to tread, it seems, questions concerning the salience and ultimate potency of Nietzsche's thought nevertheless abound: does, as Nietzsche seems to suggest in *The Gay Science*, the cycle repeat everything including the smallest of details? If so, does this entail epistemic access to previous cycles? The worry here is that the burden of unbearable responsibility seems to be found in the notion that one knows one will have to endure their loneliest loneliness innumerable times. Yet, if one has no knowledge of the previous cycles (given that if two cycles are to be identical, a latter cycle cannot contain any knowledge of previous one) then it follows that the threat of identically recurring cycles becomes as Soll claims "a matter of complete indifference (Soll, 1973, 339). Aaron Ridley puts the point in this way:

> [I]f the demon is coming to me *now*, he has presumably also been to me at precisely the same point of my life innumerable times before—and what difference has *that* made? If I passed the test I passed, if I failed I failed, and I'll go on doing whichever I did infinitely many more times, without changing a thing. The thought of the eternal recurrence, then, should be a matter of the deepest indifference. Why *care*? (1997, 20)

As Ridley makes clear, the worry is simply that with no recollection of a previous cycle the thought of the eternal return seems to lose its bite.

In an attempt to bypass concerns of this order Deleuze offers a rather nuanced reading of Nietzsche's thought of the eternal return. Deleuze claims, for instance: "We misinterpret the expression 'eternal return' if we understand it as 'return of the same'" (2005, 45). That is, Deleuze's Nietzsche can be read *prima facie* as denying that every specific event must return in its same order and sequence. Rather, and put positively, Deleuze suggests: "*It is not the same which returns, it is not the similar which returns;* rather, the Same is the returning of that which returns, *—in other words, of the Different*; the similar is the returning of that which returns, *—in other words, of the Dissimilar*" (1994, 93). In the following section, I shall reconstruct Deleuze's tripartite typology of historical cycles in the hopes of, in the first instance, unpacking this seemingly cryptic remark and, in the second, of making sense of Deleuze's suggestion that the thought of the eternal return should serve as a principle of selection: to lay bare, in other words, "Nietzsche's secret," namely "*the eternal return is selective*" (2001, 88). Or, in other words, the eternal return may create the conceptual space to engender revolutionary actions by selecting, for instance, aspects of the past worthy of emulation.

Deleuze on That Which Returns: The Intracycle, the Cycle, and the Eternal Return

Deleuze, in *Difference and Repetition*, suggests "Nietzsche is profoundly correct to oppose 'his' conception to every cyclical conception" (Deleuze 1994, 93) and this is particularly significant because, as R.G. Collingwood has noted, many such approaches are on offer:

> Since Plato announced that the course of history returned upon itself in 72,000 years, since Polybius discerned a 'circular movement' by which the history of states came back, over and over again, to the same point, the

> theory of historical cycles has been a commonplace of European thought. (1965, 56)

Deleuze highlights three such cyclical conceptions of history: the intracycle, which he teases out from the work of the twelfth century scholar Joachim of Flora; the cycle, which he notes in the work of Giambattista Vico; and the eternal return, found in the work of Nietzsche; in an attempt to make clear the manner in which the final third mode is both "most important and mysterious" (Deleuze 1994, 93).

Deleuze describes the first mode of repetition, the intracycle, in the following way:

> An intercycle repetition, which involves the manner in which the first two ages repeat one another—or rather, repeat one and the same 'thing,' an act or event yet to come. This above all is the thesis of Joachim, who establishes a table of concordances between the Old Testament and the New; but it is a thesis which cannot go beyond simple analogies of reflection. (1994, 93)

That is, in the first instance this mode of historical repetition is differentiated from the other cyclical models of history insofar as Joachim's interpretation "establishes a table of concordances" and, accordingly, trades in analogies of reflection. Morton Bloomfield teases out the nuance of Joachim's conception of historical repetition, so I quote him at some length:

> The general historical theory of Joachim is well known. Arguing on the basis of analogies between the Old Testament and the period of the New Testament down to his time. . . [he] assumed a coming new period in the future. The 'trinity' of ages would correspond to the divine Trinity . . . Concordance is the key-word for Joachim's method. The period of the Old Testament was primarily the age of the Father, the period from the time of Jesus down to roughly Joachim's time was that of the Son, and the third age, which is a naturally completing period, would be that of the Holy Ghost . . . each subsequent age follows the pattern of the first age . . . the Old Testament, then, is the key to the meaning of history. (Bloomfield 1957, 264)

Joachim's notion of historical repetition moves by noting the concordances and proceeds analogously to tease out that the "coming times of the Holy Spirit are successively prefigure[d] in the first and second epochs of the Father and Son, which are strictly concordant, for each figure and event of the Old Testament, if understood spiritually, is a promise and signification of a corresponding figure and event of the New Testament" (Löwith 1949,

149). Hence, Joachim's intercylic mode of repetition does not venture past the simple analogies of reflection because each period both repeats and prefigures one and the same thing.

Deleuze frames his discussion of the second cyclical conception of history as follows: "A cycle repetition in which it is supposed that, at the end of the third age and at the end of a process of dissolution, everything recommences with the first age: here, the analogies are drawn between two cycles (Vico)" (Deleuze 1994, 93). Two points are worth making: the cycle, according to Vico, culminates in a process of dissolution, and, as such, the cycle recommences with a return to the first age. In other words, according to Vico, history transverses through cycles, or ages, which are distinguished from one another by certain generally recurring features, and this process is, as Collingwood notes, "not a circle but a spiral; for history never repeats itself but comes round to each new phase in a form differentiated by what has gone before" (Collingwood 1994, 68). Second, the distinguishing feature of Vico's version, which serves to demarcate this mode of cyclical history from the intercycle, is that this mode trades in "analogies drawn between cycles" whereas the former mode deals solely in "analogies of reflection" (Deleuze 1994, 93).

In considering the first point, it may be useful to make a start by noting Vico's account of the three ages of history:

> These are: (1) The age of the gods, in which the gentiles believed they lived under divine governments, and everything was commanded them by auspices and oracles, which are the oldest institutions in profane history. (2) The age of the heroes, in which they reigned everywhere in aristocratic commonwealths, on account of a certain superiority of nature which they held themselves to have over the plebs. (3) The age of men, in which all men recognize themselves as equal in nature, and therefore, there were established first the popular commonwealths and then monarchies, both of which are forms of human government. (Vico 1968, 20)

A cycle, which consists of the three ages, ends in a process of dissolution, a period labeled the barbarism of reflection. Vico's description of the latter mode of barbarism is certainly worth quoting at length:

> [N]o matter how great the throng and press of their bodies, they live like wild beasts in a deep solitude of spirit and will, scarcely any two beings able to agree since each follows his own pleasure or caprice. By reason of all this, providence decrees that, through obstinate factions and desperate civil wars, they shall turn their cities into forests and their forests into dens and lairs of men. In this way, through long centuries of barbarism, rust will consume the misbegotten subtleties of malicious wits that have turned

> them into beasts made more inhuman by the barbarism of reflection than the first men had been made by the barbarism of sense. For the latter displayed a generous savagery, against which one could defend oneself or take flight and be on one's guard; but the former, with a base savagery, under soft words and embraces, plots against the life and fortune of friends and intimates. Hence peoples who have reached this point of premeditated malice, when they receive this last remedy of providence and are thereby stunned and brutalized, are sensible no longer of comforts, delicacies, pleasures, and pomp, but only of the sheer necessities of life. And the few survivors in the midst of an abundance of the things necessary for life naturally become sociable and, returning to the primitive simplicity of the first world of peoples, are again religious, truthful, and faithful. (Vico 1968, 424)

In an attempt to leave to one side hyperbole, it seems the cure for maladies of the mode of barbarism of reflection is the *ricorso*, the reoccurrence of the simplicity of primitive barbarism—the reoccurrence, in other words, of the religious, the truthful and the faithful. Or, to put the point another way, Karl Löwith for instance notes: "The real end of it is decadence and fall, after which the whole course (*corso*) begins anew from a new barbarism in a recurrence (*ricorso*) which is, at the same time, a resurgence" (Löwith 1949, 132). As such, this mode of repetition is demarcated from the other modes of historical repetition insofar as the end of a cycle of history recommences with the first age.

In turning to the second point, to determine how the cyclical mode may deal in analogies between cycles, we may note for Vico, at least in *The New Science*, one can argue analogically from one occurrence to the recurrence insofar as two different periods may have the same general character. So of heroic barbarism, for instance, we may note the occurrence of "a government by a warrior-aristocracy, an agricultural economy, a ballad-literature, a morality based on the idea of personal prowess and loyalty and so forth" (Collingwood 1994, 67). Such recurrence, *ricorso*, is not the exact repetition of the former cycle. Rather, one may, and in following Collingwood, note: "the Christian barbarism of the Middle Ages is differentiated from the pagan barbarism of the Homeric age by everything that makes it distinctly an expression of the Christian mind" (Collingwood 1994, 68). Hence, the two periods share a set of particular features, a general character, which is teased out by the historian. It is for this reason Deleuze notes the cyclical mode trades in analogies drawn by the historian between cycles.

The third conception of history, the eternal return, is according to Deleuze, "precisely what is needed both to correct the intracyclical hypothesis and to contradict the cyclical hypothesis" and, as such, this mode trades neither in an analogy of reflection nor in the attempt to trace an analogy between cycles (Deleuze 1994, 93). Rather, according to Deleuze:

> the repetition in the first two moments no longer expresses analogies of reflection, but the conditions under which eternal return is effectively produced by means of some action or other; on the other hand, these first two moments do not return, being on the contrary eliminated by the reproduction of the eternal return in the third. (Deleuze 1994, 93)

Here the third mode of historical repetition is distinguished insofar as within the mode of the eternal return the analogous does not reoccur. Hence, Deleuze's contention is that it is precisely the Different or the Dissimilar, which cannot take the form of the analogous, which returns.

If the foregoing is an accurate reconstruction of Deleuze's thought on that which returns, a word exploring the cogency of this position is certainly in order. David Couzens Hoy has recently put forward a set of concerns questioning the persuasiveness of Deleuze's proposal:

> Independent of the issue of whether what repeats is the same or different, there are two problems with any interpretation, including Deleuze's, that holds that what recurs is the repetition of *patterns* rather than *specific events*. First, patterns are not as regrettable as specific instances. One might very well not want to repeat a particular action over and over again, even while granting that the pattern is very likely to recur. . . . Second, the doctrine is potentially vacuous if all it means is that the general patterns recur. All events are alike in some basic respects, and if these basic respects are what recur, then any event could substitute for any other event. (Hoy 2009, 84)

Leaving aside the vexed issue of determining the scope of that which repeats, it does seem, as Hoy argues, that patterns are of a different order than events. That is, under this schema, events are subsets of patterns. Granting this formula, the subsequent claim, "one might very well not want to repeat a particular action over and over again, even while granting that the pattern is very likely to recur . . ." nevertheless seems, at best, rather tenuous (Hoy 2009, 84). For instance, one might suggest the archetypal disposition of despair or dread of life is at least as pernicious as a particular event, for example, one's particular love lost and one's own broken heart. If this is right, it does not seem clear that trading in patterns, general despair or dread, is decisively better than peddling specific events, heartbreak. That is, *contra* Hoy patterns may be at least as regrettable or unfortunate as specific instances.

Given that "all events are alike in some basic respect," Hoy's second worry is that Deleuze's reading of the thought of the eternal return is potentially vacuous if all it means is the general patterns recur (Hoy 2009, 84). This claim, namely that all events are similar at some initial or rudimentary level, is certainly trivially true, as in, for instance, two commands, which, even if they pull in opposite directions, share the common feature of being commands. Yet, it seems Deleuze in invoking Nietzsche's thought of the eternal return is after something more than claims of this order, namely the trivially true and the potentially vacuous. And, an attempt to make clear how Deleuze's rendering of Nietzsche's fundamental insight may circumvent these sorts of worries will be the subject of the final section.

Conclusion: Towards "the Condition of Historical Action Itself"

At the start of this essay, I posed the following interpretive query: "Does [the eternal return] stand for factual repetition, for the repetition of the past which should be willed as it was, or a . . . return-reactualization of that which was lost in the past of occurrence, of its virtual excess, of its redemptive potential?" (Žižek 2006, 193). I have attempted to argue that the notion of the eternal return in the hands of Deleuze is certainly not the repetition of the past just as it was. Rather, it is the Dissimilar or the Different that Deleuze contends returns. Further and in an attempt to elide the charge that this interpretation renders Nietzsche's thought either vacuous or trivially true, we may do well to recall another of Deleuze's claims:

> Marx's theory of historical repetition, as it appears notably in *The Eighteenth Brumaire of Louis Bonaparte,* turns on the following principle which does not seem to have been sufficiently understood by historians: historical repetition is neither a matter of analogy nor a concept produced by the reflection of historians, but above all the condition of historical action itself (Deleuze 1994, 91).

If the central tenets of my argument have been compelling we can note why historical repetition in the hands of Deleuze is not solely a matter of analogy. Nor, Deleuze claims, is historical repetition properly understood as "a concept produced by the reflection of historians." Deleuze puts the worry in this way:

> the question is asked whether the second displays sufficient resemblance with the first to be identified as the Same again: a question which can be answered only by the establishments of relations of analogy within judgment, taking into account the variations in empirical circumstances (is Luther the analogue of Paul, the French Revolution the analogue of the Roman republic?) (Deleuze 1994, 294-295)

Here the worry is that cyclical repetition becomes a mold pressed upon the past by the historian who notes likenesses between phenomena and forms a particular judgment: that, for instance, the French Revolution is the analogue of the Roman republic. Rather, historical repetition properly understood engenders the "condition of historical action itself" and aims to tap the redemptive potential of the past, as in, for instance, the manner Marx lays out in *The Eighteenth Brumaire of Louis Bonaparte*:

> The tradition of all the dead generations weighs like a nightmare on the brain of the living. And just when they seem engaged in revolutionizing themselves and things, in creating something that has never yet existed, precisely in such periods of revolutionary crisis they anxiously conjure up the spirits of the past to their service and borrow from them names, battle cries and costumes in order to present the new scene of world history in this time-honoured disguise and this borrowed language. (Marx 1963, 15)

We misunderstand the eternal return if we take it to be akin to the cyclic mode, as evidenced in the work of Vico, by presenting a *post facto* interpretation of historical phenomena by noting, for instance, the reoccurrence of battle cries and costumes. We misunderstand the eternal return if we take the borrowed names and language as strictly concordant with the first instantiation, as in the case of Joachim. Rather, in the work of Deleuze, the eternal return captures the virtual excess, the trapping of that which previously went without instantiation, and creates conditions for revolutionary actions. In other words, it is, when properly understood, an "expression of overflowing energy that is pregnant with a future" (Nietzsche 1974, 329).

Works Cited

Badiou, Alain. 2000. *Deleuze: the Clamor of Being*. Translated by Louise Burchill. Minneapolis: University of Minnesota Press.

Collingwood, R.G. 1965. "Oswald Spengler and the Theory of Historical Cycles." In *Essays in the Philosophy of History,* edited by William Debbins, 57-75. Austin: University of Texas Press.

—. 1994. *The Idea of History*. Oxford: Oxford University Press.

Daniel, E. Randolph. 1980. "The Double Procession of the Holy Spirit Joachim of Fiore's Understanding of History," in *Speculum* 55: 469-483.

Deleuze, Gilles. 1994. *Difference and Repetition*. Translated by Paul Patton. New York: Columbia University Press.

—. 2001. *Pure Immanence: Essays on a Life*. Translated by Anne Boyman. New York: Zone Books.

—. 2005. *Nietzsche and Philosophy*. Translated by Hugh Tomlinson. London: Continuum.

Hallward, Peter. 2006. *Out of this World: Deleuze and the Philosophy of Creation*. London: Verso.

Kundera, Milan. 1984. *The Unbearable Lightness of Being.* Translated by Michael Henry Heim. New York: HarperCollins Publishers.

Löwith, Karl. 1949. *Meaning in History*. Chicago: University of Chicago Press.

Magnus, Bernard. 1978. *Nietzsche's Existential Imperative*. Indianapolis: Indiana University Press.

Marx, Karl. 1963. *The Eighteenth Brumaire of Louis Bonaparte*. Translated by C. P. Dutt. New York: International Publishers.

Nietzsche, Friedrich. 1968. *The Will to Power*. Translated by Walter Kaufmann. New York: Vintage Books.

—. 1968. *Twilight of the Idols.* Translated by R.J. Hollingdale. Harmondsworth: Penguin.

—. 1974. *The Gay Science*. Translated by Walter Kaufmann. New York: Vintage Books.

—. 1978. *Thus Spoke Zarathustra.* Translated by Walter Kaufmann. New York: Penguin Books.

—. 1989. *On the Genealogy of Morals and Ecce Homo.* Translated by Walter Kaufmann. New York: Vintage Books.

Owen, David. 1994. *Maturity and Modernity: Nietzsche, Weber, Foucault and the Ambivalence of Reason*. New York: Routledge.

Ridley, Aaron. 1997. "Nietzsche's Greatest Weight." *Journal of Nietzsche Studies* 14: 19-25.

Risse, Mathias. 2009. "The Eternal Recurrence: A Freudian Look at What Nietzsche Took to Be His Greatest Insight." In *Nietzsche on Freedom and Autonomy*, edited by Ken Gemes and Simon May, 223-246. Oxford: Oxford University Press.

Soll, Ivan. 1973. "Reflections on Recurrence," in *Nietzsche: A collection of Critical Essays*, edited by Robert Solomon, 322-342. New York: Doubleday.

Vico, Giambattista. 1968. *The New Science of Giamattista Vico*. Translated by Thomas Goddard Bergin and Max Harold Fisch. Ithaca: Cornell University Press.

Žižek, Slavoj. 2006. *The Parallel View*. Cambridge: MIT Press.

Women and Becoming-Woman: Deleuze and Feminism

Anupa Batra

Deleuze and Guattari consider themselves part of the conversation in feminist theory, though not all feminist scholars would agree.[1] Deleuze offers a theory to account for the unequal relation of men and women and what would be required to overcome this inequality. The notion of becoming-woman is key, and it has been the source of much of the controversy. Deleuze is, in general, a philosopher of becoming, and it is from this direction that he takes up the issue of gender inequality. More broadly, Western thought has been structured around the focal point of the white rational man. Deleuze argues that this is repressive at a number of levels: at the socio-political, the individual, and the sub-individual. It is an ethico-political motivation that leads him to argue for a movement beyond this structure. However, he claims that this movement must begin by the exit out of gender. Deleuze's reason is because of the central role of sexual difference in the constitution of Western thought. The issue of the inequality of women, for Deleuze, is one aspect of a more general way in which our thinking and our modes of life are oppressive. Thus, the inequality of men and women can only be overcome by a more fundamental change in the way our existence is structured. Becoming-woman, then, must be viewed as an essential step toward a transformation of who we are and, thus, as revolutionary.

---

[1] I rely primarily on *A Thousand Plateaus: Capitalism and Schizophrenia*, which is co-authored by Deleuze and Guattari (1987). However, I only use Deleuze's name, and this for several reasons: First, my argument is related in an essential way to Deleuze's theory of becoming; second is for simplicity's sake; and third is that this has come to be common practice in the scholarship.

Becoming, Event, and Time

All becoming, in Deleuze's sense, is virtual rather than actual. Becoming comes to be congealed in the actual, but it always goes beyond the actual. To understand the notion of becoming, it is important to grasp its temporal structure. Deleuze has always been concerned with how change occurs, including historical change, and his notion of becoming is central to his theory. The other, related notion I take up here is the event. Deleuze's claim is that historical change cannot be explained by reference to simple linear time (referred to as *Chronos*), that is to say, as the sequence of actual events. The event is the object of philosophy in that we conceive it only through an extraction from the actual. Indeed, the virtual event occurs in excess of actual, historical events. The event has a non-linear temporality (*Aeon*) and it provides him with a way to theorize the way in which change occurs. We cannot, of course, provide a historical date at which virtual events occurred, as they elude the present. In terms of linear time, the event always seems to be before or after the present. Instead, the ways in which these virtual events become embodied in actual bodies and states of affairs are through the historical, actual events that occur at specific dates and locations. To be clear, we must distinguish between the virtual event, which is always incorporeal, and the actual event, which occurs as a particular determination of the virtual event. The "instant" of the virtual is always available and thus "alongside" the linear time of the actual. In addition, we distinguish between bodies and states of affairs which are actual, and becoming, which is virtual. Becomings always correspond or belong to an event: Becoming is always movement out of bodies and states of affairs. In addition, becomings are always too fast or too slow for perception. They always, in principle, exceed the categories of recognition.

The identity of subjectivity as sexed must be viewed as a virtual event. It articulates for us a transition between one historical structure to another. This framework of understanding the subject as sexed obviously cannot be localized in linear time. Thus, we cannot point to a historical date at which subjectivity became indexed on sexual difference. For Deleuze, all events are made up of other events and intersect with others besides those. The sexualization of subjectivity is no different in that regard. For each actual girl, we must always speak of a time before a history and gender norms have been imposed on her and a time after, when she is a woman. The event of gendered subjectivity has already happened and continues to occur for women and will happen again for every actual girl.

Let me turn to the examination of becoming-woman and its relation to becoming-molecular. Becoming is always molecular for Deleuze. The molar designates identifiable forms, like persons or things, while the molecular are the particles that escape the sedimentation of these structures and escape any representation by the subject. For this reason, they can rupture and open up the identity of the molar. However, molecular becomings can become codified and are thus the subject matter of the molar. Molar traits are simply the codified or segmented lines of desire or of becomings. The notions of the molar and molecular provide another way to understand how identifiable forms are constituted through the more fundamental processes of becoming, that is to say, how identity is constituted in and through difference.

Becoming-woman is thus set apart from other becomings in that it entails those molecular becomings that escape the dualistic economy of gender. What I examine here is the way in which becoming-woman can overcome the gendered structure of thought and subjectivity. To return to the point made by Deleuze, that becoming-woman is the opening onto all becoming in Western thought: "Although all becomings are already molecular, including becoming-woman, it must be said that all becomings begin with and pass through becoming-woman. It is the key to all becomings" (Deleuze and Guattari 1987, 277). Rosi Braidotti has engaged in a significant way with Deleuze's work, and one important criticism she makes is that he implies that the exit from personal identity will be the same for everyone such that becoming is sexually undifferentiated (Braidotti 2002, 82). Braidotti argues, and rightly so, that the relationship between the masculine and feminine is not a symmetrical one and, consequently, we need to speak of gender-specific becomings.[2] Her notion of the "politics of location" is a way of taking up the concern with differentiated becomings. Thus, the exit from identity could only begin from the identity one has: man, woman, black, white, Latina, gay, etc. In addition, Braidotti's view is that there is no possibility of the erasure of sexual difference—indeed, that is not even desirable—only the displacement of it such that it is less repressive.

My analysis combines the view of Braidotti with that of Deleuze. On the one hand, with Braidotti, becoming could only be gender-specific *initially*. On the other hand, it is possible to go beyond gendered subjectivity and, indeed, desirable. In fact, if a complete exit from the gendered structuring of subjectivity and thinking does not occur, this

[2] At least at one point Deleuze suggests there are two different but complementary movements, one from the majority and one from the minority group, but the significance of this is unclear (Deleuze and Guattari 1987, 291).

structure will continue to be repressive, but in new ways. The elimination of the gender structure would truly be revolutionary, as it would mean a transformation in Western thought and being at the deepest level.

In this essay, I examine a paradox that seems to occur when considering Deleuze's notion of becoming-woman. On the one hand, men (in the sense of molar persons) are in the privileged position in Western society and, in fact, almost every society right now. On the other hand, there is a way in which women are in the privileged position in terms of change and transformation. In the end, the overcoming of gendered subjectivity would necessarily be a transformation of all persons, since the subjectivity of all is constituted as gendered. However, I will argue that women form the initiation point. To be clear, the individual molar subjects known as women are in the privileged position for becoming-woman. There is a danger of romanticizing the suffering of a minority group, which I seek to avoid. Deleuze's theory forms the basis for my claim. While he does not elaborate on the point, he does suggest it as a possibility: "There is a becoming-woman, a becoming-child, that do not resemble the woman or the child as clearly distinct molar entities (although it is possible—only possible—for the woman or child to occupy privileged positions in relation to these becomings)" (Deleuze and Guattari 1987, 275).

The Center vs. the Periphery

We must first understand Deleuze's account for the gendering of subjectivity. We can ask the following question: Who is considered the subject? The answer is, of course, the rational, heterosexual white man. This is the center around which everything is structured. All others are at the periphery and are constituted *as others* in relation to the center. (While the racial aspect of subjectivity is significant, my concern in this essay is with gender). According to Deleuze, all activity occurs at the periphery, while the center remains still. Thus, the temporality of the center is the dead present. The activity occurring at the periphery is molecular, as it escapes the coding of molar forms and is beyond any system of representation. The division of gender, masculine and feminine, is not symmetrical. If it were, then there would be two poles, not a center. "The central point, or third eye, thus has the property of organizing binary distributions within the dualism machines" (Deleuze and Guattari 1987, 292). The feminine is constructed as the non-masculine. This does not imply absence, since the feminine becomes reterritorialized (that is to say, indexed in a rigid way) onto other traits. For example, submissiveness, passivity, and weakness (or delicacy) have been constituted as feminine traits.

Genuine change for Deleuze does not occur at the molar level, only at the molecular. Louise Burchill argues that parody or subversive repetition of molar feminine traits will not lead to transformation (Burchill 2010, 90-91). We must recall that for Deleuze gender is not simply socially constructed. Their claim is much stronger: Gender is a codification of the body or of desire. "The question is not, or not only, that of the organism, history, and subject of enunciation that oppose masculine to feminine in the great dualism machines. The question is fundamentally that of the body—the body they steal from us in order to fabricate opposable organisms" (Deleuze and Guattari 1987, 276). If gender were simply a social construction, change would be possible at the molar level, and the only difference between the genders would be in their molar traits, like personality traits. The most important difference between repeating feminine traits ironically or in parody rather than with sincerity is that of intention. For example, a comment can be ironic only insofar as the audience grasps the intention of the person making the comment. A comment made in sincerity appears to be the same as a comment made ironically; that is to say, the two comments are the same on the surface. In this way, the two occur as a repetition that is the same. In contrast, intention does not occur at the surface but, instead, forms a kind of depth to the comment. An audience can confuse the ironic with the sincere comment because the difference is not a genuine one. For Deleuze, real change must be a change of surface, or a change in the movement of desire (that is to say, a change in the body, which is the congealing of desire). Depth has to do with the molar subject, and a change at this level is not a real change. Burchill's point is that parodic repetition does, indeed, extract a difference, but one that is indexed on the past and thus cannot truly be new.

Therefore, change cannot result from voluntaristic choice, which can only come from a molar subject. It occurs only when it is forced to occur; hence Deleuze's emphasis on experimentation. My claim is that the exit from the gender structure that is initiated in women will occur not by anything they consciously choose to do, but only because of what they are forced to experience.

Experimentation

Experimenting for Deleuze always occurs through the interaction of two or more elements, never by an isolated one. The notion of deterritorialization is important here. The deterritorializing of an organ, form, or a function occurs when it becomes freed of its previous framework or rigidified path to connect and interact with other elements in new ways. Since for Deleuze there is no essence apart from what something does, the deterritorialized organ becomes something genuinely new. Deterritorialization never occurs for an organ or form in isolation, but in the encounter with another element. The danger, of course, is that whatever has been deterritorialized can be reterritorialized onto new molar identities, and then become commodified.

In experimenting, one does not strive for a particular result. The outcome remains unknown until it actually occurs, including whether the deterritorialization of an organ will occur and lead to something new. Only in linear time is the future a variant of the past: It occurs as repetition of the same. The temporality for teleology and, hence, all planned activity is linear. In terms of linear time, transformation or change always appears as accident or chance. All becoming is precisely the undermining of a plan. Those particles that are molecular escape any plan. This notion of becoming, which can only come from experimentation, depends on the notion of temporality proposed by Deleuze. Transformation into a particular state of affairs cannot be achieved as a goal. This means, too, that the liberation of women at the molar level cannot be the goal of this experimentation, if the goal is for women to retain their identity as women. Deleuze does say that a molar politics is necessary, meaning that it is important to fight for women's rights, for example. However, the achievement of these goals will not truly be revolutionary because it will not change the fundamental structure of gender. It is a kind of concession, not a real transformation, and it is even possible that such a concession could further entrench the gender structure.

However, I do not think the disarticulation of gendered subjectivity can be viewed as a goal in the sense of the achievement of a particular state of affairs. Instead it is an issue of loosening and rupturing molar boundaries, of freeing the movements of desire from certain pathways (based on whether one is feminine or masculine), with no expectation of the paths that desire would then take. The question I believe we *cannot* ask is this: *What* will woman (and man) become, or what new form of subjectivity will come about? The two points of a becoming are not significant, according to Deleuze, only the movement between them. It is the *outcome* that we cannot determine in advance, as it is simply another dead present. Certainly new molar forms will come about, as becoming always becomes congealed into molar forms. What is problematic is when that molar form is so repressive that it makes lines of flight out of it very difficult, when movements of desire along any other pathways become impossible. Becoming-woman, which both men and women must ultimately undergo, though asymmetrically, will not leave intact either the masculine or feminine.

Most importantly: Change (understood as the deterritorialization of organs, forms, and functions and the subsequent new connections that are formed) is forced to occur to a greater degree or with greater intensity in women than in men for two related reasons: First, women's identity is more fluid because their subjectivity is not constructed as the center pole. Second, again for the reason that women's subjectivity is not constructed as the center pole, their subjectivity gives them less pleasure. Our social and political structures in the West are built on male subjectivity, so men, in principle, have a greater chance for success and pleasure. (Pleasure should be understood broadly, such that it includes satisfaction and fulfillment). For these reasons, women are in the privileged position for the transformation out of gendered subjectivity. However, there is correspondingly a greater danger for women than men. More about this shortly.

The evolution beyond gender for humans is no different from other cases of evolution. So, for example, the Deleuzian account of the colonization of land by the first lobe-finned fish would state that it was due to the deterritorialization of the fins from their function in water when the fish encountered land. A new kind of relation was produced between the two elements, fins and land. The fins became feet: a genuine transformation. In addition, new possibilities for habitats opened up. It is not simply a space that was pre-marked and was not previously used, but a real creation of space-time because the marking out of the space and time could only occur in and through the movements of the fish, the first element, in relation to the land, the second element. Evolution is not a simply a linear change, or a change that can be measured in reference to some previous state. It is a radical break from the past.

Deleuze argues that even a small detail can force becoming to occur: "We can be thrown into a becoming by anything at all, by the most unexpected, most insignificant of things" (Deleuze and Guattari 1987, 292). The important feature here is the unexpected. However, it seems to me a subject would be more likely to undergo a becoming when the experience is not merely unexpected, but is uncomfortable or painful, thus providing greater impetus. Women are in the better position for experimentation precisely because they receive less pleasure from their identity as it has been defined in otherness to man. In contrast, since the greatest chance of success and pleasure for men is in accordance with the possibilities given by gendered subjectivity, there is nothing to impel the movement in another direction. Women are forced to seek alternative paths for pleasure and, therefore, forced into new relations. More precisely, their bodies are forced to become deterritorialized and to enter into new relations. That social institutions are structured according to the right and power of man means that individual men achieve greater success and receive greater fulfillment within these institutions. For women, the options are these: One, the adoption of the traits that are constituted as feminine in and through the relation to the standard, man. The subjectivity constituted in this way offers a certain amount of pleasure or satisfaction in that it reinforces the dualistic structure. In other words, women who have these traits to a high degree are viewed as being "real" women. Two, women can also be "male." This is in the sense of having traits that have been indexed onto the masculine. However, since these traits are coded as masculine, an individual woman with these traits will either fall short in comparison with a man or the expression of the trait will appear as a kind of perversion of masculinity.

Third is the path that will lead to the transformation of gendered subjectivity. This occurs when the first and second alternatives fail. As a result, new kinds of relations with other people, institutions, and events (events in the ordinary sense) *may* be formed, that is to say, relations that do not reproduce the gendered structure. However, there is no guarantee deterritorialization will lead to new connections and to transformation. If a new relation occurs, it will open up and constitute a new sphere which did not exist before, and which would bring with it new possibilities and entirely new kinds of relations.

Let us consider a specific example. We have said that problems or events intersect with one another. In Western culture, with its history of individualism, there is a corresponding problem of the isolation of individuals. Community does not have the same function as it does in other cultures. It would take another essay to fully develop and justify these claims. However, my interest is in the way this event intersects with that of gender. One "solution" to the problem of the isolation of the individual is the nuclear family.[3] However, this serves the needs of men better than women and, so, is more fulfilling to them. When career is a driving factor as well, individuals must be willing to relocate in order to advance their careers. Since women, more than men, index their own success on the success of their (male) partners, women are more likely to uproot their own lives and move with their partners. In contrast, men are less likely to do the same for a female partner. The result is that men who are successful in their careers are more likely to have a family and, thus, to not be isolated, while women in a comparable position are more likely to be alone and isolated. This can be the impetus for experimentation. However, there are three possibilities. (To be clear, these possibilities arise only if there is a problem of isolation). One, women can simply take the secondary role in the relationship with the man and avoid pursuing a career that might demand relocation. The problem of isolation is avoided altogether in this case. Two, a woman can pursue a demanding career and, assuming it requires relocation, she can accept her isolation. These two possibilities center on the traditional heterosexual relationship as the primary relationship. The third possibility is the creative moment. It resembles the second possibility, but the established path of the heterosexual relationship is replaced with other social encounters.[4] These need not be sexual or with men. Although it cannot be determined in advance what connections will occur and what roles each individual would play in these connections, it would mean that these women's interpersonal relationships would not center on the nuclear family. The transformative or revolutionary moment occurs as the deterritorialization of desire from the way it flows in the nuclear family so that new connections are established.

My second point is this: It seems to me that the feminine is less rigidly codified, precisely because it is not the center. Through experimentation, the freeing of desire is more likely for molar women. In this way, molar women are more able to enter into the time that is "not yet" or, in other words, be more future-oriented. A quote from Paul Patton, regarding the nature of lines of deterritorialization, is helpful: "[The lines of flight] are the imperceptible cracks in a person's make-up" (Patton 2009, 47). The

---

[3] Solution here does not mean the correct resolution to a problem. A solution, in Deleuze's sense, is the way in which an Event becomes embodied in a state of affairs. Thus, there are many solutions, each of which "resolves" the problem of the Event in a different way.

[4] I do not want to suggest that anyone can choose a homosexual relationship over a heterosexual one. My point is that the heterosexual relationship is part of the structure of gendered subjectivity.

cracks are the imperceptible movements, the movements that occur outside of the rigid pathways that have become fixed. It seems to me that there are more "cracks" in this sense in the make-up of molar women. If this is true, then there are more lines of flight out of the strictures of molarity. In other words, if it is true that feminine traits are less sedimented, then the possibility for something new is greater. It is more likely that the movement of molecular particles that escape the coding of gender will be greater in women than in men.

Let me indicate very briefly the danger for women. Obviously at the molar level, the situation is worse for women, since at least the way in which desire is codified in men allows them more pleasure and satisfaction, more movement, and more connections. Deleuze clearly states that a molar politics for women is necessary, in the sense of fighting for equal rights, for example, and in general fighting for their subjectivity as their own (Deleuze and Guattari 1987, 276). However, the danger for women is greater at the molecular level too. It is the danger that corresponds to the greater possibility for transformation. If there are more "cracks" in the make-up of molar women, the possibility for the cracks to open up without producing something new is greater. To put it another way, the greater possibility of freeing desire from its constrained paths means, correspondingly, there is a greater possibility for total breakdown or total dissolution of subjectivity. This would manifest itself in psychological breakdown.

I conclude by noting two points. First, certain formations of subjectivity are certainly more repressive than others, and going beyond those would be revolutionary. The subject-forms constituted in opposition to the Western white rational male adult certainly fall into this category. Second, the strength or force of creativity of life comes from becoming, not from the subjectivity that subdues that life. "It is dangerous to confine oneself to such a subject, which does not function without drying up a spring or stopping a flow" (Deleuze and Guattari 1987, 276). The reification of subjectivity in general is what makes certain exploitative relations possible. Subjectivity is imposed on each individual, that of man or woman, white or black, etc., and it is the construction of how one can act and what one can feel and think. This also allows us to *take* an individual as being such-and-such a subject. In this way, it becomes predetermined what relations can occur between individuals, what can be demanded of another individual, and what rights and privileges an individual has. I have been discussing the issue of woman apart from other minorities, like the non-white, the child, and the animal, but it cannot in actuality be disentangled from these.

Returning to Braidotti's argument, at this point in Western history, with its current ethico-political configuration, it is necessary to be recognized as a subject and, for example, to be recognized as having certain rights in relation to others. Deleuze is not advocating the obliteration of subjectivity. Rather, his point is that subjectivity, like other forms of molarity, belongs to one abstract pole, while pure becoming belongs to the other pole. Nor is he claiming that some sort of subjectivity or molarity must be maintained, but rather that these must be understood as abstractions. Movement and creativity occur only between the poles. In the end, it is desirable, first, that gendered subjectivity is overcome and, second, that the strictures of subjectivity in general be loosened.

Works Cited

Braidotti, Rosi. 2002. *Metamorphoses: Towards a Materialist Theory of Becoming.* Cambridge: Polity.

Burchill, Louise. 2010. "Becoming-Woman: A Metamorphosis in the Present Relegating Repetition of Gendered Time to the Past." *Time & Society* 19 (1): 81-97.

Deleuze, Gilles and Felix Guattari. 1987. *A Thousand Plateaus: Capitalism and Schizophrenia.* Translated by Brian Massumi. Minneapolis: University of Minnesota Press.

Patton, Paul. 2009. "Events, Becoming and History." In *Deleuze and History*, edited by Jeffrey Bell and Claire Colebrook, 33-53. Edinburgh: Edinburgh University Press.

NEW EARTH, NEW PEOPLE: DELEUZE, DEMOCRACY AND THE POLITICS OF THE FUTURE

BRYAN NELSON

> The question of the future of the revolution is a bad question because, in so far as it is asked, there are so many people who do not become revolutionaries, and this is exactly why it is done, to impede the question of the revolutionary-becoming of people, at every level, in every place.
> —Deleuze and Parnet, Dialogues II

The Untimely: When Philosophy Becomes Political

It is well known that the philosophy of Deleuze organises a thought exterior to the categories of history. For Deleuze's work is populated by problems of events, becomings and multiplicities, that which eludes a historicism—a logic of origins, necessity, interiority, revelation—which according to Deleuze, has prevailed over philosophy from Hegel to Husserl to Heidegger.[1] As Deleuze suggests in dialogue with Claire Parnet, this demands a reconsideration of the kinds of questions posed to philosophy, of the structure of the questions themselves:

> Questions are generally aimed at a future (or a past). The future of women, the future of the revolution, the future of philosophy, etc. But during this time, while you turn in circles among these questions, there are becomings which are silently at work which are almost imperceptible. We think too much in terms of history, whether personal or universal. (Deleuze and Parnet 2007, 1-2)

For Deleuze, it is the task of philosophy to think these becomings, to express and *effectuate* them through concepts which are constructed with

[1] See Jay Lampert's assessment of this position in *Deleuze and Guattari's Philosophy of History* (2006).

the unique capacity to chart or map, much like a cartography, their deterritorialising forces, lines of transformation and potentials for new relations and powers. Becoming-woman, becoming-revolutionary are examples of such concepts which should not be confused with women and revolutions in themselves. As a concept, becoming-woman (which has troubled so many) never intends to replace a notion of woman as subject or subject position, nor will it disregard or discredit the significance of women's struggles for the autonomy of their bodies, their history and subjectivity (Deleuze and Guattari 1987, 275-6). But without compromising the integrity of the macropolitical, Deleuze claims that philosophy is realised to consider a very different set of problems.

At least since *Nietzsche and Philosophy*, Deleuze has maintained that philosophy is the practice of concept formation. Philosophy is creation, a self-positing, a building and forging of new concepts, an exercise in radical constructivism. And yet, when *What is Philosophy?*, written many years later, adopts this argument as its principal thesis, Deleuze and Guattari explain in far more detail that the concepts that philosophy creates will correspond neither to historical conditions nor to states of affairs but are rather engineered to "diagnose our actual becomings" (Deleuze and Guattari 1994, 112). Philosophical concepts do not arrange regimes of representations and cannot be evaluated in terms of true and false; they are rather tools to encounter, express and counter-actualise the virtual becomings which both flow through us at every moment and escape history altogether. Perhaps this is why, late in life, Deleuze described himself unashamedly as a "pure metaphysician" in the tradition of Bergson and Whitehead (Villani 1999, 130). Thus, if there can be no proper history of becoming-woman, it is not because it remains too abstract or merely ideal but because the *actual becomings* which the concept intends to express—those vital transformations which may pass through women and men alike, deterritorialising both what it means to be a woman and what woman means—remain irreducible to a history of women or women's struggles, however detailed. For Deleuze, philosophy does not think according to a model of historical times but in terms of a metaphysics of the "infinite Now" of our becoming: a becoming-woman which cannot be framed according to the past, present and future of women, a becoming-revolutionary which refers not to any particular historical revolution but (following Kant) grasps the intensity or "enthusiasm" of the revolutionary *event* immanent to all people, in all places.[2]

It is with this rather peculiar term—the *infinite Now*—that philosophy arrives at a very different conception of the future: "[T]he future is not a historical future, not even a utopian history, it is the infinite Now, . . . the Intensive or Untimely, not an instant but a becoming" (Deleuze and Guattari, 1994, 112). But perhaps more important than a new concept of the future which is no longer historical but a becoming, is philosophy's adoption of an entirely new orientation towards that future. When philosophy dislodges itself from history and is no longer confined to questions regarding a past or future time, it not only begins to conceive of the future in terms of our immanent becomings but initiates a thought which itself summons forth these becomings through its *untimely* relation to the present. Throughout his authorship, Deleuze constantly references what Nietzsche understood as philosophy's critical or untimely vocation. In the foreword to the second essay of his own *Untimely Meditations*, Nietzsche qualifies the untimely as "acting counter to our time, and thereby acting on our time, and let us hope, for the benefit of a time to come" (Nietzsche 2011, 60). Deleuze cites this passage repeatedly.

According to Nietzsche, if philosophy does not strive to represent the world in thought, it is because its first task is to organise a thought which acts upon the world, which counters the world in its present form and thinks against it, in the hope of changing it, of bringing forth a new and better world to come. Thus, the nature of philosophy is not communication or contemplation or reflection but *resistance*. And what it resists is the present time, the time in which we find ourselves, *our time*, so that we may become something else altogether. In *What is Philosophy?*, Deleuze and Guattari envision the very activity of concept creation as a kind of call for resistance in itself: "We lack creation. *We lack resistance to the present.* The creation of concepts in itself calls for a future form, for a new earth and people that do not yet exist" (Deleuze and Guattari 1994, 108). Thus, if Deleuze rejects a notion of philosophy as the recognition or justification of existing states of affairs and forms of life, it is in order to pursue philosophyas an untimely mode of thinking, an absolute deterritorialisation of the present in thought (Patton 2000, 12). For it is only when philosophy renders itself untimely, when the concepts it generates are structured to counter, to challenge the conditions of the present time, for the benefit of a time to come, that philosophy itself

[2] Deleuze and Guattari reference Kant who, in *The Contest of Faculties*, draws a distinction between the historical events of the French revolution and the *idea* or *spirit* of revolution itself (see Deleuze and Guattari 1994, 110 and 112).

becomes political. This is precisely what we discover with Deleuze. Political philosophy is not a genre or stream of philosophy; it is not the philosophy which ponders the topics of politics. *Philosophy must become political.* How is this achieved? Philosophy becomes political when it exceeds itself, when it touches the outside, when it becomes *nonphilospohical*, when its absolute deterritorialisations in thought intersect with the relative deterritorialisations in a given historical milieu, when it participates in creating new possibilities for thought and new possibilities for action, in order to help create a future that is different from the past and different from what is today.[3] Likewise, the value of political philosophy, its merits and virtues, may only be appraised and evaluated by the capacity of its concepts to engage with and assist in this resistance to the present and the making of a new and better future. In this respect, while Nietzsche is rarely considered a political philosopher himself, Deleuze could very well affirm that he represents the source and model for all political philosophy, past, present and future.

Therefore, philosophy does have a unique politics all of its own: a politics of the future which is entirely distinct from the question of the future of politics. When philosophy adopts an untimely orientation, when it functions to deterritorialise the present in thought—*when philosophy itself becomes political*—the future will no longer appear as something to be appropriated as an object of contemplation or reflection but rather as something to create, something to be made. For Deleuze, this is philosophy's critical task, its revolutionary potential: to render thought active, aggressive, *violent*, to raise thought to a force or power capable of encountering the present and altering it:

> This is why philosophy has an essential relation to time: it is always against its time, critique of the present world. The philosopher creates concepts that are neither eternal nor historical but untimely and not of the present. The opposition in terms of which philosophy is realised is that of present and non-present, of our time and the untimely. (Deleuze 1983, 107)

An untimely philosophy is one which realises its capacity to *think against*, to critique its time, to create forms of resistance in thought which may intersect with transformative movements in a social field, informing coordinates and orientations, inspiring possibilities, calling forth a revolutionary becoming. But this is a task which cannot be achieved once and for all. The project of philosophy, as a critical or political philosophy, must be taken up again and again, epoch after epoch, the image of the

[3] See also Patton 2010, 155.

philosopher constantly renewed and regenerated, in his own time and against his time (Deleuze 1983, 107).

For Deleuze, philosophy is intrinsically bound to a politics of the future, but a future which is no longer historical or utopian but once again, "Intensive or Untimely, not an instant but a becoming"(Deleuze and Guattari 1994, 112). The present time may very well remain a privileged position for philosophy, its point of departure and site of critique, the object of its *being against*. But it is through this specific relation to the present—its resistance to the present—that philosophy summons forth a future form, not a utopian future, a transcendent time eternally broken from the present, but an immanent future of our becomings, a future produced only through our creative resistances to the present, through the effective displacement of the conditions of our own present time, so that we may become something else entirely: *a new earth and a new people*.

Diagnosis: Becoming-Democratic

In Deleuze and Guattari's final collaborative work *What is Philosophy?*, this cluster of ideas—the untimely, resistance to the present, becoming-revolutionary, new earths and new peoples—appears alongside and inextricably bound to the elusive concept of *becoming-democratic*, a term raised in the text but never developed. On first encounter, the concept may seem somewhat perplexing given its position in the work. It appears only once and quite unexpectedly following a series of remarks offering a rather disparaging assessment of the sad condition of our own modern democratic States. Throughout, Deleuze not only condemns democracy as the keeper of capitalism and justification for Western imperialism but also vehemently denounces what is often taken as some of its most characteristic features: majorities, consensus, circulation of opinion, the rights of man.[4] And yet, the idea of democracy itself will not be abandoned. For immediately following this critique we are asked to consider a new concept, one which attempts to conceptualise democracy

[4] We should assess Deleuze's criticism, particularly of human rights, with care. What Deleuze takes issue with is not so much rights in themselves but the claim of their *universality*, which not only returns the philosopher to the ancient task of defending abstract universals but far more problematically, dictates a relation to those rights as one of reference and application. Rather, for Deleuze, we should think about the legality of politics in terms of creating and inventing forms, to which Deleuze elsewhere gives the name *jurisprudence* which he opposes to rights in this regard.

not as a State or form of governance or set of institutions but as a *becoming*.

The concept appears interwoven with a number of topics which have already been introduced above. The passage reads:

> [T]he object of philosophy is not to contemplate the eternal or reflect history but to diagnose our actual becomings: a becoming-revolutionary that, according to Kant himself, is not the same thing as the past, present, or future of revolutions. A becoming-democratic that is not the same as what States of law [États de droit] are, or even a becoming-Greek which is not the same as what the Greeks were. The *diagnosis* of becomings in every passing present is what Nietzsche assigned to the philosopher as physician, 'physician of civilisation,' or inventor of new immanent modes of existence. (Deleuze and Guattari 1994, 112-13)

Although democracy is clearly not a theme in Deleuze's writings, this passage will function to raise the question: does his philosophy contribute anything to a democratic politics? This has certainly been subject to doubt.[5] But if it does, it is most directly via this concept of becoming-democratic. And while we are given little more in the text regarding how we should proceed to diagnose democracy's becoming and construct its concept, Deleuze will certainly offer some direction. We know becoming-democratic is not the same thing as the constitutional State (État de droit) and is in no way reducible to a logic of majorities, consensus and rights. Rather, the concept compels us to consider democracy in an entirely different way, as the movement or experience which both resists the present state of democracy (our sad "democratic" States) and strives towards the creation of a new mode of existence, a different, hopefully more democratic future. Thus, it helps forge a new meaning of democracy by resisting the hegemony of what democracy means today. But at the same time, the concept will not provide an ideal schema for democracy in any way. It does not present a model to be realised or projected into the future. Rather, like other untimely concepts, it is practical, created to be engaged and used in the present, to help us think against what passes for democracy in our own time in order to assist in the drawing forth of democratic potentials in all domains and at all levels of social life. Indeed, in *What is Philosophy?*, Deleuze and Guattari's principal criticisms of democracy are not organised around any essential argument against

---

[5] Both Philippe Mengue (2003) and Nicholas Thoburn (2003) identify a certain hostility or antipathy towards democracy throughout Deleuze's work. Many of Paul Patton's essays are structured as a response to this position. See particularly "Deleuze and Democracy" in Patton 2010.

democracy in itself but rather the observation that in our society, democracy is structurally restrained in two directions: by its nationalisation by the State and by its subjection to the requirements of a global capitalist axiomatic (Deleuze and Guattari 1994, 104-6).[6] Therefore, we may initiate an analysis of becoming-democratic as Deleuze's term for a deterritorialised democracy, a democracy no longer shackled, arrested or captured by the vast networks of the capitalist and State machines, a democracy which opens up sets of relations, social bonds and institutional practices for their subsequent reformation—not sometime in the future but in the now of its becoming.

What Deleuze is gesturing to, what becoming-democratic attempts to conceptualise, is democracy as a virtual movement, as a dynamic transformative process, democracy as *event*. We have already seen that Deleuze always opposes history and becoming (a history of women and becoming-women, historical revolutions and becoming-revolutionary). So the concept is already arranged so that it may never simply be corresponded to a particular regime, institution or organisational structure. It references nothing constitutional, legal or procedural. Indeed, as Paul Patton notes, there is something of democracy's becoming which is always betrayed by its actual historical manifestations (2005, 64; 2010, 171). Rather, the becoming of democracy is what passes through institutions, challenging their limits, pushing them in new directions. It is the carving out of new spaces for democracy, discovering locations where there is potential for more democracy and confronting the points where democracy is obstructed, mitigated, channelled away or denied altogether. Becoming-democratic is thus the Deleuzian concept which is not content to approach democracy as a static entity, a completed project, a sanctioned politics, its limits and boundaries carefully delineated and governed by official State institutions and fiercely defended by the police. It rather incites a conception of democracy as the persistent process of democratisation, the extension or deepening of democracy, the introduction of democracy to new social fields precisely where it "does not belong."

Becoming-democratic is the concept which expresses the powers or capacities of a democracy at its most relentless, its most unyielding, a democracy which is constantly in the process of discovering its own relative limits and then transgresses those limits. It is the concept which expresses the potentials for democracy immanent to every social sphere, every institution, every set of relations where modes of domination and exclusion are discovered and experienced. And it is the concept which

[6] Also see Paul Patton's "Becoming-Democratic" (2008, 187).

itself summons forth the radical democratising impulses inherent to all people and in all places where that domination is confronted and disputed, where rigid hierarchies and concentrations of power are challenged, dismantled and replaced with more participatory, egalitarian arrangements. And yet, as a term which attempts to think democracy's becoming not historically but as a pure event, becoming-democratic cannot simply be understood as the struggle or challenge or confrontation in itself, however spontaneous, but as something *inside* the struggle, its potentiality, what provides the struggle its intensity, its libertarian character. The concept considers what democracy is capable of becoming and how democratic those becomings can be.

As democracy deterritorialised, the becoming of democracy remains irreducible to any form, strategy or model. It is creative, self-differentiating, constantly in the process of transforming and reinventing itself. At times it may appear merely dormant and quiescent, only to manifest vibrant revolutionary forms and subjects without warning. In this regard, Deleuze's becoming-democratic will approximate or compose certain affinities with Claude Lefort's concept of *savage democracy* and Miguel Abensour's concept of *insurgent democracy*. Like Lefort's savage democracy, it recognises the indeterminacy of democracy, that it has no ultimate foundation or reference, that all bearings of certitude are lost. It recognises that democracy has no formula, that it cannot be mastered or contained; there is something of democracy which always escapes only to return more militant, more democratic. This is democracy independent of any principle, authority or established rule, a democracy which evades ideology and cannot be codified into a doctrine. This is a democracy which, despite Tocqueville's dream, resists domestication, resists submission to an established order, a democracy which can never be possessed or tamed, democracy at its most *anarchic* (Abensour 2010, 105-13).[7] And like Abensour's insurgent democracy, it is a democracy which is never an ally of the State but appears as a permanent insurrection against the State, which rises up against the unifying, integrating State-form which functions to neutralise democracy, to incorporate its powers into a "democratic" State (Abensour 2010, xxxiii-xxxiv). Therefore, becoming-democratic is indeed, as Deleuze remarks, a democracy that is not the

[7] Here I cite Miguel Abensour's (2010) lucid analysis of Lefort's *savage democracy* (a concept which remains almost as enigmatic in Lefort's work as becoming-democratic is in Deleuze's) in his essay "'Savage Democracy' and the 'Principle of Anarchy,'" from which Abensour develops his own concept of *insurgent democracy* as an immanent critique. Abensour himself completed his *Doctorat d'État* under the supervision of Deleuze.

same as what the State is. It is what counters the State, both the autocratic State and the democratic State alike. And for this reason, it always discovers itself, much like the war machine of *A Thousand Plateaus* (Deleuze and Guattari 1987), exterior to the State apparatus.

It is in this respect that becoming-democratic discovers the register in which democracy intersects with everything that is revolutionary. It constitutes the forces which charge and animate democracy, from below, from the outside, from a minoritarian position: blacks, women, refugees, proletarians. Becoming-democratic is a becoming-minoritarian. It is a process which challenges the constituted standard or measure of a majoritarian logic, the principle which has already decided who counts and who does not, who is included and who is not, who may participate and who may not and thus has already produced and distributed a set of minority subsystems or determined states (whether it be women, workers, children, the mentally ill, etc.): "Majority implies a state of domination, not the reverse. . . . The majority in a government presupposes the right to vote, and not only is established among those who possess that right but is exercised over those who do not, however great their numbers" (Deleuze and Guattari 1987, 291).[8] Even in democratic Athens, as is well documented, the Athenian-adult-male-citizen, those included in the assembly, were never the *majority* of the city in terms of their numbers. Indeed, although Athens represents a remarkable and inspiring experiment in direct democracy, a state of domination returns to haunt the city of Athens as soon as that citizen body who self-legislate as equals in the assembly return home to govern absolutely those classes who have no right to participate in democracy (women, children, slaves, non-Athenian residents). Thus, as a "state of domination," and not a reference to quantity, the majority itself is never something embodied but represents an empty form, an abstract rule (it is not adult white men themselves but the principle which distributes a different status, role or position to adult white men than to women, blacks, children). It is this rule or measure, this a priori principle of distribution which becoming-democratic, as a becoming-minoritarian, will always resist and inevitably seek to transform: problematising its authority and legitimacy, eroding its distribution of roles, forms, appearances and capacities and creating new possibilities for autonomy, participation and transversal relations.

Deleuze's concept of becoming-democratic helps render democracy's becomings thinkable. We know that for Deleuze, philosophy is not

[8] Compare this with Jacques Rancière's description of the "distribution of the sensible" and the emergence of the *demos* from their own minority status in *Disagreement: Politics and Philosophy* (1999).

concerned with universals, fixed states or static forms (perhaps these are already abstractions). Rather, philosophy builds concepts which are engineered to conceptualise flows, virtual processes, becomings, transformations. By rendering democracy's becomings thinkable, philosophy can assist in the actualisation of its immanent potentials everywhere. This is what makes the concept political. This is what calls forth that new earth and new people to come. And yet, the concept of becoming-democratic should not be confused with the *democracy to come* that we discover in deconstruction. Although this distinction requires far more extensive argumentation than can be offered here, it is worth noting that the politics of deconstruction have always relied on certain transcendent ideas or have composed a certain transcendent relation to its ideas. And accordingly, when deconstruction considers the question of democracy—which is very much consistent with its ethics—what we discover very quickly is that the *idea* of democracy or the *truth* of democracy is posited as an ideal horizon, a transcendent field projected into a future which can never be realised. This is why it is often said to represent a messianic promise. For deconstruction, this is the paradox of democracy, both its necessity and its impossibility. But at the same time, it is precisely this paradox of democracy, the promise of a true democracy which can never be, to which deconstruction will appeal in order to incite our own initiatives to move towards it. Thus, what deconstruction discovers is the truth of democracy located perpetually in a future time—a *democracy to come*—which is set out against us as a standard to be acknowledged, an ideal to ascribe to but always with the knowledge of the atheist, the knowledge that true democracy is a promise which can never be truly fulfilled.

However, becoming-democratic organises an entirely different temporality. As we have seen, Deleuze's politics of the future can only be engaged through the present and through a resistance to the present. Of course, something of democracy's becoming, something of its potentiality, will inevitably escape any historical actualisation; this is the fate of every becoming. But becoming-democratic will never orient itself according to an impossible future, a transcendent democracy to come but may only be initiated in the present of our resistance to what is, our conditions of today. And it is precisely through this immanent resistance to the present that we may create something entirely other, a future which otherwise would not have been, a future which is, hopefully, more democratic. And yet, analogous to deconstruction, Deleuze's becoming-democratic can offer no promises, no guarantees. There is always an element of risk in every becoming. This is why, following Nietzsche, I wish to emphasise the word *hope*.

With Deleuze, it is never simply a question of an open future or a closed future. It is a question of the infinite Now of our becoming. To speak of a Deleuzian future (which seems so much more noble than Foucault's over-quoted reference to the twentieth century as *Deleuzian*) means to call for the now of our resistance to the present. This resistance to the present inevitably demands that we ourselves become something else: a new earth and people. Deleuze believes philosophy can have a role in this. By creating new ways to think about the forces which shape our future, philosophy may contribute to producing new possibilities for thought and new for action. This is the nonphilosophical value of philosophy. Recall that for Deleuze, Spinoza is the philosopher who teaches philosophers how to become nonphilosophers (Deleuze 1988, 130). This is what makes philosophy political. It is the task of philosophy to make the future different from the past and different from what is today. This is philosophy's political vocation, its revolutionary potential and what renders its concepts untimely.

Works Cited

Abensour, Miguel. 2010. *Democracy Against the State*. Translated by Max Blechman and Martin Breaugh. Cambridge: Polity.

Deleuze, Gilles. 1983. *Nietzsche and Philosophy*. Translated by Hugh Tomlinson. New York: Columbia University Press.

—. 1988. *Spinoza, Practical Philosophy*. Translated by Robert Hurley. San Francisco: City Lights Books.

Deleuze, Gilles and Félix Guattari. 1987. *A Thousand Plateaus: Capitalism and Schizophrenia*. Translated by Brian Massumi. Minneapolis: University of Minnesota Press.

—. 1994. *What is Philosophy?* Translated by Hugh Tomlinson and Graham Burchell. New York: Columbia University Press.

Deleuze, Gilles and Claire Parnet. 2007. *Dialogues II*. Translated by Hugh Tomlinson and Barbara Habberjam. New York: Columbia University Press.

Lampert, Jay. 2006. *Deleuze and Guattari's Philosophy of History*. London and New York: Continuum.

Mengue, Philippe. 2003. *Deleuze et la question de la démocratie*. Paris: L'Harmattan.

Nietzsche, Friedrich. 2011. *Untimely Meditations*. Translated by R. J. Hollingdale. Cambridge: Cambridge University Press.

Patton, Paul. 2000. *Deleuze and the Political*. London: Routledge.
—. 2005. "Deleuze and Democratic Politics." In *Radical Democracy: Politics Between Abundance and Lack*, edited by Lars Tønder and Lasse Thomassen, 50-67. Manchester: Manchester University Press.
—. 2008. "Becoming-Democratic." In *Deleuze and Politics*, edited by Ian Buchanan and Nicholas Thoburn, 178-195. Edinburgh: Edinburgh University Press.
—. 2010. *Deleuzian Concepts: Philosophy, Colonization, Politics*. Stanford: Stanford University Press.
Rancière, Jacques. 1999. *Disagreement: Politics and Philosophy*. Translated by Julie Rose. Minneapolis: University of Minnesota Press.
Thoburn, Nicholas. 2003. *Deleuze, Marx and Politics*. London: Routledge.
Villani, Arnaud. 1999. *La guêpe et l'orchidée. Essai sur Gilles Deleuze*. Paris: Belin.

NOSTALGIA, PAST AND FUTURE HOPES

ROCHELLE GREEN

Introduction

Were it not for the possibility of the future, social and political change, revolutionary change, would forever remain a foregone conclusion. However, understanding the very richness of the possibility the future presents to us raises concerns about the relation of hope to contemporary life. Nevertheless, in order to consider the constitution of possibility in the future, the very quality of time itself becomes a focus of inquiry. What can the future be without recourse to past and present? Furthermore, how can either past or present be considered without the aid of memory, of history, and of hopes and projections toward a future yet to become? This essay brings focus to these questions through an examination of the nature of nostalgia and its relation to revolutionary hope thematized in the work of Ernst Bloch.

Nostalgia often brings to mind images of days long passed, days to which we would like to return, days better than the days in which we live now. Etymologically, conceptions of nostalgia take their content from the combination of a desire to return *home* and of the pain caused by that desire, a bittersweet pain induced by an insatiable desire. As Geoghegan has summarized:

> The term 'nostalgia,' although it has come to be used to refer to time, was originally coined to refer to place, for it meant the agony of being separated from home. The word, as its ending suggests, was used to describe the illness of desire for one's own homeland, quite literally homesickness. The temporal dimension was, however, implicit, since it was the depressed state brought on by the memory of (recollection of) home. (Geoghegan 1997, 16)

Thus, we can think of nostalgia as a kind of painful memory. Problematically, however, the relationship between this act of memory and agency is unclear. Obviously, a *real* return to what has been or what used to be is impossible. The past can only emerge as a time to which we can literally return through fantasy or imagination. Thus, we may wonder what good it is to wallow in this impossible desire to re-live a time now long passed. As Robert A. Reeves has noted, "nostalgia is realism giving birth to fantasy" (Reeves 1992, 97). That is, nostalgia is a form of memory that induces a comparison between what was past to what is present, and judges the past to be more desirable than the present. In this sense, nostalgia makes a clear assessment of the hardships of this moment. Under this spell of longing for what used to be, however, it is far less clear if the assessment made of the present is quite so judicious. Reeves further describes nostalgia as a kind of *Weltschmerz*. Nostalgia emerges in relation to our sense of world-weariness and as a fantasy-driven palliative for our experiences of alienation and discontent given the social and political circumstances in which we find ourselves. The agency of nostalgia, then, is the directed creation of a wish that can never come to fruition.

Because nostalgia forms the affective basis or a means of thinking relegated to an intractable and unproductive emphasis on "what has been," I argue in what follows that a conception of history and of memory capable of illuminating the ideals of the future is necessary for a truly revolutionary conception of what is possible, and for grounding hope. A sense of the past, of history, and of memory with such illuminating power lends itself to a refiguring of the past as such, and most importantly, also of what we imagine to be possible in the future. Rather than regarding "what has been" as a petrified archive of factual events, this view of history suggests a conception of memory as a series of images of a past future, a series of hopes for the future in a present now gone. As history presents itself as this kind of illumination of the future within the present, our interpretive horizons of history and memory must serve as tools for opening up our own futures rather than as stagnant and petrified productions of generations now expired.

This essay begins with a brief account of Bloch's use of history in order to illuminate what I consider to be the more pernicious side of nostalgia. By juxtaposing a sense of nostalgia to what Bloch describes as Utopian Consciousness, I identify four key dimensions of nostalgia: 1. Nostalgia is an emotion deeply integrated with what is present in our social and emotional lives and is a reflection of aspects of life already present to hand. For Bloch, this relation to what is present in our social and emotional lives is what signifies a "full emotion." 2. Nostalgia, like other modes of "remembering" involves a specific kind of excess. I argue that nostalgia is founded upon a saturation of what Bloch describes as an "eternal now." Consequently, it is inextricably conjoined to similarly situated full emotions like abandonment, regret, and guilt. 3. Nostalgia can be and probably is laden with ideology that can occlude an actual hopeful rendering of the past, which would usher forth a front of consciousness, upon which a revolutionary future could emerge. 4. Nostalgia is the creative impulse facing backward, in the direction of "what was" or, even worse, toward a "what should have been." The second section of this essay is informed by a discussion of Bloch's distinction between anamnesis, which is the mode of remembering endemic to the Western intellectual heritage, and particularly harmful to Utopian Consciousness, and anagnorisis, a kind of insightful recollection that allows for the past to be used in constructing a future yet to become. Relating nostalgia to anamnesis, I argue nostalgia further impedes the kind of future-oriented appropriation of the past Bloch advocates. Consequently, nostalgia is linked with despairing dispositions and precludes genuine expressions of hope.

Utopian Consciousness and an Outline of Nostalgia

In an interview discussing Bloch's controversial relationship to orthodox Marxism and his potential threat to the German Democratic Republic, Wolfgang Harich commented, "Bloch in actual fact was considered philosophically harmless, because nobody could understand his writings—they were esoteric; his stylistic idiosyncrasies did not appeal to the masses—[The GDR had] no reason for concern in that respect" (Ernst and Klinger 1997, 9). Bloch's style is idiosyncratic at best. He regularly plumbs centuries of history and cultural productions, myriad dimensions of the Western intellectual tradition for traces of hope. His intellectual heritage speaks to an impressive and erudite background, while his personal commitment to political emancipation and freedom from exploitation speaks to a great courage of conviction regardless of whether one agrees with his heterodox Marxism or not. Bloch traverses an odyssey of hope through culture and history all the while insisting upon the creative potential of human consciousness to recognize a future yet to become.

Nostalgia as Full Emotion

Bloch's conception of hope is a response to what he describes as the "learned" emotions of fear and anxiety. Both fear and anxiety comprise "full emotions," meaning they are intentionally directed toward objects already present in social life. While Bloch's focus in the opening sections of *The Principle of Hope* (1996) is clearly directed toward a description of fear and anxiety, his *Heritage of Our Times* (1991), in which he investigates the formation of National Socialism, opens a space for understanding nostalgia as an even more pernicious "full emotion". In fact, it is in this text rather than in *The Principle of Hope* that Bloch most closely identifies a political manipulation of memory as a shared cultural phenomenon or as heritage, described as a creative use of the non-contemporaneous, arguably in the form of nostalgia. Here, he suggests that the German people experienced a series of full emotions, most of which were promoted and incited by a variety of non-contemporaneous contradictions. It was, according to Bloch, precisely because these non-contemporaneous contradictions were available for exploitation that the National Socialists were able to gain a foothold with a variety of different groups and classes of people from the youth to the peasantry, to the petite bourgeoisie. Fascism was capable of tapping into this non-contemporaneity in a variety of ways in order to invite people to reimagine a longing for national identity tethered to archetypal constructions of blood, soil, and home. Further, Bloch's conception of this manipulation of non-contemporaneity rivals leftist interpretations of fascism and puts orthodox Marxism on guard for complicity in the emerging formation of a National Socialist ideology. As David Kaufmann has suggested:

> Heritage of Our Times indicts the left for being partially responsible for the success of the National Socialists. In Bloch's account, the left misunderstood the appeal of the Nazis by dismissing all the atavisms and irrationality of the fascist movement as mere childish regression from the superior ratio of Marxism. Socialists and Communists failed to see that Nazism had a strong emotional appeal, a strong draw on archaic desires and cultural outcroppings . . . (Kaufmann 1997, 34)

Such expressions of deep faith in a national identity, for instance, incited a sense of nostalgia founded upon any number of immediate concerns, worries, anxieties, and fears. In this context, nostalgia operated as a full emotion much in the way Bloch describes fear and anxiety in *The Principle of Hope*.

Full emotions are oppositional to "expectant emotions," of which hope is the exemplar. That is, expectant emotions for Bloch have two components: intentional states and indefinite objects. Because they are directed toward indefinite objects, expectant emotions orient us toward the future. Hope involves being directed toward an object we can only vaguely grasp. Further, Bloch conceptualizes hope, not as an emotion in the colloquial English sense of the term, but rather in terms of the German, *Affekt*. *Affekt* implies a much more active state of becoming. Richard Roberts remarked:

> What Bloch calls the 'emotion [Affekt] of hope' precipitates activity that opens, broadens and frees; we are to throw ourselves into what is becoming [ins Werdende] and not merely be passively thrown in 'what is' [ins Seiende]. (Roberts 1990, 27)

Bloch relates hope to a form of social consciousness, or utopian consciousness of history as a container for a cultural inheritance yet to be tapped. This inheritance, yet to be realized, is what Bloch refers to as the utopian content of the past. The utopian content of the past harbors the potential of allowing for a better, more emancipated future. At the same time, however, this content also allows for both reactive manipulation and ideological construction. It is due to the very fact that this cultural heritage can speak to a people's sense of themselves and sense of culturally available tools for the resolution of present concerns that it is good fodder for ideological propaganda. In fact, ideology is remarkably more effective when it incorporates, accounts for, and responds to such a cultural inheritances. Accordingly, Utopian Content can speak to both progressive and reactionary purposes.

Nostalgia as Excess

For Bloch, the worlds of the past are not yet finished. However, he does not mean to imply that there is something always left to be completed in the productions of peoples now long gone. Rather, it is the notion that contained within these cultural productions is something excessive, something left over, yet to be exposed, and yet to be mined for its relevance to present and future projects. According to Bloch, we ought to approach the past to find what within it can be utilized to direct us toward a future. Further, this utopian consciousness is that which undergirds all social constructions, cultural products, and experiences of hope, and hence is responsible for the emergence of new things. In other words, it is the basis of social and cultural creativity. Roberts describes it as an "open process" in which "succeeding ages 're-function' the material of the past to suit their ideological requirements, whether reactionary or progressive" (Roberts 1990, xxvi). In light of his suggestion that past cultural productions, artifacts, and achievements involve a kind of excessive dimension, it follows that nostalgia as a contemporary production of the past similarly involves excess. However, this nostalgic excess is somewhat distinct from the positive excesses Bloch wishes us to cultivate in past productions.

Rather than indicating a surplus from the past that would, ideally, be projected forth into the future, nostalgia remains blind to this positive cultural excess of desire toward the future. Nostalgia counter-intuitively revolves around a dynamic but nevertheless redundant affect bound to an eternal now. Nostalgia neglects the inevitability of change in the present all the while it repeatedly tries to locate the present in various ways in the past. In its more despairing forms, nostalgia may even suggest acts of frenzied remembering in efforts to re-constitute glory days and galvanize those images of good fortune like a fortress against the present. In many respects, as is true with many differing forms of memory, veracity to actual occurrences is of little relevance here. Rather, the act of remembering itself serves a function of placating or of re-constituting desire. Just as I previously suggested that nostalgia be understood in Blochian terms as full emotion, I further suggest here that part of its excess is its complicity with any number of emotional dispositions that can be assumed in relation to past events. However, because of its relation to affects deeply embedded with the present, nostalgia involves an excess of backward looking abandonment, regret, discontent, and perhaps even guilt.

Nostalgia as Ideology

Utopian thinking is what is carried over from the progressive minded thinking of the past. However, Bloch identifies progressive thinking as having the potential to lie dormant depending upon ideological circumstances. Ideology can occlude the progressive dimensions of cultural artifacts and constructions of the past, though it can never obliterate them altogether. A hopeful consciousness, according to Bloch, assumes an interpretative stance regarding the past in order to exploit ideological counter-points and ultimately to irritate fissures within oppressive cultural and political ideology.

Of interest in the context of ideology's capacities to obfuscate the traces of utopian content in the past is the relation of nostalgia to ideology. If the foregoing two points regarding the qualities of nostalgia are compelling, nostalgia can be a prime social and cultural nexus of ideology. This would be the case because of nostalgia's rootedness in the present as well as because of its complicity with other backward full emotions. In order for ideology to hold sway, it must resonate with the cultural and social expectations and needs of a contemporary moment. Consequently, impregnating memories of the past with an interpretative structure coding the past as a gilded age and fostering a desire to return, has the potential to ground ideological claims in a sense of tradition, a sense of heritage, and imbue them with a sense of legitimation. Further, if it is the case that nostalgia is an excessive and dynamic conjunction of backward-looking full emotions like regret, guilt, or disappointment, these full emotions become a prime locus for exploitation and for manipulation. Thus, nostalgia remains in the ideological domain, capable of fortifying any number of specific ideological claims.

This, of course, stands in marked contrast to the utopian content discussed above. The actual utopian content of the past can be disguised and hidden by such ideology but as Bloch warns, this utopian content can never be obliterated totally. Consequently, the ideological push of nostalgia, or of cultural practices and traditions that commemorate, perpetuate, and exhort nostalgia, may be a source of forewarning. That is, in a culture permeated by images of what "once was" or by what "could have been," we may be alerted to a variety of ways in which ideology is functioning in contemporary cultural productions. Similarly, if nostalgia maintains a deep ideological dimension, such widespread collective remembering which recasts specific eras of national history, for example, may also be an indicator of what kinds of affective responses are readily available within contemporary social contexts. Consequently, nostalgia may act as an inadvertent harbinger for a revolutionary consciousness, provided such a sensibility were capable of escaping the eternal now, shrewdly veiled by backward looking and secretly forlorn idyllic images of a past barely recognizable as fictive. However, this revolutionary consciousness is, no doubt, Bloch's conception of the utopian consciousness already at work in the traces of utopian content accessible in canonical cultural productions of the past. Nevertheless, as Kaufmann has pointed out, "The past is not an open treasure-trove. It has to be claimed. Progress is not the self-actualization of the world—it is, at best, a product of human choice and bravery in the face of terrible odds" (Kaufmann 1997, 41).

Nostalgia as the Creative Impulse Looking Backward

Once the utopian consciousness has located traces of utopian content, it becomes aware of a not-yet-conscious, which is that which carries utopian content from the past through the present to instantiate the future. "The Front," as Bloch describes the temporal location of hopeful insights, is the result of bringing to explication and actuality latent moments of hope left coded in ideological cultural artifacts and historical constructions. The front is the temporal place at which the ideological occlusion of the hope contents of past creations gives way to a creative, hermeneutical impulse driving us forward into the future, that which is "Not-Yet." This exposition of hope contents is only found through the proper activity of thought: "venturesome thinking." This is thinking that goes beyond the confines of contemporary social and political hurdles and is capable of creatively utilizing as of yet untapped potentialities of history. The present colliding with the future creates a rupture allowing for the experience of venturesome thinking to come to the fore as a dawning hopeful consciousness.

Unlike the hopeful venturing and wandering of thought toward the future that signifies utopian consciousness, nostalgia maintains a disposition of looking backward. However, what is crucial about nostalgia is that it engages the same creative tendencies found in hopeful consciousness. That is, it utilizes a kind of venturesome or creative modality of consciousness, but rather than directing it toward the open horizon of futurity, it harnesses its potential and grounds it in the past before it is able to take flight in an actual manifestation of hope. As Reeves has pointed out, "[Nostalgia] surveys the past and the present and wants the past in the present's place; it says, "If only that could happen now!" (Reeves 1992, 96). As such, and rather unlike the disposition necessary disposition for hope in Bloch's thinking, nostalgia maintains a desire to move back, to return to the past. In so doing, nostalgia neglects a careful or considerate reckoning of the present. Instead, it imposes a feeling of dread toward the present, promoting a turning away, a denial, a refusal. Thus, we can see in the nostalgic frame of mind glimmers of nihilism.

Nostalgia's response to this disavowal of the present in lieu of a romanticized past "tries to establish the unreal in place of the actual" (Reeves 1992, 96). This is in contrast to the nature of hopeful venturesome thinking. The priority of fantasy over reality relegates nostalgia to an impotency in contradistinction to the activity and the viability of forward facing creative thought. For Bloch, the impulse that is captured by nostalgia remains at the root of the hope: that impulse is fundamentally creative. Nevertheless, when ensnared by wishes and yearnings for what has already been or what is imagined to have already been, those longings are subverted from their rightful place as sources of revolutionary consciousness and truly future-oriented possibility.

As Bloch repeats in multiple places within his corpus, one cannot predicate revolutionary potential on a reactionary gaze backward. As Kaufmann has summed up, "The past does not return. Bloch does not want the call to an unfinished tradition to sound like an incitement to historical regression, to a fumbling nostalgia" (Kaufmann 1997, 39). Thus, Bloch maintains our indebtedness and our situatedness within historical lines of possibility but does not want us to be at the center of an historical flight of fancy. However, this does raise some concerns regarding how Bloch understands our relationship to the past, concerns to which I shall now turn.

Nostalgic Anamnesis

The hopeful disposition relies on an abdication of the desire to assert epistemological axioms in order to predicate a world of stasis and petrification. This philosophical predilection to "fix" reality in this way is a tendency Bloch identifies reaching back to Plato. As Bloch suggests, knowledge as backward looking begins in the Platonic conception of anamnesis. Remembering what was shown to an individual knower becomes the manifest model of knowledge despite the multifarious variations this theme takes throughout the history of philosophy. "Being-ness simply coincides with Been-ness, and the owl of Minerva always begins its flight only after dusk has fallen, when a form of life has always become old" (Bloch 1996, 8). As such, creativity is stripped from conventional articulations of epistemology, thereby limiting what is hopeful in canonical theories of knowledge.

Because nostalgia involves a kind of remembering inextricably bound to a representation of the past as "better," or as a locus of potential, this form of memory is particularly anamnetic. That is, it renders the present as a lesser version of the past. In nostalgic remembering, the past is invested with either the possibility of what could have been or the actuality of what was better. In both cases, it moves the gaze of the social subject from a future possibility to a past now dead. Further, rather than seeing the past as a repository of utopian content capable of moving the present into the future, nostalgia sediments the subject *in the past*. It is a kind of epistemic remembering like that found in Plato that Bloch criticizes. However, this form further contends that the best situation for the remembering subject has actually expired. Consequently, just as was the case with philosophical knowledge entirely tethered to the past, a subject enslaved to such remembering loses a vital sense of self within a social context.

This further implicates the ideological dimensions of nostalgia in terms of alienation. If it is the case that nostalgia is a backward looking form of memory, and a kind of full emotion, it is also the case that nostalgia posits that the contemporary circumstances of anguish and discontent are resolvable in what has already now passed, not with the tools in the present or in the future. There can be no future with such a sentimental reversal of the drive toward the future. In fact, in such nostalgic formulation, one wonders if history itself would not seem dead, considering the primacy of the eternal now. The lodging of the remembering subject entirely in the past renders history familiar in a comforting way. Nostalgia dislocates the "shock" of a kind of recollection (*anagnorisis*), which Bloch recommends accompany our forays into the cultural productions of history. A sense of recollection, unlike simple Platonic remembering, invests us with a sense of familiarity that is uncanny, that leaves us feeling decentered, and consequently provides the requisite context for the generation of the truly new. Nostalgia forecloses the possibility of such shock as is found in recollection by massaging a memory into a complacent backward wish for what is imagined to be better than what exists now.

Conclusion

Nostalgia is a full emotion in the Blochian sense. It takes advantage of those objects, experiences, and qualities of life that are ready to hand. Unfortunately, nostalgia is conditioned by the judgments made regarding present dimensions of our experience. Those judgments denigrate the present so as to support an active turning away, a turning toward the past as unrealistically utopian or glorified. As Gabriel Marcel has described despair, it is an emotional state in which one is committed to a certain epistemic mistake (Marcel, 1948). In despair, we make a claim to *know* things will not change. It is in this light that we may also find nostalgia. In order for nostalgia to maintain the feeling of homesickness, the feeling of loss and a desire for return, we must in some way be simultaneously despairing of the present. Due to this sense of ineluctable despondency, the mistaken notion that the past must have been better, without trouble, without strife or affliction creatively molds the preoccupation with an intractable and deeply unsatisfactory present into a palliated, glorified past.

The relevance of this analysis to conceptions of political agency and engagement cannot go unnoticed. Bloch's aim in *The Principle of Hope* is to chart the processes and mechanisms whereby we can remain steadfast to the work of overcoming exploitation and alienation. Nostalgia as a reactive, yet creative emotion stands in direct opposition to this project of emancipation. The ideological significance of nostalgia is that it operates using a number of the same underlying currents as does hope. Consequently, nostalgia and hope emerge from remarkably similar contexts. What distinguishes them is a specific use and understanding of time itself. Nostalgia maintains a reactionary sense of time. Too afraid to look ahead, it concentrates its energy on creating a picture of what may have been. Nostalgia can pose a great danger and it can be a great obstacle to the formation of truly revolutionary hope. Hope necessitates a radical time; a sense of the temporal in which what is potential in the future can be locatable in the past and utilized in the present. Hope is revolutionary time.

Works Cited

Bloch, Ernst. 1991. *Heritage of Our Times.* Translated by Neville and Stephen Plaice. Berkeley and Los Angeles: University of California Press.

—. 1996. *The Principle of Hope, Volume I.* Translated by Neville and Stephen Plaice. Cambridge: The MIT Press.

Ernst, Anna Sabine and Gerwin Klinger. 1997. "Socialist Socrates: Ernst Bloch in the GDR." *Radical Philosophy* 84:6-21. Canterbury: Radical Philosophy Group.

Geoghegan, Vincent. 1997. "Remembering the Future." In *Not Yet: Reconsidering Ernst Bloch*, edited by Jamie Owen Daniel and Tom Moylan, 15-32. New York: Verso.

Kaufmann, David. 1997. "Thanks for the Memory: Bloch, Benjamin, and the Philosophy of History." In *Not Yet: Reconsidering Ernst Bloch*, edited by Jamie Owen Daniel and Tom Moylan, 15-32. New York: Verso.

Marcel, Gabriel. 1978. *Homo Viator*. Translated by Emma Craufurd. Gloucester: Peter Smith.

Reeves, Robert. 1992. "Nostalgia and the Nostalgic." *Southwest Philosophical Studies* 14: 92-97. Lubbock: Texas Tech Press.

Roberts, Richard H. 1990. *Hope and Its Hieroglyph: Critical Decipherments in Ernst Bloch's Principle of Hope.* Atlanta: Scholars Press.

PART III

UNTIMELY TIME

BOGAN TIME AND THE RACE THAT STOPS THE NATION

NATALIE CHURN

In 2011, Australian Prime Minister Julia Gillard was voted "Australia's biggest bogan" in Ninemsn.com.au's annual online FIX awards—a title she earned based largely on the mere fact that she sports a very broad Australian accent. It is intriguing that a Prime Minister could be described with a slang label that reflects a perceived lack of education, productivity and culture, given that her predecessor, Kevin Rudd, was partly unpopular due to his apparent inability to connect with the common voter as he was "too" intellectual and a reputed over-worker. In Australia, it seems, everyone is either too cultured or too uncultured, too hardworking or too lazy. The time is ripe for a review of so-called Australian values, which are constantly used in the media to allude to a shared Australian past and vision of the future, as this linear view of time promoted by the state and the media suppresses many of the alternative visions of Australian temporality and thus of alternative identities.

An exploration of the subculture known as bogans, often used to represent the less educated, suburban-living, "blue-collar" working masses, demonstrates that the way in which this cultural group is stereotyped as perceiving and using notions of time exposes the constructed nature of hierarchies of time in Australian culture, for which a revolution in thought and representation is long overdue. Bogans are stereotyped by anyone who considers themselves more high class, educated or deserving, whether it be economically, socially or culturally. Despite Australia's reputation as a culture of egalitarianism (which has its own problems, see Thompson 1994), such name calling exposes huge class mentalities within the nation which have long been taken for granted and which the Australian Occupy movement is finally trying to make visible. At the same time, so-called bogans, stigmatised as they are, represent a symbolic revolution on the part of "everyday" Australians in that they fight for alternative temporalities, reflecting a projection of the desires and fears of many Australians who are finding that they no longer have the time for time.

Australia provides an interesting national case study in that the time identity(ies) within Australia reflect a tension between two dominant ideologies of time. When the media both beyond and within Australia represents the Australian identity, it typically refers to a laid-back notion of time, where productivity is subjugated to leisure. Australians appear to have all the time in the world, and use time to their advantage. Phrases such as "no worries mate" and "she'll be right" are associated with the Australian way of life, and images of barbeques, early morning surfing and long outback drives do not necessarily conjure up images of a nation in a hurry, a nation concerned with progress and a lack of time.

Yet, of course, Australians *are* preoccupied with issues of time. Issues of Australian identity are tied up not only with images of where Australians came from but also questions of where they are going, what their future will be. There are discrepancies between the image of Australia that is sold to the world, of a timeless nation that takes it easy, and how some Australians might really see themselves: as hurried, stressed, hard-working, and constantly running out of time. Australia's unique history and the various aspects that make up dominant strains of the national identity have led to a mixed time-identity influenced not only by British colonialism and Aboriginal culture but also globalisation, Americanisation and capitalism as well as a sunny, hot climate which may lend itself to leisurely concepts of time.

Yet there are groups within Australia that are marginalised for favouring one or the other of these two dominant poles, and that at times even exist outside of these poles, in "third times." Among these groups is the negatively yet sometimes also endearingly portrayed group known as bogans. Like "chavs" in Britain or "white trash" in the USA, they are both mainstream and yet stigmatised for the way in which they interpret time and how they work for time or make time work for them. The following analysis of bogan time, as well as the carnivalesque form of their participation in the annual Melbourne Cup celebrations (the biggest annual horse race in Australia), focuses on the significance of bogan stereotypes and their relevance for understanding a multitude of desires, fears and stigmas regarding time in Australia. In addition, this chapter will argue for the need to stop time-stereotyping individuals and groups, and thus make room for plural temporalities.

The Bogan Stigma and Bogan Time

The label "bogan" is often directed at working-class people in Australia who are seen as being less cultured than other Australians. They are stereotyped as having strong sense of patriotism, broad Australian accents, a tacky way of dressing and being uncouth. As the bogan is difficult to define (see Nichols 2011), existing mainly in the eye of the beholder, and is a label that can be used when convenient and discarded when inconvenient, this paper will thus use the term bogan to refer to those who consider themselves bogans (as professed for instance on the website BoganAndProud.com.au), as well as the stereotype label used in popular culture.[1]

Bogans are often marginalised by other Australians as embarrassing to the national image of Australia; however in patriotic times such as Australia Day bogans find their numbers suddenly boosted as many Australians identify more strongly with their open sense of patriotism and perceived idleness. One online commenter with the pseudonym "Bogansnob" argues, "Everyone is someone else's bogan, just like everyone is someone else's snob. The real problem is when you need to put down another lifestyle to feel good about your own" (comment on Jennings 2010). In this vein, David Nichols suggests that "[t]he bogan delusion is a mirror" (2011, ch. 1), saying more about those who call others bogans than it does about the existence of bogans themselves, and he calls for the elimination of the stereotype as it only functions to denigrate a particular group of people in Australian culture. Yet one must also consider that some Australians prefer to identify with the bogan, not in order to openly portray themselves as uncultured, but rather because they celebrate a certain laid-back lifestyle and see themselves as harmless and fun. Such a view could even be constructive in the face of increasing late-capitalist values in that it emphasises quality time over quantitative (time-is-money) time.

As such a central element of the Australian culture, the bogan stereotype is important for indirectly exposing and critiquing aspects of Australian identity. Mel Campbell argues:

> Many Australians think they know exactly what a bogan is. Some say it's a socio-economic class. Some say bogans are a subculture. And others say that tastes or pop-cultural references make someone a bogan. Ultimately, bogans are none of these things. . . . And when we talk about bogans, we're really talking about national identity. Bogans pop up in the media and in

[1] This chapter will not consider certain aspects of the bogan stereotype such as bogans being racist or misogynistic, as it goes without saying that not all who associate with (or are labelled with) the bogan stereotype are necessarily prejudiced against others. See Nichols 2011, ch. 6.

> the public imagination as figures that are both embarrassing and 'un-Australian', and instantly, recognisably 'hyper-Australian.' (2010)

Bogans are marginalised because they are seen as lazy, unproductive, and damaging the reputation of Australians in general. Their time identity is often stereotyped as one of resistance to the dominant capitalist ideology of time as productive time. Yet Campbell makes a good point: "By labelling these tastes bogan, we don't have to consider the unsettling proposition that they're our tastes, too" (2010). All Australians are allowed to celebrate Australia Day overtly, but when a person practices patriotism or leisure in other contexts, he or she is a racist, lazy bogan.

Also associated with unemployment, bogans are often characterised as selfishly claiming time for themselves while the Australian taxpayer works hard to support them. E.P. Thompson suggests that acts such as sleeping in or taking excessive holidays were seen in the 1700s as leading to poverty (1967, 83), and he argues that this view has not changed much in recent times, for "the propaganda of time-thrift continued to be directed at the working people" because the "leisured classes discovered the problem of the leisure of the masses" (90). Interestingly, leisure is something that has to be "earned" or "deserved," and this does not seem to apply to many working class or unemployed Australians.

However, such stereotypes as described above have changed dramatically since the onset of the mining boom in states such as Western Australia. Mining companies need thousands of extra employees to work at mining sites several hours' flight from the capital cities. Jobs are offered not only to engineers, drillers and miners but also to truck drivers, cooks, road-workers and so on. These jobs often require work on a fly-in fly-out basis with workers living at the mines for several days or weeks while their families stay in the larger cities. While such long divisions of time between home and working life and the isolation of the mining communities can be criticised as being detrimental to the family unit (see Sharma and Rees 2007), these jobs are coveted because they pay extremely well. The popularity of these jobs, where often no degree or work experience is required, has led to a rise in a new cultural group popularly known as "CUBs" or cashed-up bogans. This group is stereotypically viewed as consisting of bogans who take on fly-in fly-out jobs in order to quickly earn a lot of money. The cashed-up bogan has become a popular stereotype within Australian culture and is slowly being associated with a new kind of time identity.

This time identity consists of working very hard over one or two weeks, and then returning home for a few days to participate in a period of intense consumerism and leisure. Time at the mines is used purely for the sake of earning money, whereas time spent back home must be seized to the largest extent possible. CUBs are often characterised by the fact that they buy very large, expensive houses and extravagant products associated with leisure time, such as jet skis and large televisions. Yet as Nichols comments, "It's ludicrous, plainly, to try and mount a case that middle- or working-class (to use the old parlance) people in outer suburban areas are more materialistic than the bohemian bourgeois" (2011, ch. 7). In terms of temporality, while they may appear to be in control of their time, their lives are in fact subjugated to the clocks and schedules of the fly-in fly-out lifestyle. CUBs have a vastly different time identity to ordinary bogan stereotypes in that they work extremely hard and have very little leisure time, with the result that this leisure time is a much more intense experience due to there being so little of it.

Interestingly, the CUBs, despite breaking free of the bogan stereotype of laziness and instead working hard in difficult conditions, are still marginalised by other Australians, as can be seen in the many disparaging comments on the website ThingsBogansLike.com.au. The fact that CUBs are able to earn twice as much as other Australians doing jobs that may require no degree could cause envy among those "white collar workers" who have spent years earning a degree in order to earn half as much money. As Campbell writes, "Put simply, the term cashed-up bogan suggests that there's something embarrassing and unsophisticated about a certain sort of spending—that some aspirations are somehow wrong or misguided" (2010). Of course, one *could* argue that the CUBs should in fact be praised within the capitalist paradigm for their hard work carried out "for the nation" and the time they invest in this difficult lifestyle, yet CUBs continue to be marginalised despite fitting well into the dominant capitalist ideology.

This marginalisation is largely based on the fact that CUBs are seen as purely consumerist from the populist point of view, working only to earn money and not actually getting any personal satisfaction out of their work. They are not seen as striving for progress, but rather as living intensely in the moment and wasting precious months or years of their youth (including motherhood or fatherhood) at the mines when they could be spending more valuable time at home. This is obviously a prejudice levelled at CUBs, and the truth may be quite different. Perhaps CUBs are scorned because they are seen as the "wrong people" to be getting rich, and they are not educated enough to spend their money wisely. This could also be a modern take on Australia's tall poppy syndrome (the tendency to criticise successful people). While such a view is regrettable, there are Australians that feel threatened by the bogan way of living. Unlike those

Australians who go for a camping trip up the coast with the family in order to forget about work, CUBs who live intensely in the moment on their days off are not given the empathy of other Australians.

This chapter does not condone excessive consumerism or spending. The issue is that CUBs are stereotyped for their consumerism, despite consumerism being an issue across classes in Australia regardless of education or background. Thus CUBs are merely a symptom of a system where money and possessions are prioritised over time. As Emma Norton writes,

> This stereotype is fostered by the establishment to silence the legitimate complaints of mining workers and to obscure the real bludgers and profiteers: the mining bosses. It affects the residents of mining towns who, having expected to reap some of the benefits of the mining boom, often end up blaming fly-in fly-out (FIFO) workers for the poor resources and development in rural Australia. The caricature of FIFO miners is also fed by elitist middle class prejudices towards the 'undeserving' workers, and is even adopted by some on the left. (2012)

She argues that "the stereotype of the 'cashed-up mining bogan' provides a convenient scapegoat for discontent over the contradictions of the resources boom" (Norton 2012). Equally important is the time-stereotyping that this involves: hierarchies of time are constructed among the working class on the part of those controlling the means of production, causing discontent and name-calling and thus veiling the true issues of ecological damage, declining living conditions in rural communities, and the perpetuation of a society that values productive time over quality time.

Looked at from a different perspective, the fact that mine workers work so many hours and do not have much time for their families could be seen as a way of deferring time. By earning lots of money now and essentially handing over their time to the capitalist machine, they may hope to have an easier life in the future where they can work less and thus be in control of their time. This is reminiscent of Michael Ende's novel *Momo* (1984) where the adults work tirelessly and put their time into a time bank to be used in the future, which they however never regain because they never stop working. Whether mine workers are able to find their own equilibrium between work and free time is an issue which is still being explored by Australian sociologists (see for instance Kaczmarek and Sibbel 2008).

The Melbourne Cup as Bogan Carnival

An Australian tradition that must be addressed in this exploration of the bogan stigma is that of the popularity of the Melbourne Cup among bogans. This race in Melbourne is popular with many Australians, not because all Australians are interested in horse racing, but rather because in numerous schools and workplaces work is stopped so that people can watch the televised race. In Victoria, the state where the race takes place, Melbourne Cup Day is a public holiday, whereas in the rest of Australia work is often only stopped for a short time for the main horse race. For this reason the race is often nicknamed "The Race that Stops the Nation."

In this way the Melbourne Cup functions as a kind of time island in Australian life where ordinary schedules are disrupted for an event that for most people has purely symbolic value in that it means they can stop working for half an hour or so. This time island is distinct from other nationally recognised time islands such as the one minute of remembrance for ANZAC Day that many workplaces and schools follow, in that it does not have sentimental value for most Australians. Instead it is largely seen as a moment where workers can break free of the drudgery of the ordinary work day and experience time differently for a few minutes.

The Melbourne Cup is seen to have popular following among so-called bogans, as identified on the website ThingsBogansLike.com.au (number 17 in their list). The event is traditionally associated with a fashion parade and It-girls wear fancy cocktail dresses and decorated hats to the race. "Ordinary" Australians sometimes take this opportunity to dress up too, with young women wearing fancy hats, "strappy little chainstore dresses, painful stilettos or streaky tans" (Griffin 2009). Griffin's language alone embodies the bogan stigma. Such attire is viewed by those who consider themselves of a higher class as a kind of performance in that bogans are seen as performing outside of their cultural group, attempting to live up to an ideal that is beyond their reach. This Melbourne Cup phenomenon is often portrayed in Australian popular culture, such as, for instance, in an episode of the television series "Kath and Kim" (cf. "The Mango Espadrille") where the cashed-up bogans Kath and Kim go to watch the Melbourne Cup, and end the day carrying their shoes in their hands and vomiting in public. These popular culture stereotypes of the bogan performance at the Melbourne Cup call attention to persisting cultural hierarchies regarding who is allowed to dress up, to get drunk, or to take a day off from work.

This portrayal of the Melbourne Cup is reminiscent of Mikhail Bakhtin's carnivalesque time. Bakhtin describes the feast-days and carnivals of the Middle Ages as embodying certain carnivalesque elements which included eating and drinking to excess, and an inversion of the social order for a limited time. Bakhtin sees carnival as closely associated with aspects of time and identity:

> Carnival with all its images, indecencies, and curses affirms the people's immortal, indestructible character. In the world of carnival the awareness of the people's immortality is combined with the realization that established authority and truth are relative. (1968, 256)

The time of the Melbourne Cup, like carnival, represents an island in time where the normal day or schedule is frozen. Describing carnival time, Bakhtin writes, "Let us stress this complete liberation from the seriousness of life. In the atmosphere of carnivalesque freedom and familiarity, impropriety also has its place" (Bakhtin 1968, 246), which is reflected in the drunkenness that often accompanies the Melbourne Cup. Bakhtin reinforces this with a quotation from Goethe's "Italian Journey," where carnival is "the suspension of all hierarchic differences, of all ranks and status; carnivalesque revelry is marked by absolute familiarity" (Goethe quoted in Bakhtin 1968, 246).

In the same way, Melbourne Cup time could be seen as a carnivalesque time because the distance is broken between socially marginalised groups such as bogans and those who take the Cup seriously as a fashion event and horse race, due to the bogans dressing up, drinking champagne and sometimes making fun of the entire event simply by participating in it. This equality and parody is what organisers of the Melbourne Cup may find threatening, as it is celebrated by the ordinary people to such an extent that the high class aspects come to be seen as ridiculous and superfluous. The Melbourne Cup, having been claimed by bogans, workers and other "ordinary" Australians as their own, is an island of time which is used for leisure, but also functions to contradict the dominant ideology in that it is not seen as productive time. Thus carnivalesque performances (and representations and stereotypes thereof) also say much about the desire to break free of schedules on all social levels.

As much as the Melbourne Cup exists as an island of time outside of the normal concept of time within Australia, it is nevertheless an officially recognised, delineated time presented to Australians as carnivalesque time, and thus exists within the dominant ideology. This reflects Bakhtin's suggestion that carnival times "uncrowned and renewed the established power and official truth" (1968, 99), because while they may upset the social order for a short time, they reinforce the dominant social order by recognising the fact that such an order exists, and returning to this order once Melbourne Cup Day is over. In this way such carnivalesque "mini" revolutions take on elements of revolution in the non-political sense, as a kind of repetition or cycle that ultimately changes nothing. Melbourne Cup Day is about creating a utopian now, seizing upon the free time given to workers by the nationally recognised event and enjoying it for the moment, whether it be for five minutes or the entire day, before life resumes its normal schedule and sense of time. One sees similar phenomena in holidays such as Halloween in the U.S. Like Homi Bhabha's (1995) concept of the third space as an in-between site of negotiation and expression of identity on the part of marginalised groups, time islands and other alternative cultural concepts of time function as "third times" that can be seized by marginal voices and thus maintain a sense of time-pluralism within the Australian identity.

A Call for a Revolution in Time-Stereotyping

Bogans may not usually be considered marginal voices; however they represent individuals exercising alternative or subversive temporalities in opposition to the dominant sense of time imposed by the state or populist media (while at the same time living within these imposed senses of time). Even the cashed-up bogans, who appear to conform to the capitalist time of productivity, nevertheless find ways to subvert its totality. Many CUBs are also subject to the control of commodified time in that they constantly defer their dreams in order to earn money today, and can celebrate their wealth only in short, intense bursts. As the example of the Melbourne Cup demonstrates, it is within such carnivalesque processes of signification that sites of temporal dissidence, or time islands, can upset the social order, even if just for a day.

As David Harvey points out, "[s]patial and temporal practices are never neutral in social affairs. They always express some kind of class or other social content, and are more often than not the focus of intense social struggle" (1995, 239). It is surprising that not more has been done to address the intolerance caused by contending concepts of time across the world. Australia's two dominant time identities (she'll-be-right time vs. productive, progressive time) are exposed by the bogan stereotype as constructed to suit the needs of those who define policy and seek to define the Australian identity in order to exclude those "values" that are not convenient to a view of the nation at a given time. Maybe the popularity of the Melbourne Cup is indeed a sign of the strain and stigmas that class differences place on workers, as well as their desire to escape from the drudgery of the everyday, suggesting that the dominant capitalist stereotypes of time in Australia are not really conducive to a happy lifestyle.

A revolution in cultural thinking is needed to prevent time stereotypes from being accepted as natural. As Australian band The Cat Empire sing, we must learn to "find the hours that the clocks cannot define" (Reibl 2003), for it is in this dimension outside of measured time that contrasting voices can be expressed and heard, and alternatives to dominant societal structures might be discovered. If the dominant classes continue to suppress or exploit liminal time identities, then the revolution in cultural thinking and the fight against time stereotyping must come from below. At the time of writing (2012), movements such as Occupy Melbourne or Occupy Sydney represent an organised voice speaking on behalf of workers' rights and equality that is making itself heard in the national media, which could also work to expose these constructed time identities. Yet the media is also going to great lengths to portray the oppressed as a splintered group with conflicting (rather than plural) interests, pitting "bogan" stereotypes against "hippie" stereotypes and claiming that bogans enjoy watching hippies be beaten by the police. The website ThingsBogansLike.com.au suggests in their entry on "Bashing Hippie Skulls" (no. 248):

> The bogan, as we have well learned by now, has an astonishingly broad vocabulary with which to insult other bogans. . . . As the political class have increasingly courted the bogan over successive generations, and the corporate world has become more sophisticated in persuading the bogan to act against its own interests, we live in a world today of growing income inequality. Meanwhile bogans remain the prime beneficiaries of the explosion in welfare payments that do little to even out income inequality. . . . [The Bogan] is comfortably ensconced in its cocoon of normalcy. . . . Thus, it is offended by the existence of those unwashed and dissimilar who insist on appearing in public places where they cannot be easily ignored.

While this analysis only functions to portray bogans as unintelligent, violent and lazy, it is correct in that although bogans are denigrated by the upper classes, for politicians the masses (which are sometimes grouped as bogans) are also their voting base, and thus it is in the interest of politicians to portray participants in the Occupy movement as hippies and the enemy of the "bogan" masses (because hippies value non-consumerism), so as to belittle the Occupy movement and further promote consumerism and fear politics.

Participants in Australian Occupy movements and any movements in the future must speak out to fight the stereotyping of any class. This can start with attempts to communicate to the media that the Occupy movement itself is not a collection of unemployed, lazy hippies but represents an alternative temporality in that it *makes time* for plural voices and concerns to be heard. Protesters must also be careful not to fall into the stereotyping trap themselves. The problem is not uneducated, lazy bogans, but rather the high cost of university education, high unemployment, and the need for a more balanced media. The problem is not consumerist seize-the-moment bogans, but rather a system that depends on consumerism to perpetuate itself. It is not the cashed-up bogans ruining rural towns and inflating prices with their high-consumption weekends but a government that promotes this culture in its reliance on mining to facilitate progress. It is not that drunken women in chain-store dresses ruin an Australian holiday, rather we must consider what such cultural portrayals really say about how time, leisure and work are valued and who is allowed to participate in escaping from the drudgery of the everyday. Time stereotypes are a symptom of a greater, prejudiced, self-perpetuating system that needs to be challenged at its very base, and thus any movement such as Occupy must be seen as part of a possible solution.

Works Cited

Bakhtin, Mikhail. 1968. *Rabelais and His World.* Translated by Helene Iswolsky. Cambridge: MIT Press.

Bhabha, Homi. 1995. *The Location of Culture.* London: Routledge.

Campbell, Mel. 2010. "Perhaps There's a Little Bogan in Everyone." Accessed Aug. 25, 2011. http://melcampbell.com.au/archives/perhaps-theres-a-little-bogan-in-everyone/

Ende, Michael. 1984. *Momo.* Hammondsworth: Puffin.

Griffin, Michelle. 2009. "Bogans Seem to Have More Fun at the Races." *The Age,* November 2. Accessed August 26, 2011. http://www.theage.com.au/opinion/bogans-seem-to-have-more-fun-at-the-races-20091101.hrpe.html.

Harvey, David. 1995. *The Condition of Postmodernity.* Oxford: Blackwell.

Jennings, Joshua. 2010. "Suburbia: The Hits and Myths." *Domain,* May 22. Accessed My 8, 2012. http://news.domain.com.au/domain/real-estate-news/suburbia-the-hits-and-myths-20100521-w1ba.html.

Kaczmarek, Elizabeth and Anne Sibbel. 2008. "The Psychological Well-Being of Children From Australian Military and Fly-in/Fly-out (FIFO) Mining Families." *Community, Work and Family* 11.3: 297-312.

Nichols, David. 2011. *The Bogan Delusion* (Ebook). Mulgrave: Affirm Press.

Norton, Emma. 2012. "The Myth of the Cashed-Up Mining Bogan." *Socialist Alternative*. Accessed Mar. 11, 2012. http://www.sa.org.au/index.php?option=com_k2&view=item&id=7224:the-myth-of-the-cashed-up-mining-bogan&Itemid=392

Reibl, Felix. 2003. "How to Explain." *The Cat Empire* (album). EMI. Performed by The Cat Empire.

"The Mango Espadrille." 2004. *Kath and Kim.* Gina Riley and Jane Turner. Australian Broadcasting Commission.

Thompson, Elaine. 1994. *Fair Enough: Egalitarianism in Australia.* Sydney: University of New South Wales Press.

Sharma, Sanjay and Susan Rees. 1997. "Consideration of the Determinants of Women's Mental Health in Remote Australian Mining Towns." *Australian Journal of Rural Health* 15.1: 1-7.

Thompson, E.P. 1967. "Time, Work-Discipline, and Industrial Capitalism." *Past and Present* 38:56-97.

SEPTEMBER 12TH: FROM HEROES TO TERRORISTS

DALBIR SEHMBY

> While in the last century or so Sikhs had been becoming intricately woven into the fabric of Western societies, 9/11 questioned whether individuals with such a peculiar and religious-based identity should be part of those societies—in short, Sikhs were not only being treated differently on account of their appearance; their identity marked them as different, and therefore a suspect element in the West.
> —Sidhu and Gohil

> When the story of the siege was told to the House of Commons by the Secretary of State for India, every member rose spontaneously to his feet in tribute to the men of the 36th Sikhs.
> Each one of the defenders received the immediate posthumous award of the Indian Order of Merit.
> —James Lunt

Introduction

Over a century ago, during the height of colonial rule, a small regiment of self-sacrificing, turbaned, bearded Sikhs roused the British House of Commons to a spontaneous ovation. In the new millennium, when the world has become supposedly more pluralistic and open-minded, the hairy-faced, turbaned individual is treated with xenophobic suspicion. Such cultural paranoia requires inquiry and change, especially in our current global climate. Early 21st century protests and social movements the world over are only a part of the solution. Indeed, a revolution rests on the idea of progress. Although we have lived through civil rights movements and although we live in a time of human rights, stereotyping and cross-cultural conflicts persist. More than progress based upon conflict, there needs to be a serious re-evaluation of contemporary assumptions of time, re-evaluations which lead to greater awareness of the way our relationship to time is far from natural, scientific, or fixed.

Realizing that our view of time can vary, we may explore and perhaps even develop alternative relationships to time. This study hopes to offer some insight into the link between the dangerously blinding assumption of western chronological progress and the process of stereotyping, an act that relegates a target group to an assumed past and situates that same group as a future threat, leading to a conflict-ridden present. Western societies benefit, ideologically, from the position of assumed progress, for it situates the West in a reductionist, hierarchical binary with those cultures deemed regressive.

In particular, post-9/11 hate crimes against turbaned Sikhs indicate a violent lack of cross-cultural understanding. Despite their recognizable headgear, or maybe because of it, little is known of the Sikh concept of time and oneness, or their history, including dates like September 12th. As the anniversary date of the Battle of Saragarhi and as the day after the tragedy of 9/11, for contemporary Sikhs September 12th exists as a loaded signifier of polarizing associations: the turbaned Sikh as heroic defender or the turbaned Sikh as mass murdering terrorist. September 12th marks an inspiring last stand, when, in 1897, twenty one turbaned Sikhs held the fort of Saragarhi against thousands of Afghans. Sikh bravery resulted in praise from the British, and ever since, the Battle of Saragarhi has inspired Indian Sikh (and non-Sikh) regiments. However, because it is the day after 9/11, September 12th signals another meaning, one which conflates the Sikh turban (*daastar*, *paghri*, or *pagh*) with the terrorism of Osama bin Laden. The day after 9/11 signaled change: when the post-Soviet world replaced the red menace with a brown one, and when the turban-wearing Sikh became stereotyped as outdated, primitive, and tyrannical. However, as this paper will illustrate, stereotypes surrounding the turbaned male were already lingering, due to a perspective rooted in the chronology of western progress. Thus, due to its historical association with the Battle of Saragarhi and its contemporary association with the era after the 9/11 attacks, September 12th exists as a date of superimposed binaries for Sikhs, overwhelming the Sikh heroic past with contemporary fears of the turbaned male and all he is assumed to represent.

First, to move beyond the West's self-positioning at the lead of progress, I will offer insight into the Sikh implications of time. In particular, during these fast-paced times, the Sikh pursuit of meditative oneness may serve as an alternative to the contemporary urbanite's servitude to time as a marker of productivity. Second, instead of treating stereotyping solely through the lens of prejudice, this paper views stereotyping as intricately linked to a Western chronology which results in a reductionist material narrow-mindedness that fosters a difficulty to view cultural difference outside of the temporal framework of progress versus those deemed regressive—in this case, the turbaned Sikh. Third, due to the alluring narrative of chronology associated with Western progress, the complicated histories of Sikhs—exemplified by moments like Saragarhi—require recovery.

Time, Turbans, and Oneness

For Sikhs, time is simultaneously finite and infinite, as no moment is more important than the now. As opposed to the either-or dilemma of Hamlet's famous inquiry of "To be or not to be," a Sikh answer may be "To be and not to be," for being at one in the flow of the moment is a state that one pursues (to be) and abandons into (not to be). The intellectualizing goal-oriented "To be" is ego-driven and misleading due to the control it implies, so such decisive action requires a sensitive balancing with the humility of chance, the unanticipated, and uncertainty. The meditative life of a Sikh is dance of active pursuit with the faith of letting go. Pursuing/allowing for such moments of oneness is an important path to peace of mind. Sikh time can be understood through two predominant metaphors: oneness and the double-edged sword.

Beyond self-doubt and overcoming inaction, being at one at any particular moment is crucial to Sikhism. The opening line of the Sri Guru Granth Sahib, "*Ik Onkar*" is often translated as "God is One." However, the term is more than a monotheistic declaration. *Ik Onkar* connotes "One Being" and all that implies: one mysterious force in the universe (the "God" concept for Sikhs) and a state, a certain oneness of being. Oneness can be achieved through meditating and raising awareness in everyday activities. Such oneness is an experience of the divine in life. Along with oneness of being, there is "oneness of beings," in the sense that the oneness of soul—*Ik Onkar*—lacks a clearly delineated space and boundaries of the divine, spirit, and human beings—and between human beings (regardless of race, gender, class, caste, or religion).

In Sikh perception, the God concept is transcendent, yet immanent; moreover, such a conception is experienced in the oneness of time. As the first line and the goal, the opening line "*Ik Onkar*" of Guru Nanak's Japji Sahib prayer (the first morning prayer and the opening of the Sri Guru Granth Sahib, the Sikh sacred text) marks the beginning and the end, the infinite: "God is One." This oneness differs from other monotheistic religions in that this oneness does not signify a spiritual entity as such, but rather this entity's encompassing and inclusive being-ness. For Sikhs, the divine is a mystery, even though it can be experienced. Because of this, no one can have exclusive claims to the divine—a nameless, shapeless, and timeless entity which is indescribable. As reiterated through poetic repetitions in Guru Gobind Singh's composition, Jaap Sahib (the second morning prayer), God "is the cause of worldly beauties and is present everywhere," and "is present in nether regions, the skies, and in the space [beyond Earth]" (*Sacred Nitnem* 2002, 91). In other words, oneness is the Sikh vision of the divine in all, the Sikh concept of a common humanity, as well as the state and a goal of the Sikh sense of being. The "oneness" that opens the Japji Sahib marks the developmental spiritual goal of a Sikh that one must achieve through continuous (and unrestricted to a particular space or time) personal meditation on the nameless name of God.

Oneness also refers to the state of balance of truthfulness that is indefinable and indeterminate. In contrast to traditional views of truth that are defined by religious authorities and are meant to be conclusive, liberating, and satisfying, in Sikhism, truth is symbolized by a multifaceted image of a double-edged sword. On one level, the two blades of the curved swords (used in the *Khanda*—the emblem of Sikhi) symbolize the spiritual and secular realms that must be balanced through and condensed in the moment of unity and oneness symbolized by the *Chakkar* circle (timeless, limitless, and all inclusive) and the central double-edged sword—the tip of which provides a site for and a reminder of the difficulty of attaining a state of graceful balance. Grace here is an important concept as well. While one can make an effort, one must also be effortless, and through such a sense of faithful abandon, one will experience moments of balancing: flowing oneness through grace. As described in the fifth morning prayer, Anand Sahib from Guru Amar Das, the spiritual journey of a Sikh is a journey along "the path which is sharper than a sword and finer than a hair" (*Sacred Nitnem* 2002, 177). Thus, the symbolic representation of spiritual development, as well as time, is an unresolved balancing act—atop the tip of a sword or along a strand of hair—that must be experienced in the present.

The Sikh notion of time is difficult to define using the traditional categories of either linear or circular time, for time is a paradox: "According to the Gurus, it is neither real nor illusory, yet both; neither subjective nor objective, yet both. On the one hand, time is the matrix in which the self is trapped . . . On the other hand, given that the locus of one's existence in time is the body, time is a gift." Hence,

> the Guru teaches that attachment and suffering result only from our habitual obstruction of the natural flow of time. By accepting time's impermanence as our own essence and seeing every attempt to control time as ultimately illusory like the waves of the sea, which are here one

> moment, gone the next, one can be released from suffering. (Shackle and Mandair, 22)

In Sikhi, hair becomes a reminder of the trap and gift of time.

Alongside other reasons for hair growth, moving with the flow of life is reflected in the growth of hair that is protected by the turban (where the winding cloth itself is an active daily reminder of this timeless state of flow). Instead of controlling and cutting hair, Sikhs are encouraged to manage their hair, as a reminder of how time cannot be stopped, that the flow of life should not be denied. Just as time should not be denied, or "cut," the hair (and time) should be managed. Cutting hair resists nature's flowing gift of time, analogous to the wish of stopping time, a wish that communicates insecurity, fear, and anxiety. Metaphorically, cutting hair mimics tyranny, where one practices an aggressive control over things, instead of working with them. Fretting over lost time and control sparks a vicious cycle, until one realizes how in each moment, time is lost forever, but also available. Simultaneously, one lives and dies in each moment—rebirth arrives in every breath. Remembering the divine gift of life and learning happens in each breath (literally "Sikh" implies "to learn"), with the inhale (*Wahe*—an expression of wonder or awe) and exhale (*Guru*—movement from darkness to light) of *Wahe Guru.* Like riding the wave that is *hukam*, or God's will, the Sikh should not fight against time, but move with it, striving (without forcing) to live at one in each moment of every thought and action for peace of mind, body, and soul.

Mapping Regress: Stereotyping and the Turban

Stereotyping, as well as the lack of differentiation and the discrimination that arises from it, stems from a linear, chronological view of time, where the culturally progressive is contrasted with the primitive—and all this is mapped into easily identifiable visible markers, such as race, gender, or, in the case of Sikhs, the turbaned male. Before the word meme became an Internet phenomenon, stereotyping may have been a persistent type of cultural meme, allowing communities to build collective identities through the difference guaranteed by the biases of making relatively efficient generalizations of outsiders (or those deemed as outsiders). By drawing sweeping conclusions, stereotyping's hasty generalizations save time, binding a community together through imagined fears.

If a stereotype has existed for some time, then the act of stereotyping breaks down barriers of time, linking one to the biases of the past. Stereotyping can even make the history of relationships (between insiders and outsiders) for one community become the history of another community, as can be observed in the way formerly disparate (even feuding) communities become more homogeneous. A Russian and American at odds with one another a generation ago can bond over a collective terrorist threat. Contemporary political efficiency, rather than accuracy or fairness, makes stereotyping practical for a unified (or unifying) community. Stereotypes allow one to make sense of others in a falsely-informed manner, by enabling the perceiver to categorize with both ease and (perceived) peer support. Even though none of the 9/11 hijackers were Sikh, wore a turban, or had long beards, from the timeline of progress, the turbaned male nonetheless already fulfilled the scapegoating fantasy of a barbaric threat.

Along the line of chronological progression held by the perceiver, the Sikh male is easily associated with the culturally backwards past: the uncivilized, hairy, and oppressive primitive who contrasts the modern, civilized, "clean" democratic citizen. In "Locating the Sikh *Pagh*," Kalra explains:

> Whether American modernity or Indian modernity, the turban seems to now stand for an absolute symbol of opposition—eighth century for Muslims and eighteenth for Sikhs—to any tendencies of modernity . . . A binary opposition of almost phantasmagoric dimensions between the modern West, medieval Islam; between the clean-shaven white man and the bearded and turbaned fundamentalist, begins to take form. (Kalra 2005, 77-78)

The differences propositioned by prejudice function through a binary hierarchy, situating one community as more culturally and politically advanced than another.

Space functions in this imagined chronology as well, for the turban is associated not only with another time, but with another imagined space, an exotic or even brutal landscape, a space which itself is implicitly characterized as out-of-date or as a space of lack—without modernity. After the tragedy of 9/11, the Sikh turban also suggests the patriarchal extremism of the Taliban and al-Qaeda. Along with time and space then, politics are mapped onto the turban, a contemporary, antagonistic politics which associates the turban with tyranny, non-democratic values, and violence. With the turban being out of synchronicity (and thus space and politics) with the contemporariness of the western or modern perceiver, the turbaned hirsute individual is associated with primitiveness, or a lack of grooming.

Indeed, contemporary beauty nomenclature (trimming, plucking, and shaving) draws a parallel between managing an open, natural, and potentially wild space (a field, yard, or garden) and maintaining one's face. Hair growth is likened to the growth of grass, or weeds in the garden. If language carries culture, then the linguistic (and cultural) implication is that the civilized manage the natural space by taming the wild through technology (rakes, pitch forks, shovels, saws, knives, and shears). Correspondingly, one's measure of civility can be located upon one's face. Through the technology (combs, brushes, scissors, blades, tweezers, combs, and brushes) of grooming, one must manage all those wild hairs sprouting upon the face of the civilized male.

In an almost contradictory way, civilized maleness is associated with some degree of facial hair, and perhaps most tellingly, less with hair itself, than with the capability of growing hair. The capability to grow facial hair is both a biological and a social indicator of masculine maturity. With burgeoning sprouts, civilized men are always on the verge of going wild. Facial hair is a signifier fraught with loaded cultural meanings, some more playful than others. A bit of five o'clock (after the work day) shadow makes one rugged. Too much makes one into a hippie, or with age, Santa Claus. More seriously, hair connotes a reactionary lifestyle choice (the hippie as counter to "culture"); the hippie is stuck in time, the 1960s. Or hair connotes a state of time (the advanced age of a bearded elder); in a youth-oriented culture especially, hair growth communicates a lack of grooming or the passage of time. Of interest here though is the relationship between the notions of temporal progress, the civilized, and the wild.

Masculinity is associated with the capability of growing hair and the management of such hairs, paralleling the way (patriarchal) civilization is always on the verge of collapsing back into the wild that surrounds it, if it were not for the efforts of those institutions (judicial, political, educational) and forces (military, police) who can pluck the weeds of devolution. The contemporary, civilized male then, wears a face that embodies the relationship that contemporary society has with nature: one of exerting dominance and control through excising or managing the wild. With the morning shave, one engages in a ritualized manifestation of treating hair as a distraction to (and detractor of) the civilized clean.

In contrast to the apparent cleanliness of "clean-shaven," unshorn hair implies dirtiness. Dirtiness also indicates a state of temporal weight. Dirtiness, as it relates to the civilized and progress, is extremely important, for even racial differences can be associated with cleanliness and the apparent lack of it. If whiteness is clean, then anything of colour is sullied whiteness. For the narrowest and most conservative mindset, skin colour variations have less to do with melanin and more to do with dirtiness, an imagined natural reminder of the darker individual's fundamental inability to become truly cleansed by civilization.

Clearly, with the notion of civilized progress, a hierarchy is at work, one that undermines the various mechanisms of civilization. If one's lack of civilization is intrinsically linked to the apparent dirtiness of one`s skin colour and quantity of hair, then no amount of education, culture, and religion can completely redeem the imagined primitive. Moreover, such racist reluctance to accept the bearded non-white individual is also steeped in the superstition of prejudice, or a sense of managing the temporal future. Scapegoating provides the prejudicial community with a rationalizing narrative, where the blamed group bears the fears of the stronger community and the implied elimination of the prejudiced group is believed to guarantee a problem-free future. In other words, if the hairy-faced turbaned individual becomes contemporary and thus "civilized," by cutting his wild hair, then all may be well for the larger community's continued progress. The turbaned, bearded Sikh evokes fear because he not only represents difference, but also because he reminds the "clean-" shaven community of its own potential for regression.

Lost Histories

Unfortunately, after 9/11, there have been hundreds of reported attacks against Sikhs, ranging from verbal abuse and bullying to physical attacks and even murder. Occurring four days after the tragedy of the towers, Balbir Singh Sodhi's death marked the first post-9/11 murder (Sidhu and Gohil 2009, 76). Three years later, in 2004, Rajinder Singh Khalsa and his brother were beaten by five attackers in Queens, New York (Sidhu and Gohil 2009, 78-79). In 2006, in California, Iqbal Singh was stabbed in the neck. In these two incidents, the victims survived. However, as reported on MSNBC online, ten years after 9/11, in California, two elderly, turbaned Sikhs, while taking a stroll, were shot dead in what is assumed to be another hate crime (Hoang 2011). Targeting of turbaned Sikhs brings western misunderstandings to the fore. Although twenty-first century North American cities are pluralistic, contemporary nations have difficulty comprehending turbaned Sikhs as multi-faceted and varied personalities—let alone as heroes.

Unlike the 9/11 era's imagined chronology that situates turbaned Sikhs as backwards, savage, and undemocratic, Sikhs have a contested history, one replete with memories of egalitarianism, individualism, and democratic values. September 12th is telling because, perhaps in search of ways to counter the post-9/11 stereotyping, in recent years there has been a resurgent interest in the Battle of Saragarhi, with Jessi Thind's novel *Saragarhi* (2007), a recent graphic novel, and talk of a mainstream Indian film.

Saragarhi remains a testament to the unflinching bravery of twenty-one turbaned Sikhs on September 12th, 1897. Their bravery may not be causally attributed to their turban, their situation, or even their Sikhness, for bravery exists in many individuals of different appearances and beliefs. Donning a turban and becoming a Sikh does not automatically make one brave or heroic. Nevertheless, despite the variety of Sikh personalities and events in history, by the time of Saragarhi, Sikhs were already building a reputation as heroic defenders. Guru Nanak, the first Guru, challenged prevailing biases against caste and gender; however, Nanak's methods were peaceful. With the increasing tyranny of Mughal rulers, violent altercations arose. In Sikh memory, the martyrdom of Guru Tegh Bahadur prompts the eventual development of the *Khalsa*. Prompted to stand up for the religious freedom of Hindus and Sufis, Tegh Bahadur peacefully challenged the conservative, power-hungry Muslim Emperor Aurangzeb (who even jailed his father Shah Jahan). Refusing to convert to Islam through financial coercion and then torture, Tegh Bahadur died a martyr for individual freedoms in 1675. Tegh Bahadur's son, Guru Gobind Singh, the tenth and final living Guru, formed the *Khalsa* in 1699, establishing an *amrit* ceremony to initiate followers into a distinctive dress. *Khalsa* Sikhs upheld/uphold the five gifts (the five *kakkars*: hair, comb, long undergarment, steel bracelet, and dagger) from Gobind Singh. Fuelled to fight against tyranny and for the burgeoning democratic vision of the *Khalsa*, Sikh armies helped ensure the survival of multiple faiths in India.

Because of their willingness to stand up for principles at great cost to themselves, the examples of Guru Tegh Bahadur and Guru Gobind Singh (and others) inspire Sikhs. The early eighteenth-century *Khalsa*'s

> insistence on military training, on spirited resistance to tyranny, on individualism, on equality and democratic decision-making, on the creation of the Khalsa itself, were milestones on the journey to the Sikh state. (Patwant Singh 2002, 70)

In the face of persecution, Sikhs were resolute. Patwant Singh continues,

> No one persecuted the Sikhs with greater zeal than Zakariya Khan . . . His columns combed the countryside for Sikhs, and those captured alive were brought to Lahore for daily public executions. He fixed a reward of fifty rupees for every Sikh head brought to him. (Patwant Singh 2002, 83)

Battle-hardened, by the time of Maharaja Ranjit Singh, in the early nineteenth century, Punjab's Sikhs developed into a formidable force, one that impressed the British during the two Anglo-Sikh wars. A mutual respect developed early into the annexation of Punjab, leading to comments such as the following: "the British 'Raj' has to-day no supporters more capable or more loyal than the Sikhs" (Gough and Innes [1897] 2005, 2). Unlike the pre-colonial period, the turbaned *Khalsa* (initiates who have taken *amrit* and adopted the five *kakkars*) and the Sikhs increasingly would "become interchangeable terms" in the twentieth century (Dhavan 2011, 12). Promoting the *Khalsa* Sikh image, the British provided the first distinctive definition of Sikhs in the 1891 Punjab census and that distinction was visual: long-haired, turbaned, and bearded (Takhar 2005, 15). Into the twentieth century,

> Sikhs formed about one-fifth of the imperial army in the First World War . . . The Sikhs won 14 of the 22 military crosses awarded to Indians for conspicuous gallantry. From Shanghai to Mesopotamia, Sikhs heroically guarded the outposts of the Empire (Nikky-Guninder Kaur Singh 2011, 143).

As this brief survey should indicate, contesting histories are at work: a post-9/11 demonization versus Sikh heroics. While post-9/11 stereotyping works on imagined binaries, the historical recognitions of Sikh achievements serve as a "reminder of the limitations of any Saidian approach which suggests that a simple 'self versus other' dichotomy was utilized by British Orientalists to distance themselves from Indians" (Ballantyne 2002, 116). While Said's contributions have been extremely valuable, the relationship between Britain and its colonies was complicated and often contradictory, especially in different periods of the Raj.

Such contradictions and changing perspectives come to the fore when India gains independence. "After the first world war, the amicable Anglo-Sikh relationship underwent a major change" (Nikky-Guninder Kaur Singh 2011, 150). In the mid-nineteenth century, a precursor to Gandhi, Ram Singh Namdhari had already "launched a non-violent and civil disobedience movement in the Punjab" (150). So, alongside Sikh service to the Raj, there was a peaceful movement advocating self-rule. However, with increasing racial propaganda, double-standards, and violence (such as the *Jallian Wallah Bagh* massacre), Sikhs once again stood up against blatant abuses of power—this time for Indian independence. Citing Puri, Patwant Singh provides the following numbers:

> Out of 2,175 Indian martyrs for freedom, 1,557, or 75 per cent, were Sikhs. Out of 2,646 Indians sent to the Andamans for life imprisonment 2,174, or 80 per cent, were Sikhs. Out of 127 Indians who were hanged 92, or 80 per cent, were Sikhs . . . And the Sikhs comprise only 2 per cent of India's total population! (Quoted in Patwant Singh 2002, 254)

With memories such as these floating in the Sikh consciousness, the Battle of Saragarhi functions as more than just a reference to a particular date, but as defiance in the face of stereotyping and lost memories (for the West and for those Sikhs unaware of such a heroic history).

Unfortunately, the Indian, British, and Sikh accounts of inspirational Sikhs fall victim to varying degrees of cultural amnesia brought about by the politics of independence, Sikh extremism, and imagined 9/11 associations. The service and sacrifice of Second World War veterans (Sikh and non-Sikh) become lost between one time, the ending of the Raj, and another time, the birth of an independent India. England saw Indian soldiers as unfaithful for no longer being a part of the Raj, while India saw their soldiers as unfaithful to the new India, for having served the British. For the turbaned Sikh, this state of limbo lingered until 9/11, after which the turbaned Sikh male identity merged with mass media images of a non-Sikh, Osama bin Laden. Before 9/11, in the 1980s, Sikhs had already flirted with terrorism through the Khalistani movement for an independent homeland. Certainly, the contributions of Sikhs fighting for Indian independence were also terrorists—albeit ones that became Indian heroes. Before 9/11, terrorism was a less negative term in the West. France's Reign of Terror voices an important moment in western democracy. Since then, national independent movements and Civil Rights movements the world over have resorted to acts of terror. In western thought, terror also forms the basis of Marxist revolution. Terror signaled a time for change in the face of perceived injustice. After 9/11, terror becomes definitively negative, or anti-civilization. With this change, notions of the wild turbaned foreigner are also revived—notions that link back to Imperialistic stereotypes of the savage who requires colonization and who is without an admirable history.

Conclusion

Contemporary North American society views turbaned Sikhs through an imagined chronology of linear western progress that fuels a cultural amnesia, where dates like September 12th, 1897, once important for even the West via Britain, become lost in time. Despite living in a period when protests shake nations and annoy Wall Street investors, there remains a lack of inter-cultural inquiry into the link between change, time, and culture, prompting the proliferation of ethnic stereotyping and the initiation of hate crimes; as a result, the chronology of hate includes moments of cyclical violence. Chronology as progress dominates contemporary conceptions, but time is more than the clock of progress chasing an imagined future through the elimination of superstitious fears. Slipping into the habit of justifying hate seems too recurrent a compulsion in human history, one that requires a shift. Perhaps this paper's inquiry into time will help us with future navigations. With multiple views of time (as linear, cyclical, regressive, progressive, or paradoxical), time is perhaps better referred to as times—in the plural. This paper's glimpse into some Sikh thoughts on time is but a minor contribution to a burgeoning discussion on the intersection of time, culture, belief, and perspective. Let us hope that inquiries such as these will help ease misunderstanding and quell the momentum of hate.

Although not a Sikh, with his beard and turban, Osama bin Laden's post-9/11 videos amplified the fear of patriarchal regression already a part of contemporary societies. Even though Islam and Sikhism are differing faiths, and despite the fact that the majority of Muslims are not al-Qaeda members, India's Sikh turban becomes confused with the Taliban. With crowded, dirty streets and infamous poverty and with cave-dwelling gunners, India and Afghanistan fit the popular imagination of the uncivilized, more easily than the clean, rich, more temporally progressive spaces of Saudi Arabia or the United Arab Emirates—despite Middle Eastern dress or the fact that the groomed 9/11 hijackers were from there. Rather, in the twisted rationalizations that form stereotyping, the turbaned Sikh male (and hijab-wearing Muslim female) become visually convenient targets. Such faulty reasoning is analogous to blaming the moustache for Hitler and Stalin. Of course, the issue harbors less reasoning than it does prevailing cultural fears, embedded in the imagination of temporal progress and selective memory.

While the hair and turban of the Sikh embodies the philosophy of moving with the flow of time, the "civilized" view is to cease and control time. Nevertheless, the issue is more than masculinity and wild hairs, for there exists an underlying fear of an imagined future. Terrorism evokes a collapse into brutality worse than experienced in the world wars, for terrorism implies the very final minutes of civilization. Wars have rules; terror does not. With cave-dwelling men primed to attack at any moment, civilizations are threatened by barbarians immune to the rules of warfare. Most frighteningly, time may repeat itself, for the memory of Rome's collapse echoes in contemporary imaginings of western nationhood. Rome's past meets with the West's present in a fear of the future evoked by the sight of the East's turbaned Sikh: civilization may crumble once again into the dark—and hairy—ages.

Works Cited

Ballantyne, Tony. 2002. *Orientalism and Race: Aryanism in the British Empire*. New York: Palgrave Macmillan.

Dhavan, Purnima. 2011. *When Sparrows Became Hawks: The Making of the Sikh WarriorTradition, 169-1799*. New York: Oxford University Press.

Gough, Charles and Arthur D.Innes. (1897) 2005. *The Sikh and the Sikh Wars: The Rise Conquest and Annexation of the Punjab State*. New Delhi: Kaveri.

Hoang, Lien. 2011. "Mistaken for Muslims, Sikhs hit by Hate Crimes." Associated Press. Accessed Feb. 13, 2012. http://www.msnbc.msn.com/id/41962756/ns/us_news-crime_and_courts/t/mistaken-muslims-sikhs-hit-hate-crimes/#.T0mY7vWROSo.

Kalra, Virinder S. 2005. "Locating the Sikh Pagh." *Sikh Formations: Religion, Culture, Theory* 1 (1): 75-92.

Lunt, James. 1977. "Saragarhi: Sikhs and Pathans, 1897." *History Today* 27 (4): 226.

Sacred Nitnem: Containing the Divine Hymns of the Daily Prayers by the Sikhs. 2002. Translated by Harbans Singh Doabia. 23rd edition. Amritsar: Singh Brothers.

Shackle, Christopher and Arvind-pal Singh Mandair. 2005. *Teachings of the Sikh Gurus: Selections from the Sikh Scriptures*. New York: Routledge.

Sidhu, Dawinder S. and Neha Singh Gohil. 2009. *Civil Rights in Wartime: The Post-9/11 Sikh Experience*. Surrey: Ashgate.

Singh, Nikky-Guninder Kaur. 2011. *Sikhism: An Introduction*. London: I.B. Tauris.

Singh, Patwant. 2002. *The Sikhs*. New Delhi: Rupa.

Takhar, Opinderjjit Kaur. 2005. *Sikh Identity: An Exploration of Groups among Sikhs*. Burlington: Ashgate.

Thind, Jessi. 2007. *Saragarhi*. Victoria: Trafford Publishing.

STRANGERS TO LIFE ON EARTH: THE GENDERED POLITICS OF NOSTALGIC NATIONALISM AND RIGHT-WING POPULISM

AMANDA CONROY

> The loss of the dimensions of earthly inheritance goes hand in hand with the neglect of Hestia in favor of the male gods, defined as celestial by philosophy from Plato onwards. These extra-terrestrial gods would seem to have made us strangers to life on earth . . .
> —Luce Irigaray

Introduction

The ways in which we inhabit time are intimately connected to the ways in which we matter. By matter I refer not just to the meanings attached to us as individuals or as members of a group, but to our relationship to ourselves as matter. To enter History is to transcend time; we come to mean something, and we avoid the spatio-temporal limitations of our bodily matter. When we suddenly find ourselves exiting History—pulled off stage with a vaudeville hook or forced to play a bit part in another person's drama—we stop meaning something and start mattering again. We materialize on Earth again, as a stranger and suddenly and painfully incarnate. Occupying the privileged position of the protagonist of History offers transcendence and significance—a kind of de-mattering. This privileged position is disrupted by the introduction of an alternate narrative, a competing temporality.

I say "we," but the invocation of a collective subjectivity is a temporal politics of my own: an injunction to you, the reader, to join me on the page—at least for the length of a paragraph—in occupying imaginatively and temporarily a temporally-induced subjecthood in which, I contend, a significant number of our fellow denizens of the present dwell. The aim of this chapter is to explore this element of the psychic drama that goes hand in hand with social change and revolution. The problematization of dominant, Western-centric, and patriarchal notions of time, with their emphasis on progress, modernity, and civilization, has been and remains a critical element of feminist, post-colonial and other progressive movements. However, as this essay will explore, right-wing, conservative, and nationalist movements mobilize around a politics of time as well. This chapter will challenge the assumption of coevality between different social groups, even those operating in the same (traditionally-defined) "chronological" time. In particular, the ways in which "nostalgic melancholy . . . may be deployed as a specifically masculine defense against loss" (Radstone 2007, 159) will be considered. Right-wing populist politics of time revolves around nostalgia for the past and apocalyptic visions of the future, temporalities which serve to give legitimacy and meaning to the dominant identities—those of white men—which have been challenged.

I begin with a discussion of the challenges to Modern, colonial, patriarchal visions of time–as-progress and with the white male as protagonist. I then discuss the sexual politics of right-wing populist movements, linking the fact that men are overrepresented in such movements with the nostalgic nationalism and myth-making around which their political ideology is oriented. Of particular interest are the ways in which nationalist mythologies serve to create and justify a mission, a supreme battle of good versus evil, us versus them, survival versus utter dissolution. In creating such narratives, right-wing populist groups invoke an alternative historical trajectory in which they are the protagonist-saviors, thus reversing time, and reclaiming a place in History. Time, as the essays in this volume explore, is intimately connected with social change; these alternative temporalities—revisions of the past, new futures, or, simply, new relationships to the present—both make possible and are themselves made possible by new constellations of relationships between social groups. However, the opening up of alternate temporalities for some groups may lead to the retrenchment of old temporalities for others. Throughout the chapter I explore nostalgia as not only a melancholic reaction to loss—a mourning for a future that is gone forever—but a reaction that is intimately bound up with a gendered and racialized subjecthood.

Whose Time is Our Time?

However one conceives of the causal mechanism that moved us into a post-modern moment, what is almost indisputable is the fact that such a move has involved a disruption to temporality that entails a concomitant disruption to Modern selfhood. Modernity and its faith in progress, civilization and time itself is racialized and gendered, as in order to conceive of itself as advanced, the Modern West invented Others who were not, for it could not be farther along the time line of human progress if there was nothing behind it (Yegenoglu 1998, 96).[1] The Modern Subject is rational, transcendent, and male. It is detached from the messiness of everyday life, and the body's spatio-temporal concerns (Seidler 1989); it is an "ahistorical, disinterested and disembodied subject" (Yegenoglu 1998, 109). It moves transcendently and celestially, orbited by its overly-embodied feminized and racialized others.

The post-modern self, on the other hand is "nomadic" (Bauman 1992; Melucci 1989), "depthless" (Jameson 1991), or "pastiche" (Gergen 1991). For Jameson the post-modern self is schizophrenic: "the links in the signifying chain snap, then we have schizophrenia in the form of the rubble of distinct and unrelated signifiers [since] personal identity is itself the result of a certain temporal unification of past and future with the present before me" (Jameson 1984, 72). But just because we can no longer unify the past, present and future does not mean we do not try. If one takes a closer look, we might be able to see that perhaps we are not all post-moderns. A reaction to the "ontological insecurity" of post-modernity can result in over-identification with stable entities that provide meaning and order, such as nationalism (Giddens 1985): "Two psychological portraits are contained in such visions of the post-modern world: the portrait of the depthless psyche, bobbing long on the postmodern tide; and the regressive psyche, struggling against the flow" (Billig 1995, 138).

Social change is not just the result of spontaneous, de-temporalized collective action; rather, the long-term transformation of society destabilizes traditional power centres and creates new spaces for the evolution of organizational structures to fill in the power vacuum left by the hollowing out of such centres (Tilly 2004, 2005). But grievances are not either there or not there to provoke action; structures, events, and situations arise and then undergo a process of interpretation in which they are framed as grievances requiring action (Snow et al. 1986). Because the interpretation of events is as multitudinous as the number of reflexive actors engaging in meaning-making activities, the number of different interpretations of the present moment and its meaning is similarly multitudinous.

It is the contention of this chapter that, if one looks at the ways in which the present is interpreted, in which fractured access to or desire for a fragmented post-modern selfhood is distributed amongst groups, certain

---

[1] See Osborne 1995 and Whitrow 1988 on modernity and the emergence of linear time, Terdiman 1993 on modernity and the idea of progress and McQuire 1998 on the rise of History as a way of measuring progress.

patterns emerge—patterns which refract the gendered and racialized politics of Modern time. Right-wing populist groups in Western societies are both frequently populated overwhelmingly by white men and orient their political ideology around nostalgic nationalism. Nostalgia is key in struggles over the creation of collective memory and, thus, in "political conflicts over modernity" (Smith 2000). They are "an important weapon in the debate over whose memories count and what kinds of desires and harms are politically relevant" (Smith 2000, 507). These politically-relevant "desires and harms" are oriented around narratives of divine battles between good and evil, us and them, which extend from the past into the present and lend themselves to apocalyptic possible futures. Nostalgic nationalism and narratives of cosmological battles serve a purpose: "Unable to handle ambivalence, authoritarians need unambiguous truths and clearly demarcated hierarchies. They seek the security of a clear world-view in which evil 'others' can be hated and a pure 'us' loved" (Billig 1995, 137). These narratives are intimately connected to the temporal politics of the "post-patriarchal," "post-colonial" moment; I would like to suggest that that they offer a way for (some) men to create an alternative, counter-historical trajectory in which they are again the protagonist.

Masculine Nostalgia and Right-Wing Populism

The beginning of the 21st century saw a rise in right-wing populist movements in Western Europe, North America, Australia and New Zealand (Rydgren 2007). Such movements are oriented around "ethno-nationalism" and mythical reconstructions of the past (Rydgren 2007). This section will discuss current right-wing populist movements in the UK and the U.S.

Right-Wing Populism in the US and UK

In the U.S., the 1990s saw a resurgence in citizen's militias, a resurgence that morphed into nationalist/nativist groups such as the Minutemen and, more recently, the Tea Party. In the 1990s militia activity in the US in general saw a sharp increase (Abanes 1996; Diamond 1995; Dyer 1997; Kimmel and Ferber 2000; Van Dyke and Soule 2002). In the late 1990s and early 2000s, groups of American citizens calling themselves Minutemen began patrolling the U.S.-Mexico border for the purpose of stopping undocumented immigration. Made up primarily of white men

(Navarro 2009), this movement is part of the long tradition of militia activity in the United States; increasingly, there is greater overlap between Minutemen and other, older militia groups. David Holthouse of the Southern Poverty Law Center reports that:

> growing numbers of Minutemen and their fellow travelers now perceive immigration as merely a glaring symptom of a much broader problem. The larger problem, they believe, involves shadowy conspiracies threatening American sovereignty, unwelcome demographic changes polluting American culture, and a potentially totalitarian government. (Holthouse 2009)

At the same time, more and more Minutemen are swelling the ranks of the Tea Party, another populist right-wing party that emerged on the scene in the late 2000s. Many Tea Party leaders are former or current Minutemen, one result being the swelling of nativist and anti-immigrant sentiment in the Tea Party and the rise of Tea Party nationalism (Burghart and Zeskind 2010).

In the UK, the British National Party has emerged from the ashes of older, fascist organizations to become a legitimate player in electoral politics. The BNP is the most successful extreme-right wing party in Britain today. Coming out of entities such as the National Front, in the 1990s the BNP began to "modernize" by "adopting populist-style themes such as 'freedom,' 'identity' and 'democracy'" (Goodwin 2010, 173), and has begun to achieve electoral success. It sent its first representatives to the European Parliament in June 2009. In 2008, BNP candidate Richard Barnbrook won 130,000 votes (5%) to become the first BNP member of the London Assembly and almost 70,000 Londoners voted for him for mayor (Ford 2010).

Nostalgic Nationalism

In both contexts, nostalgic reconstructions of the past play a key role in narrating political ideology. Citizen's militias and Minutemen draw on the importance of "citizen soldiers" in the Revolutionary War in order to justify their right to organize and to bear arms (Churchill 2009). The name Tea Party evokes America's revolutionary past. Populist groups frequently identify with the "frontier values" of independence and community self-policing (Mulloy 2004). The courageous posse makes a frequent appearance as well, taking the law into their own hands to defend civilization against the savagery of Native Americans and Mexican bandits (Mulloy 2004). Citizen's militias reify the owning of guns as a way to connect to the past

(Churchill 2009, Mulloy 2004), as guns "embody the spirit of self-reliance that Americans cherish" (Strain 2010, 73). Like its counterparts across the Atlantic, the BNP draws on national myths and symbols in order to justify their actions and ideology and to lay claim to a given heritage. Of particular importance are examples of British resistance and fortitude in the face of foreign invaders. They reference the "spirit of the Blitz" in describing their reactions to the threat posed by immigration and "Islamification" (Goodwin 2011), seeing themselves as operating in the same spirit as the "million and more brave lads who were killed or maimed fighting in every battle to see off every would-be foreign invader" (Goodwin 2011, 165).

Right-Wing Populism as "Men's" Movements

The extreme right overall is overwhelmingly populated by men (Kimmel 2003). In the European context, a much larger proportion of men than women support extreme right-wing organizations (Mudde 2007). Similar trends can be found in the UK: Ford and Goodwin (2010) have found significant gender discrepancies amongst British right-wing populists. Men are similarly over-represented in extreme right-wing groups (including citizen's militias) in the U.S. (Berlet 2005, Churchill 2009; Kimmel 2007). Threats to masculine identity or privilege play a crucial organizing principle in articulating regional, national, religious, ethnic, or class-based conflicts (Fine et al. 1997). Right-wing extremist and white supremacist organizations are populated by young men whose "inheritance"—whose right to male privilege—has been undermined by the political enfranchisement of minorities and women and the global re-structuring of labour and capital. Such movements "deploy 'masculinity'" as an "ideological resource", a "rhetorical device" and a "recruitment device" (Kimmel 2005, 416).

Ghost Armies, Reconquistas and Wars at Home: Hypermasculinity, Nostalgic Nationalism and the "Mission"

These movements are overwhelmingly masculine movements, not just because of the gender of their activists, but because of the ways in which they go about their politicking. The nationalist narratives they promote are not just nostalgic but "hypermasculinized." For example, the mythic west invoked by militia members and white supremacists uses the genre of the classic Western to "re-construct" white masculinity (Ferber and Kimmel 2000, 193). The modern American militia movement is oriented around the notion of a "war" between its members—who see themselves as "ordinary," or proper Americans—and the enemies of society who threaten "Americanism" (Gibson 1994). They create a masculinized, warrior mentality (Gibson 1995, Kimmel and Ferber 2000) in order to justify action.

Hypermasculinity is a reactionary "glorification of those traits conventionally identified with men or masculinity" (Ling 2001, 1089). In particular, it can be thought of as an effect—or, even, affect—of the colonizer/colonized relationship, which constructs the colonized in the moment that they enter the non-Western gaze. Hypermasculinity is the deployment of gendered ideologies and hierarchies which traps both colonizer and colonized; the colonized constantly tries to legitimate his manliness to the colonizer while the colonizer overemphasizes traditional masculine traits in order to justify the "civilizing" mission (Nandy 1988). The white man/master/colonizer, when he encounters his other, sees his own image reflected back, naked and over-embodied (Chow 2003). He sees himself to be

> brutal, false, all-too-human, acting on whim and fancy and power. And he has nowhere to hide. The hollowness of his proclaimed ideals of Beauty, Truth, Perfection, Rationality, and Science, derived from the supposedly superior legacy of the European Enlightenment, confronts him daily and intimately. (Ling 2008, 16)

He is no longer the protagonist, the observer. He senses that he is being observed by his Others and he seems himself in their eyes. The result? Thc colonizer's masculinity goes into hyperdrive to justify the "mission" (Nandy 1988).

The "mission" here is the historical trajectory that gives its protagonists meaning. Narratives result that stress dichotomous, survivalist thinking: good versus evil, us versus them. "The populism of the extreme right," Goodwin writes, "dichotomizes issues into good and evil and aims to position the majority against minority groups on sensitive issues such as asylum or human rights" (Goodwin 2011, 6). Militia members believe that the government, minorities, feminists, liberals and foreigners have declared "war" on Americans, a war which threatens the very survival of American life and necessitates a hypermasculinized warrior mentality (Gibson 1994). They rely on narratives of the old West, narratives which provide "clear oppositions between good and evil, valuing order, permanence, stability, and strength," in which "white men are the pcrpetual heroes, rescuing white women from evil men and safeguarding civilization" (Ferber and Kimmel 2000, 195). This logic is "characterized by a series of seemingly clear alternatives: order versus chaos, man versus woman, civilization versus savagery, illusion versus truth, words versus action" (Kimmel and Ferber 2000, 190).

The British National Party orients around similar Good versus Evil thinking. The indigenous peoples of the British Isles are threatened by "a ghost army of Islamic terrorists" which have assembled in the UK in order to "wage war and inflict murder upon us until we surrender to them and an Islamic Fascist State is imposed upon us" (in Wood and Finlay 2008, 712). Members of the BNP see themselves as part of an epic battle to save Britain; members interviewed by Goodwin "framed Muslims as threatening political institutions, claiming they seek to 'Islamify' Britain, to 'fly the flag of Islam over Westminster'" (Goodwin 2011, 144). BNP activists cite "survivalism"—the survival of the British people and their culture—as reasons for joining the movement.

If the BNP is convinced that Muslims and asylum-seekers are attempting to colonize or occupy Britain bit-by-bit, Minutemen (and increasingly Tea Party nationalists) believe that Mexicans are seeking to conquer the Southwest. This "Reconquista" is a "Trojan Horse" plan to take over the United States in bits and pieces. The goal, write Minuteman leader Jim Gilchrist is "for illegal aliens to get citizenship for themselves and their children so that they can eventually vote to return to Mexico large sections of the American Southwest" (Gilchrist and Corsi 2008, 246). Mexico is, they believe, through "illegal" immigration, attempting to reconquer the territory lost to the United States in the course of U.S. western expansion (Beirich 2009; Holthouse 2009).

Nostalgic nationalism, narratives of good versus evil and hypermasculined responses are the result of the challenge to white, patriarchal temporalities and the identities these temporalities make stable. Identity is searched for in times of extreme and rapid change, times when, as Bauman puts it, "new forms of life emerge too fast to be absorbed and domesticated by the old mechanisms of control and ancient mental frames" (Bauman 1992, 680). Unsure of the part they play in the present, white men try to resuscitate a past that they recognize as their own. It is not an accident that these narratives are being reached for now. As Kellner writes,

> the turn to white male (and in some cases black male or Muslim male) identity politics represents in part a rejection of modernity . . . a pervasive mindset of white male identity politics is extreme paranoia, projecting confusion and anxiety over the tumultuous changes of the present age onto

> cosmic forces (evil, the Devil) or scapegoated villains, such as other races, 'the new world order', or, in the Bush-Cheney years, 'liberals' and 'secularists'. (Kellner 2008, 94)

The "mission" around which they orient is a nostalgic search for meaning and cairological fulfillment. While Svetlana Boym describes nostalgia as a "mourning for the impossibility of mythical return, for the loss of an 'enchanted world' with clear borders and values" (Boym 2007), it might be better described as a melancholic longing for a mythical return. We mourn when we lose something we valued, we recognize that it is lost, and we recognize that we and it are separate entities: that in losing it, we have not lost ourselves (Freud 2005). Melancholia is when we cannot recognize a distinction between ourselves and what we valued. We cannot mourn its loss because to recognize its loeryss would be an act of self-dissolution. (Some) white men cannot let go of the past because to do so would be tantamount to psychic suicide. These groups drag a mythologized past into the present in order to legitimate their attempts to thwart the machinations of the "enemy." Through commemorative practices these groups use "heritage" to legitimate their fight against cosmic forces (evil-doers, the New World Order) and racialized scapegoats. By occupying a place in history they become transcendent and ahistorical. They need this fight, this "mission." Feeling as if they are being written out of history, they grasp at ways in which to mean something again.

Conclusion

Temporally-delimited events, themselves located in large-scale processes, require interpretation in order to manifest social change and revolution. Interpretive action belongs to no single social group and the number of possible interpretations attached to a time, a place, or an event is limited only by the number of collectivities who engage in meaning-making. It is not a given that we all inhabit the same epochal moment, and individual groups may experience history in different ways. Some of "us" may have jumped ship, aligned ourselves with a different historical trajectory, claimed a different future than the one "our" past provides. But we should not forget that, in claiming a different future for "ourselves," we are taking a future away from others—namely the white, Western man. No longer in the vanguard of the Grand March of History, they are relegated to the realm of over-embodied sideliners.

However, as Billig writes, while the playful postmodern and the authoritarian may seem to represent polar opposites, "the middle" does not disappear. Right-wing populist and nationalist groups—and the mythological and polarizing narratives they deploy—may represent an extreme (or, at least, noticeable) example, but it is not so different from some so-called mainstream reactions. I may have limited my discussion to such groups and such narratives, but the sexual politics of time I have discussed can be observed in the upper echelons of politicking as well. The U.S. has been described as beset by a "paranoid patriotism" that disavows the temporality of the present—how it came to be—and "freezes the fluidity of time into eternal ideals of Good and Evil" (Oliver 2010, 198). The result is security and economic regimes in which "one hypermasculine source (e.g., U.S. foreign policy) provokes another (e.g., Al Qaeda) to escalate with iterative bouts of hypermasculinity (e.g. 'jihad'/'war on terror')" (Agathengelou and Ling 2004, 519). When Bush or Blair declare that "history will prove them right" (Oliver 2010, 198) they are doing so from the same "God's Eye view"—the view of the rational, disembodied, masculine observer—of the Modern, Enlightenment subject. It is a view from a place outside of temporality and linear history and thus free from judgment; it is "eternal and therefore cannot be questioned" (Oliver 2010, 198).

But while progressive politics may breed regressive reactions, I doubt whether we can (or should) have a feminism without a rejection of "nation time" (Oliver 2010, 198). "As a woman," Virginia Woolf has famously written, "I have no country. As a woman I want no country. As a woman my country is the whole world" (Woolf 1938, 129). However, I have always been moved by how, even in the depths of anger towards the way her Chicano culture "cripples" women, Gloria Anzaldua finds psychic time to witness, to abhor, how it "makes macho caricatures of its men" (Anzaldua 1987, 43). It seems to me that we need an ethico-feminist engagement which is not limited to the achievement of rights and equality for female subjects, but rather acts as a "sieve or cipher" for never-ending struggles for new ways of "becoming-other" for all (Grosz 2005, 167; c.f. Irigaray 1993a). We need a feminism oriented around the proliferation of endless futurities and fecundities, instead of a temporally-limited, zero-sum game in which a different future for some is a lost future—a future of loss—for others.[2]

[2] This essay would not have been possible without the help of Cathy Conroy and her constant injunctions to stop worrying and start writing. Her financial and emotional support has finally allowed me to do so.

Works Cited

Abanes, Richard. 1996. *Rebellion, Racism and Religion: American Militias*. Downers Grove: Intervarsity Press.

Agathangelou, Anna M., and L.H.M. Ling. 2004. "Power, Borders, Security, Wealth: Lessons of Violence and Desire from September 11." *International Studies Quarterly* 48 (3): 517-538.

Anzaldua, Gloria. 1989. *Borderlands/La Frontera: The New Mestiza*. San Francisco: Aunt Lute Books.

Bauman, Zygmunt. 1992. "Soil, Blood and Identity." *Sociological Review,* 40: 675-701.

Beirich, Heidi. 2011. "The Year in Nativism: 2010." *Southern Poverty Law Center Intelligence Report*, Spring, 114. Accessed Nov. 18 2011. http://www.splcenter.org/get-informed/intelligence-report/browse-all-issues/2011/spring/the-year-in-nativism.

Berlet, Chip. 2004. "Mapping the Political Right: Gender and Race Oppression in Right-Wing Movements." In *Home-Grown Hate: Gender and Organized Racism*, edited by Abby Ferber, 19-47. New York: Routledge.

Billig, Michael. 1995. *Banal Nationalism*, London: SAGE.

Boym, Svetlana. 2003. "Nostalgia and Its Discontents." Appeared originally in Boym, Svetlana. 2002. *The Future of Nostalgia*, New York: Basic Books. Accessed 15 Oct. 2011. http://www.iasc-culture.org/eNews/2007_10/9.2CBoym.pdf.

Burghart, Devon and Leonard Zeskind. 2010. "Beyond FAIR: The Decline of the Established Anti-Immigrant Organizations and the Rise of Tea Party Nationalism," *Institute for Research and Education on Human Rights Special Report*. Accessed 17 Jan. 2012. http://www.IREHR.org/images/pdf/BeyondFAIRreport.pdf.

Chow, Rey. 2003. "Where Have All the Natives Gone?" In *Feminist Postcolonial Theory: A Reader*, Edited by Reina Lewis and Sara Mills, 329-343. New York: Routledge.

Churchill, Robert H. 2009. *To Shake Their Guns in the Tyrant's Face: Libertarian Political Violence and the Origins of the Militia Movement*. Ann Arbor: University of Michigan Press.

Diamond, Sara. 1995. *Roads to Dominion: Right Wing Movements and Political Power in the United States*. New York: Guilford.

Dyer, Joel. 1997. *Harvest of Rage: Why Oklahoma is Only the Beginning.* Boulder: Westview.

Eatwell, Roger and Matthew J. Goodwin, eds. 2010. *The New Extremism in 21st Century Britain.* London and New York: Routledge.

Ford, Richard. 2010. "Who Might Vote for the BNP?: Survey Evidence on the Electoral Potential of the Extreme Right in Britain." In *The New Extremism in 21st Century Britain*, edited by Roger Eatwell and Matthew J. Goodwin. London and New York: Routledge.

Ford, Richard and Matthew J. Goodwin. 2010. "Angry White Men: Individual and Contextual Predictors of Support for the British National Party." *Political Studies* 58 (1): 1-25.

Ferber, Abby, ed. 2004. *Home-Grown Hate: Gender and Organized Racism*. New York: Routledge.

Ferber, Abby and Michael Kimmel. 2000. "Reading Right: The Western Tradition in White Supremacist Discourse." *Sociological Focus* 33 (2): 193- 213.

Fine, Michelle, Lois Weis, Judi Addelston and Julia Marusza. 1997. "(In) Secure times: Constructing White Working-Class Masculinities in the Late 20th Century." *Gender & Society* l (l): 52-68.

Freud, Sigmund. 2005. *On Murder, Mourning and Melancholia*. Trans. Shaun Whiteside. London: Penguin Books.

Gibson, James W. 1994. *Warrior Dreams: Violence and Manhood in Post-Vietnam America*. New York: Hill & Wang.

Giddens, Anthony. 1985. *The Nation-State and Violence*. Cambridge: Polity Press.

Gilchrist, Jim and Jerome Corsi. 2006. *Minutemen: The Battle to Secure America's Borders*. Los Angeles: World Ahead Publishing.

Goodwin, Matthew J. 2010. "Who Votes Extreme Right in Twenty-First-Century Britain?: The Social Bases of Support for the National Front and British National Party." In *The New Extremism in 21st Century Britain*, edited by Roger Eatwell and Matthew J. Goodwin, 191-210. London and New York: Routledge.

Grosz, Elisabeth. 2005. *Time Travels: Feminism, Nature, Power*. Durham and London: Duke University Press.

Hesford, Victoria and Lisa Diedrich, eds. 2010. *Feminist Time Against Nation Time: Gender, Politics, and the Nation-State in an Age of Permanent War*. Lanham, MD: Rowman & Littlefield, Inc.

Holthouse, David. 2009. "Nativists to 'Patriots'." *Southern Poverty Law Center Publications*. Accessed 6 Apr. 2011. http://www.splcenter.org/publications/splc-report-return-of-the-militias/nativists-to-patriots.

Irigaray, Luce. 1993a. *An Ethics of Sexual Difference*. Translated by Carolyn Burke and Gillian C. Gill. Ithaca: Cornell University Press.

—. 1993b. *je, tu, nous: Toward a Culture of Difference*. Translated by Alison Martin. New York: Routledge.

Jameson, Frederic. 1984. "Postmodernism, or the Cultural Logic of Late Capitalism." *New Left Review* 146: 53-92.

Kellner, Douglas. 2008. *Guys and Guns Amok: Domestic Terrorism and School Shootings from the Oklahoma City Bombing to the Virginia Tech Massacre.* Boulder: Paradigm.

Kimmel, Michael. 2003. "Globalization and its Mal(e)contents: The Gendered Moral and Political Economy of Terrorism." *International Sociology* 18 (3): 603-620.

Kimmel, Michael and Abby Ferber. 2000. "'White Men Are This Nation': Right-Wing Militias and the Restoration of Rural American Masculinity." *Rural Sociology* 65 (4): 582-604.

Ling, L.H.M. 2008. "Borderlands: A Postcolonial-Feminist Alternative to Neoliberal Self/Other Relations." International Affairs at the New School Working Paper. New York: The New School Working Papers Series.

—. 2001. "Hypermasculinity." *Routledge International Encyclopedia of Women's Studies*. London: Routledge.

McQuire, Scott. 1998. *Visions of Modernity: Representation, Memory, Time and Space in the Age of the Camera*. London: SAGE Publications.

Melucci, Albert. 1989. *Nomads of the Present*. London: Hutchinson Radius.

Mudde, Cas. 2007. *Populist Radical Right Parties in Europe*. Cambridge: Cambridge University Press.

Mulloy, D. J. 2004. *American Extremism: History, Politics and the Militia Movement*. London: Routledge.

Nandy, Ashis. 1988. *The Intimate Enemy: Loss and Recovery of Self Under Colonialism*. Delhi: Oxford.

Navarro, Armando. 2009. *The Immigration Crisis: Nativism, Armed Vigilantism, and the Rise of a Countervailing Movement*. Lanham, MD: AltaMira Press.

Oliver, Kelly. 2010. "Perpetual War." In *Feminist Time Against Nation Time: Gender, Politics, and the Nation-State in an Age of Permanent War*, edited by Victoria Hesford and Lisa Diedrich, 185-201. Lanham: Rowman & Littlefield, Inc.

Radstone, Susannah. 2007. *The Sexual Politics of Time: Confession, Nostalgia, Memory*. London and New York: Routledge.

Rydgren, J. 2007. "The Sociology of the Radical Right." *Annual Review of Sociology* 33 (1): 241-262.

Smith, Kimberly K. 2000. "Mere Nostalgia: Notes on a Progressive Paratheory." *Rhetoric and Public Affairs* 3 (4): 505-527.

Snow, David A., E. Burke Rochford Jr., Steven K. Worden, and Robert D, Benford. 1986. "Frame Alignment Process, Micromobilization and Movement Participation." *American Sociological Review* 51:464-481.

Terdiman, Richard. 1993. *Present Past: Modernity and the Memory Crisis*. Ithaca: Cornell University Press.

Tilly, Charles. 2005. *Popular Contention in Great Britain*. Colorado: Paradigm Publishers.

—. 2004. *Social Movements, 1768–2004*. Colorado: Paradigm Publishers.

Van Dyke, Nella, and Sarah A. Soule. 2002. "Structural Social Change and the Mobilizing Effect of Threat: Explaining Levels of Patriot and Militia Organizing in the United States." *Social Problems* 49 (4): 497-520.

Whitrow, Gerald James.1988. *Time in History: Views of Time from Prehistory to the Present Day*. Oxford: Oxford University Press.

Woolf, Virginia. 1938. *Three Guineas*. New York: Houghton Mifflin Harcourt.

Wood, C. and William M. L. Finlay. 2008. "British National Party Representations of Muslims in the Month after the London Bombings: Homogeneity, Threat, and the Conspiracy Tradition." *The British Journal of Social Psychology* 47 (4): 707-726.

Yegenoglu, Meyda. 1998. *Colonial Fantasies: Toward a Feminist Reading of Orientalism*. Cambridge: Cambridge University Press.

THE SLOW FOUNDATIONS OF HYPER-SPEED: (MIS)CONCEPTIONS OF TIME IN CHINA'S MANUFACTURING REVOLUTION

ALISON HULME

Western perceptions of China in recent decades have focussed on the immense speed of its development and the sheer scale of its production. The great manufacturing dragon is seen as somehow conducting its industrial revolution bigger, bolder and faster than any other nation state ever has. It is presented as the epitome of fast capitalism. Yet underneath this layer of hyper-speed are older, clunkier facets which provide the foundations for such velocity. Entrepreneurial traditions as old as Confucianism have informed the history of present-day small-scale business and the allegiances gained from communist work-units form the basis of many working practices. The speed of internet trading in manufactured goods must be viewed contemporaneously with the slow heaving of the container ports; the clean efficiency of wholesale commodity markets with the dirty danger of the manufacturing quarters.

There is one scene in Jennifer Baichwal's film *Manufactured Landscapes* (2006) in which a tracking shot over two minutes long captures the longest production line in the world. It is clean, bright, and most of all, it is *fast*. Nimble fingers are almost blurred due to the speed with which they go about their work. One cannot tell how they are doing what they are doing; what is clear is that the action those fingers are undertaking is one that has become second-nature to them. This is the perfect depiction of China as the factory of the world—a full-to-overflowing production fiend of epic proportions; a phantasmagorical utopia of commodities, whose subjects strive robot-like in the name of a glorious great leap to the present. It is an image that the West can find at once strange and slightly concerning; impressive and astounding. It is also an image that, with its clinical polished white floors, hairnets and strip-lighting, is as far removed as possible from the grinding, creaking, dusty realities of any industrial revolution the world has yet known. It is also perhaps, the kind of image that some Western minds may use to calm their concerns about dangerous working conditions and exploitative levels of pay.

This image, and others like it, provides (apparent) evidence for the rhetoric of those thinkers eagerly pronouncing the arrival of fast capitalism. Perhaps most typical amongst these is Thomas Friedman, whose unceasingly positive portrayal of globalization in *The World is Flat* cites countless examples of out-sourcing and flexibility, in the tone of one entirely dazzled (Friedman 2005). Moreover, his emphasis seems to be placed on the extent to which this phenomenon is democratizing, perhaps forgetting that Marx's original use of the words "flat world" was to describe the spreading of the *bourgeoisie* across the surface of the globe.

Others make far more sanguine (and nuanced) arguments on the nature of speed. In his initial work on fast capitalism Ben Agger attributed the erosion of serious critique to the fact we can no longer see how a book is independent from the world, and therefore how its contents could oppose the dominant world. For Agger, people were insufficiently distant from the world, lacked perspective to form critique, and were inculcated with conformist values due to the way in which the world appeared to have no outside and thus no exit. Most crucially, for Agger, fast capitalism was capable of speeding up the rate at which people lived out the historical possibilities presented to them (1989, 17-20). In his subsequent works, Agger has broadened his analysis to include the impact of information technologies, arguing the Internet has quickened the pace of everyday life and created a society characterized by administration, in which revolt is subsumed within the everyday and thus harder to access (2004, 5-6).

In a similar vein, Thomas Hylland Eriksen's concern is that slow time—time when we are able to think without interruption—is becoming extinct, as information technology permeates every area of our lives, causing a culture to lose a sense of its past and future and so become static (Eriksen 2001). Meanwhile, James Gleick points to our biological, psychological and neurological limits and the extent to which technology pushes these to the extreme (Gleick 1999).

In different ways, all four writers are perhaps typical of what John Hutnyk sees as clichéd writing on speed; that which accentuates electronic flows and global linkages and sees everything as a blur in capitalism. For him this belies the non-happening of reality: "it is not uncommon to find gee-whizz declarations of hyper-intensity of capital flows that seem only to lead to stasis and quietism" (Hutnyk 2004, 59). Certainly, all too often, speed is emphasized at the risk of neglecting rupture, stultification, apathy and the cumbersome nature of the labour of unseen others. However, the rhetoric has been particularly difficult to critique as it cuts across traditional political divides: speed as both a negative and positive impulse has been used by both the left and the right—Gramsci was an early admirer of the Futurists[1] and Lenin enthusiastically embraced the speed-regulation of Taylorism.

For our purposes here, it is not so much that depictions of fast capitalism do not grasp a phenomenon that is most certainly at play, and will continue to be as we move further into the twenty-first century; rather, it is that underlying this hyper-speed are older, slower forces which cannot be underestimated. These include structural elements which continue to play their hand strongly (such as corruption and tradition with the communist party in China), economic elements (such as the continued difficulty and time spent in attracting state loans and the reliance therefore upon small informal loans), as well as social and cultural elements. The squeaky-clean, whizzy development of the scene in Baichwal's film is only one side of the story.

In Shanghai, at around 5pm, a long line of workers, at least 150, trudge towards the docks to begin the evening shift. There, huge, creaking containers are hoisted by an immense slow-moving crane onto a ship half a kilometre long. Later, it will heave itself away from the dock and move eerily out into the harbour to begin its perilous journey to Europe—a floating island of iron, which will require one mile to come to a stop. The tiny figures of men on the dockside, dwarfed by the ship's enormity, seem sublimely ridiculous—men out of time and place, running to keep up with things larger and less stoppable than they have ever experienced. The scene is a twenty-first century version of such early industrial spectacles as the first great train rides or the churning cogs of giant factories; despite computerisation it is cumbersome, creaking and heavy like those earlier miracles.

Scenes such as these underscore the clear polarities between the realities of fast and slow in current-day China's manufacturing revolution; polarities that the Chinese authorities are tackling with equally polarised political rhetoric. On the one hand, they determinedly draw upon "slower" historical traditions in order to ensure continuity between past and present. This reflects, of course, political expediency, designed to prove the continued relevance and efficacy of the communist party, as much as

---

[1] Filippo Tommaso Marinetti announced the *Futurist Manifesto* in 1909, two years before Taylor's *Scientific Management*. The movement was very varied and often internally ruptured due to the differing takes on its aims by its various proponents. Most remembered however is Marinetti's proto-fascist prose which proclaimed the glorification of war and the fight against moralism, feminism, and "intellectualism" (in the form of libraries, museums, etc).

nationalistic fervor. On the other, they whole-heartedly embrace the necessity to present themselves as fast and efficient; seeking to eradicate (or at least make invisible) areas that appear "backward," and create new, fast image-spaces in an attempt to gain "world city" status for key urban metropolises.

For example, in the run-up to Shanghai's World Expo in 2010, the Shanghai authorities began to display artists' impressions of the extensive planned regeneration of certain inner-city areas. The images showed wide, tree-lined streets, with Western-style mock-Tudor houses, and emphasized clean, green efficiency, fast infrastructural links and rapid information. In addition, Shanghai's cartoon mascot "Haibao" was to be seen across the city's billboards accompanied by the slogan "better city, better life," and extolling the virtues of enhanced "wenming" (which is loosely translated as "civilization") by teaching the Shanghainese to avoid doing things the "international community" may find uncouth, such as spitting, slurping food, or not queuing. Haibao was part of a literal and cultural clean-up which attempted to erase those spaces which did not fit with the clean, efficient image of the new, speedy Shanghai.

The treatment of the city's waste peddlers was indicative of the authorities' determination. The peddlers, with their cobbled-together barrows, are seen as the image of old China—antiquated, trundling, festering, and subsumed in intricate networks of allegiances based on clan and favours rather than merit and socio-legal rationality. Just as the Beijing peddlers were banished to the suburbs during the Olympics, so their Shanghai counterparts underwent similar treatment during the Expo. The production of space cleared produc*tive* spaces, in order to make way for spaces that showed success by emphasizing the extent to which they were removed from the mechanics of production. Shanghai was presented as a temporal phenomenon, defined by speed, yet its spaces were designed to show just how relaxed and pleasant this hyper-speed could be.

This is typical of the way in which China recognizes the importance of the image of hyper-speed and markets its key cities based on the concept. However, there is another layer to China's industrial revolution which is specifically concerned with emphasizing tradition and history. This can be seen further inland along the Yangtze River Delta (YRD) in the "small commodities city" of Yiwu. Yiwu's manufacturing prowess has made it a key driver of a huge regional economy, not only in Zhejiang province, but also in the YRD economic area as a whole. Its prime location has made it an important transport hub, with express trains to and from Shanghai as well as inland flight routes, and it is home to several immense wholesale markets.

Yet, Yiwu is also keen to promote itself as having a rich history stretching back to the Tang dynasty (AD 6244). Its promotional literature draws upon great historical figures such as Tang-era poet Luo Binwang, Zong Ze—a general during the Northern Song Dynasty, and Zhu Danxi—one of the four famous doctors of the Jin and Yuan dynasties. The city park area includes a man-made lake over whose artificial islands stretch reproduction Ming bridges, linking archways and gazebos. And perhaps most poignantly, at one crossroads, a statue of a communist worker holds up his arm in a red salute to the wholesale markets that lie a few hundred metres ahead of him. These continuities with China's past are promoted domestically with almost as much fervour as the rhetoric of speed is promoted internationally.

Furthermore, there are some additional intrinsic legacies of both Confucianism and communism embedded in the fabric of Yiwu. Like many "new towns" in China, Yiwu (as we know it today) was built based on the concept of *xiaoqu* or "small districts"—planned neighbourhoods in which housing is integrated with communal facilities such as nurseries, clinics, restaurants, shops, and sports facilities. As David Bray argues, these *xiaoqu* bear a strong resemblance to the socialist *danwei* (work units) in their layout and provision of services. However, the layout of *danwei* can also be traced back to the traditional courtyard house designed to complement the ethical codes of Confucianism. Thus, it re-employed a familiar spatial repertoire in the service of socialism (Bray 2005, 35).

Therefore, the districts in Yiwu today reference slower historical times and older spatial repertoires—both Confucian and Maoist—whilst also embodying Deng's spirit of economic advancement and "speed." In fact the commercial developers of these areas even focus on creating spaces for "*linli guanxi*" (neighbourhood guanxi[2]), such as the *danwei* used to provide, proving just how deeply the ideological remnants of the *danwei* are ingrained in current practice—the "mimetic effect" Bray mentions (2005, 156). The *danwei*, born of Confucianism and appropriated for

[2] Guanxi is often translated as "relationship" or "connection," but is best explained as a combination of "ganqing" (depth of feeling within an interpersonal relationship), and "renqing" (moral obligation and "face" or social prestige). It describes personal relationships in which one is able and obliged to perform and receive favours—a long-term, obligated and heartfelt connection in which individuals have the right to demand fair return, sharing of benefits and reciprocity. It is generally accepted that the principles of guanxi come from Confucian thought. Guanxi's emphasis on interpersonal connections caused it to attract the interest of network analysts from the late 1970s onwards.

political ends, finds itself now, in the post-reform period, re-appropriated for economic ends as part of the general dominance of economic logic.

The playing out of these different times in current-day China is not without its own existential crisis. Tang Xiabing makes a case for the existence of a post-revolutionary anxiety, born out of the awareness that the post-revolutionary era has replaced politics with economics, unilateralism with pluralism, content with form, a rural romantic utopia with an urban pragmatic utopia, and perhaps most importantly, collective concerns with consumer desires (Tang 2002, 125-134). Whilst effectively creating a list of perhaps too neatly defined binary opposites, Tang's thrust here is one that poignantly pits the China of fast capitalism against that of slow agrarianism. As a result of the anxiety induced, Tang argues, participatory communal action can offer itself as an oppositional discourse, exposing the vacuity of domestic commodification and creating a nostalgic alternative to which one can return ("the heroic idea of Mao"), if not literally then through certain cultural forms or lifestyle choices (Tang 2002, 128).

What is being asserted is actually that there is a conflicted desire for the old time inside the day-to-day reality of the new time—China is being told, by its own leaders and by external commentators, that it is both speedy and steeped in slow history. Furthermore, it must be recognised that the images associated with both these types of time can be used as tools for marginalisation by the West. The ant-farm image of China as the world's factory frequently draws upon essentialist notions of a hard-working, un-questioning populace, determined to "catch-up" with the rest of the world. This image of fast revolutionary time tends to enjoy the sheer scale of China's production—what Arrighi calls its "industri*ous* revolution"—as long as China does not push its advantage in the established world order (Arrighi 2007, 86). In contrast, the rural-idyll image of China attempts to consolidate decades of communist rule by way of proving the still-backward nature of China, despite its economic success. This slow revolutionary time emphasizes the one-party system, the still-high levels of poverty, and the lack of individual freedom and rights.

Both conceptions can fall into the trap of a rhetoric which sees China as simply following an inevitable path towards capitalist democracy. This path is "our" path, the one "we" (in the West) trod during our industrial revolutions and which we now see as a prototype for any other industrial revolution in the world. Time under this logic is not only teleological, historical time, but it is already written—"we" have already lived it and a new narrative is not deemed possible. The revolution always and already belonged to "us." Such rhetoric tends to lead to assumptions regarding the inevitability of China's gaining democracy in the future, as it entails a fundamental assumption that industrial revolutions are capitalistic, and that capitalism necessarily goes hand-in-hand with democracy. There is an implicit (and unquestioned) normalisation at play here, which equates capitalism, and capitalist revolutions, to fairness, representation, transparency and other such features that most in the West insist are part of democracy. The point here is not to deny that China's revolution is capitalistic in large part, but that the West's feeling of owning the path being trodden comes from a normalising discourse in which *all* aspects of change will follow Western neo-liberal logic.

In reality, it may well be the case that in charting its own path through industrialisation, China's revolution is finding new answers to some of the key questions the West has faced since the late nineteenth century. Without doubt, China's revolution is a unique coming together of hyper-fast development and careful preservation of older, slower mechanisms of state and people which carves its own route through history. It is both the sleek efficiency of Shanghai's newly regenerated urban areas and the clunking reliability of its low-tech back-waters. It is the time of both the urban office-worker and the rural micro-entrepreneur. In many ways this is precisely the dialectic relationship with the past that China's leaders want it to have—Deng Xiao Ping's "crossing the river by feeling for stones" in order to create the "socialism with Chinese characteristics" that many have dubbed capitalism with Chinese characteristics. This ability to admit the mistakes of the past, and move forward taking some aspects of the past and disregarding others, is perhaps precisely why revolutionary time in China is made up of contrasting speeds which defy simplistic categorisations and carve a new and original path to social change.

These contrasting speeds come from both below and above—from the Party and the people, and perhaps most contentiously, from the market. The Party seeks continuity with the more picturesque parts of China's "slow" history, as well as celebration of its "fast," efficient present. For their part, some of the people seek solace in images of slow rural idyll and continue to only half-heartedly embrace credit card culture, preferring to hark back to the days of provision for all under the "iron rice bowl" policies. Others throw themselves into the fast and heroic game of catch-up that China is playing with the rest of the world, desiring not only material but cultural change in the form of democratic elections, freedom of religious practice and the tearing down of the "Great Firewall of China" to abolish media censorship. Many work in dangerous, heavy, slow industrial places, caught in the under-belly of change. Others aspire to create businesses in those sectors which ride the advantages created by the underbelly—by becoming internet entrepreneurs for example.

Finally, it must be remembered, that the market defines this revolution in a more decisive fashion than history has previously witnessed, insisting upon the absolute, unforgiving, necessity for the consumption cycle to maintain its pace from purchase to disposal/re-purchase, and utilising both clunky and speedy capitalism in order to achieve this. In many ways, revolutionary and post-revolutionary time have perhaps collapsed in China's manufacturing revolution, as part of an implosion of the old political heroism and the new economic heroism. A nationalism that attempts to embrace both is tangible. This nationalism is perhaps best described as revolutionary time running on different circuits—hyper-speedy fullness of time and space, running along on slow foundations that are still highly necessary.

Works Cited

Agger, Ben. 1989. *Fast Capitalism: A Critical Theory of Significance*. Champaign: University of Illinois Press.

—. 2004. *Speeding Up Fast Capitalism: Cultures, Jobs, Families, Schools, Bodies*. Boulder: Paradigm Publishers.

Arrighi, Giovanni. 2007. *Adam Smith in Beijing: Lineages of the Twenty-First Century*. New York: Verso.

Baichwal, Jennifer. (2006) *Manufactured Landscapes* [film]. Zeitgeist Films: Canada.

Bray, David. 2005. *Social Space and Governance in Urban China*. Stanford: Stanford University Press.

Eriksen, Thomas Hylland. 2001. *Tyranny of the Moment*. London: Pluto.

Friedman, Thomas. 2006. *The World is Flat: The Globalized World in the Twenty-First Century*. London: Penguin.

Gleick, James. 1999. *Faster: The Acceleration of Just About Everything*. London: Vintage.

Hutnyk, John. 2004. *Bad Marxism: Capitalism and Cultural Studies*. London: Pluto Press.

Tang, Xiaobing. 2002. "The Anxiety of Everyday Life in Post-Revolutionary China." In *The Everyday Life Reader,* edited by Ben Highmore, 125-135. London and New York: Routledge.

THE TRAPPINGS OF TIME: THE EFFECTS OF GLOBAL SOCIAL MEDIA UPRISINGS ON SINGAPORE'S POLITICAL SCENE

ENQI WENG

Introduction

The success of recent uprisings around the world has been greatly influenced by access to technological infrastructures, which allows for speedy communication. As witnessed in the Arab Spring and the Occupy Wall Street Movement, governments around the world face the growing problem of an increasing multitude of discontented citizens, whose natural, reflexive inclination now is to turn to the Internet to voice their political views. This paper looks at the effects of these localized uprisings and how they have impacted the political scene in Singapore, which has traditionally been conservative in its political views as well as its tolerance for diversity of political participation. In particular, it looks at how time is critical in a world that increasingly values speed and efficiency, and how, in the long haul, time is required for long-term political decision-making for worthwhile social impact to take place within any community. We will begin by looking at our human concept of and relationship with Time, followed by the effects of our changing relationship with it in how we approach democracy and online political participation in the case of controlled-media Singapore.

Our Relationship with Time

In what we can visually distinguish as a three dimensional world with its widths, lengths and heights, the fourth dimension of Time is one that we can perceive only by the physical changes around us. In this sense, our observation tells us that Time is linear. We know Time as a state of being, as a concept, but not as an object. In her contemplations, Barbara Adam asks if "[there are any] paradoxes that stubbornly remain, despite the fact that social scientists have established time as social time" (1990, 42). This social construction of time means that the speed of time is variable depending on its interaction with human agency.

The human desire to invent new machineries to take over laborious human processes led David Harvey (1989) to first write about time-space compression in reference to this effect of capitalist-driven modernization. In the capitalist aim to create a greater profit margin, production is increased so that a greater output can be formed in a shorter amount of time, thereby creating an alleged acceleration of time (Harvey 1989, 229-230). Such a culture would be ingrained in collective behavior, leading to a perceived acceleration of time which would permeate every aspect of human life and socialization. The result is a group lamentation of the shortage of time that mankind faces, an effect caused by the pervasiveness and lure of capitalism into every aspect of society. According to Adam, the "speed of movement and reaction . . . is associated with competitive advantage across species and in many cases vital to survival" (2004, 129). Klein says that "once a certain pace has been established in a community, little can be done to slow it down" (2008, 163). He goes on to acknowledge that though there are external factors driving the increasingly fast pace of society, it is ultimately still about the motivation of the individual to induce this phenomenon (163). Given the lack of motivation for the slowing down of such production processes, the hegemonic drive for an increase in speed and pace in every aspect of life then sets in. This is why the time-space compression that Harvey writes about only goes through further compression. We are able to record the speed of time by first recording time itself. In ancient days, time was recorded by the changing visuals of light in the day and the absence of it at night, thereby recording time in broad allocations of time. It is as Jacques Attali describes, where time is seen as a structure, with different structures overlapping at any one point, and this can be witnessed through "historical evolution . . . from the 'time of the Gods' to the 'time of weights', the 'time of machines' and finally the 'time of codes', the passage from one to the next is never precisely dated" (quoted in Klein 2004, 252). With the invention of the clock in modern times, we are able to record time by the hour, minute and even second. By formulating a systematic approach to calculating smaller and smaller units of time, we have come to rely on our clock pieces to dictate time. This is referred to as "clock time" and it is an invention of "capitalist industrialism" (Hassan 2007, 43) as it regulates the labor force's work time in order to measure production and output in a system of the flexible accumulation of consumer goods (Harvey 1989). For businesses, this is where the bottom-line is the fundamental goal.

The regulation of this clock time takes on a different life with the introduction of a platform such as the Internet, where different time zones collide on a common medium. It is hence a space where clock time is not adhered to and regular work time becomes diffused to accommodate an expansion of work allocation. Time-space compression is most evident on the Internet, where two parties in two completely different time zones can communicate simultaneously. The distance between two people who communicate simultaneously on the Internet collapses and there is only one perception of time at the point of interaction between the two, and this "time" is what scholars term as "real time" (Hassan 2007, 49). Hassan, however, argues for the formulation and conception of a "network time" to describe the "time" that we perceive on the Internet (2007, 48). Not only are people from around the globe able to converge at a single point in virtual space but the speed at which information can be transmitted and transferred is almost instantaneous, eliminating the need for physical time-taking transportation to take place. Increasingly, time becomes an unwanted and unvalued commodity that faces the possibility of becoming obsolete. As this paper has already addressed the fact that time is social and its perception is based on its interaction with human agency, it will now look more closely at how this "network time" has penetrated every aspect of our modern life and how that has affected our interaction with this new interpretation of time which largely takes place within a highly networked society.

Networked Society and Time

With the enabling of technological infrastructures such as high-speed broadband access, the densely networked society formed on the Internet has created a new perception of time (Castells 1996). In his research, Castells found that time in a networked society connotes a sense of timelessness because of how electronic transmissions can take place seemingly instantaneously (1996). As a global common medium and a tool for greater democracy, the Internet unites "disparate peoples through the power of the electronic message and destroy[s] geographically based power imbalance" (Warf 2008, 179). Not only is space eliminated, but power dominance in technology and democracy is flattened and, through the coming together of various local cultures, melt together to form a global hegemonic culture in this virtual space.

Improved technological infrastructures such as broadband Internet access have allowed for a greater proliferation of data to populate this virtual space. Its ease of accessibility allows for new information to be constantly uploaded to this burgeoning platform, presenting the image of machinery that runs without a pause. Where people had to initiate dialing on a modem to get onto the Internet, broadband services these days mean that we are connected to the Internet once the computer is switched on. The variety of broadband access types we can choose from also means cheaper Internet connections are made available, with lesser limits on data transfers. Such easy access to a world of databases and information can mean that all types of knowledge are readily available at our fingertips, but the sheer volume of this information presents itself as a double edged sword by bringing on a sense of loss to users when it comes to distilling what is factual from fiction. In 1987, Warren Thorngate made an observation that even before the introduction of the Internet, mankind was already facing an explosion of "content" in both "academic and non-academic areas" (quoted in Himma 2007, 259). With higher accessibility and the ease of data upload and transfer, most reputable media houses find themselves using the Internet as another means to complement their printed media form in order to reach out to more audiences. With the propagation of reputable voices on the Internet, be it reliable media owners or self-starting niche bloggers, the volume of content created makes for greater difficulty in sifting objective truth from hearsay, rumors and opinions (Himma 2007, 266). Political facts and commentaries have often been entangled in this complex heap of fact, fiction, second-guesses and commentaries.

As media companies turn to the Internet as a common platform to reach out to audiences local and global, users have easy access to local and foreign news with perspectives from various angles. On forums and blogs, users can comment on foreign political issues that affect their local concerns as well as compare political complexities from another country and use them as learning points to challenge local governmental administrations. Globalization also means that people care about political affairs that take place in another country because it means local economic consequences. Likewise, political affairs from around the world take up substantial space in local papers because of the audience's interests. In democratic countries like the U.S. and Australia, citizens eagerly criticize policies and parliamentary decisions and this receives coverage readily in the local media. In countries with limited democracy such as Singapore where the media is state-controlled, local policies are not challenged in a public forum in the same fashion as its democratic counterparts (Cottle and Rai 2008). Because of that, disgruntled citizens are made voiceless and do not have an avenue to voice their political opinions in the local media. With the availability of the Internet, political commentaries gain greater accessibility and the Internet becomes a space where citizens can comment publicly. When these commentaries gather at a speed typified by "network time" and spread through the multiplication of social shares, the bureaucratic mammoth of any government is typically unable to respond as quickly in order to manage or quench these reflexive, negative online outbursts.

Singapore and Politics

With the Internet as a new form of participatory media, citizens are able to bypass restrictive media control in their fight for greater democracy. A high-speed platform that allows for the convergence of global opinions, bureaucratic attempts to restrict and censor content online through technological means are easily overcome through the innovativeness of tech-savvy consumers. In Singapore, censorship exists for political discourse. Due to its controlled media climate, when the Internet was introduced in Singapore, fear of punishment inhibited citizens from political participation online. It was only when the Singapore Internet Community Sintercom started in 1994 and engaged in political discussions that the government intervened and instructed Sintercom to register itself as a media that will be regulated (Johal 2006, 6). With restrictions imposed on all media to refrain from controversial political discourse, Sintercom was short-lived. However, this marked the beginning of other alternative media springing up in Singapore with its participants engaging in participatory political discussions online.

The political landscape in Singapore underwent a dramatic shift between 2006 and 2011, with the Internet being one of the main contributing factors as it allowed for a greater diversity of various political opinions to be presented across different social strata. In the General Election (GE) of 2006, Singapore witnessed its first "internet election" where bloggers provided more information about what was going on at opposition parties' rallies than what mainstream media was covering (Lee and Kan 2009, 877). The strength of alternative media and political bloggers in Singapore demonstrated their influence on the election results when votes for the dominant party People's Action Party (PAP) dropped from 66.6% in 2006 to 60.1% in 2011 (Gomez 2006, 27; Chowdhury 2011). It was termed a "watershed" election (Adam and Lim 2011) and one that has not been witnessed in Singapore's history since the PAP's historical win in 1959. In the months following, the Presidential Election (PE) also witnessed a rise in opposition candidates coming forward to challenge the dominant rule of the PAP.

This changing political landscape in Singapore has been attributed to a few factors. The first is the ease with which the Internet allowed the expression of political opinions to be articulated. With a click of a button, commentaries can be submitted online, sent to fellow dissidents and can be easily shared with a wider audience almost instantaneously (Weng 2011). Building a critical mass, citizens became less fearful of the legal implications of their actions. Secondly, Singapore has been following the Western model for financial success through the enforcement of a capitalist-driven economy. This has been Singapore's motivation since its independence from Malaysia in 1959, and has been so for the past fifty years, relying mainly on the human workforce as its national resource. However, not following the democratic approach as prescribed by the Western model for success has led to divisions within the nation, as it strives instead to formulate its own version of democracy (Bokhorst-Heng 2002, 560-561). Singaporeans have been calling for a more liberal form of democracy and have been using the Internet to voice their political opinions. Lastly, the third factor is a combination of the first two factors. Increasingly, people around the world have been using social media to effect political change within their local communities and Singapore has not been sheltered from global influences in this highly interconnected environment. In what appears like a worldwide phenomena of uprisings through social media which began with the Arab Spring in 2011, Singapore's GE and PE of 2011 saw citizens reacting negatively, vehemently and immediately to election news updates, populating the online mediascape with comments and criticisms. Given that the Arab Spring events took place only a few months prior to the GE, some dissident Singaporeans were motivated and inspired to achieve the same results within Singapore (Nair 2011). The Internet has provided the leverage for power to be held in the hands of many. Barber sees the political way forward as a "strong democracy [which] does not place endless faith in the capacity of individuals to govern themselves, but it affirms with Machiavelli that the multitude will on the whole be as wise as or even wiser than princes" (2003, 151). As an interlinked platform, opinions are shared and sentiments are joined on the Internet; people find like-minded individuals like themselves. In the wake of September 11th, Parker argues that there has been an explosion of "universalist and humanist themes in left political thought" which challenges notions of "nationalism" (2003, 155). Based on John Rawls' concept of justice and fairness for all citizens, democracy forms the ideal government system where there is some form of equality among citizens to determine their life despite differences not of their own choice, such as gender (1999).

Political theorist Slavoj Žižek, however, argues against too strong a focus on democratic ideals, and says that people are misled to think that a more democratic government has greater influence over the negative effects of capitalism (2003). What people are really complaining about are consequences from the "silent weaving of the economic logic" that is embedded in the developed world which the local government has limited ability to control in a state of interconnected globalization (Žižek 2003). As the world witnesses the shaping of a global identity, how will this change communitarism-focused Singapore, which prioritizes "economic and social rights" above "political" and "civic rights", in their approach towards citizens' feedback on its governing (Chua 2004, 12)? More importantly, how can the lengthy time that bureaucracy takes to effect change stand against the quick reflexivity of citizens' criticisms without appearing tardy in responding?

In what we determine as citizens' over-reflexiveness in a heated political climate, the Occupy Movement, as part of a string of social media-enhanced activisms, exemplifies the argument here. The Occupy Wall Street movement of 2011 first started in New York as a protest against the power of financiers and later led to worldwide adoption with local Occupies focused on activism about local social causes. In Singapore, any form of public gathering requires a license and public protests are illegal. As a response to events around the world, an Occupy Raffles Place was planned, with the organizers setting up a Facebook page and making contacts to create the event at Raffles Place, the financial hub of Singapore. Despite the rallying online, nobody showed up to protest despite earlier indications that they would (Philemon 2011). The organizers were nowhere to be seen and fellow protesters who turned up were disappointed and did not appear to dare congregate at the city square to carry out what was initiated. Tng suggests that the Occupy Wall Street movement was born out of a great discontentment with the high unemployment rate in the country and people were looking for a scapegoat in the financiers to bear the brunt of their woes. Occupy flopped in Singapore possibly because the "majority of Singaporeans are not so disenchanted with corporate 'greed'" (2011). In fact, the famously uncorrupted Singaporean government is known for superior administration as well as how it replicates a business model in its operations in order to sustain and improve the country's economic progress. The political circumstances in Singapore at that point in time along with a high level of discontent with the government tempted protestors to advance their political agenda easily and quickly through the Internet but they had no long-term plan to back up their actions. Without sufficiently strong motivation or any kind of suggestion for change, Occupy Raffles Place became a symbol for Singaporeans' angst against the government instead.

Time in Politics

In 2007, Tony Blair said that "we are all being dragged down by the way media and public life interact" and that "the Internet created an environment that required media to increasingly compete for attention and forced politicians into a mode of perpetual reaction that served no good purpose" (Rosenberg and Feldman 2008, 1). With citizens' numerous commentaries online, politicians face the problem of having to silence multitudes of angry voices. Blair goes on to say that in 1997, "[the government] took an issue a day [during the period of the election campaign]" but "in 2005, [they] had to have one for the morning, another for the afternoon and by the evening the agenda had already moved on," an indication that more decisions are required to be made within a shorter frame of time (Rosenberg and Feldman 2008, 2). Protests that sprang up around the world serve as an indication of this tension between citizens' desire for change, and government's inability to administer change in its administrative system according to the citizens' timeline.

Scheuerman says that "effective legislation requires impressive capacities of foresight" (2004, 30). Increasingly, politicians no longer have the luxury of time to contemplate when they are engaging in their long-term planning and vision-casting. As the next generation is always cultivated to make quicker decisions, we face a new generation of Singaporeans who are unable to comprehend why bureaucracy takes "so long" to act (Scheuerman 2004, 12) and therefore they choose to react through social media to create pressure for instantaneous change. In western countries like the U.S. and Australia, this pressure turns the political system into one that "allow[s] for rapid communication and thus [is] subject to the whims of 'temporary considerations' inconsistent with the long-term interests of free government" (Scheuerman 2004, 44). With a short political party's life span of about four to five years depending on whether they manage to retain their voters' favor or not, this forces a political party to come up with tactics to placate citizens in the short-run so as to win their immediate vote. The time required to contemplate politically for long-term stability is virtually non-existent.

As a nation that relies heavily on human resources for its economy, residents in Singapore are acculturated through the educational system as well as in the workplace to be competitive in a global capitalist-driven economy. Under such an environment, efficiency and speed are valued in order to serve the economic purpose of the country, with the need to be constantly faster than others by having better technological infrastructure, such as being the first in the world to provide a nationwide public Internet access service in 1994 to serve its status as a highly-connected technological hub (Lee and Kan 2009, 871-872). Citizens are thus predisposed towards being reflexive rather than reflective, since contemplations have no immediate economic value. Because of its ease of use and the passive commitment required, online social activism wins popular votes with Singaporeans, who have little leisure time to spare outside of their work lives. With a click of a button, a Facebook page can be set up and sent out to communities to garner support (Shirky 2011, 38). Tactics used are "short, sporadic, episodic" and "disconnected from larger social movements" (Earl and Kimport 2011, 181). This certainly appears to be the case for Occupy Raffles Place, given its lack of connection with the larger Occupy movement and its inability to follow through with its social agenda. This incident also ultimately serves as evidence that time dedicated to decision-making is a forgotten yet required ingredient in effecting political change in the long run.

Conclusion

The history of modern civilization has provided centuries of learning lessons for our politicians. With each passing generation, history would give us valuable political advice through the successes and failures of others. However, we face the problem of the acceleration of "information overload" today (Himma 2007), which demands our attention at multiple points. Klein says that in contrast to Goethe's time, we "have all the stimulation we could possibly want" and that "in a single year, we gather more impressions than Goethe's contemporaries in a lifetime" (2008, 166). In the modern age, we advocate speed reading and even have the radical idea that decisions can be made within seconds (Gladwell 2005). Political systems have their own bureaucracies which may lengthen the decision-making process, making it appear slow and sluggish. Will Barber's strong democracy be the way forward for Singapore, where the power of decision-making lies mainly in the hands of the people? The economic success and stability of Singapore was not credited to democracy, and it would not have any incentive to have greater democracy given its pristine economic record (Rothstein 2011; Chua 1995, 655). If this new form of "wikipolitics" should take over in the struggle for greater democracy, can it have the same wisdom to see Singapore forward in the long run? If we have gone through centuries of politics to come to its bureaucratic form today yet still face the need to constantly fine-tune in accordance with changing times, what other repercussions do we need to prepare ourselves for if we make political changes within a shorter span of time (Beck 1999)? Will we end up with policies that cater to solipsism instead? Politics require constant reviewing in response to current contexts. We are confined by our biological need to take time to ponder before coming to a decision that has multiple implications for any society. This paper asks if a "wikipolitical" system will come to recognize and respect the existence and requirement of Time for governments to act appropriately and wisely on the public's feedback, instead of demanding that great democracy is the only answer.

Works Cited

Adam, Barbara. 1990. *Time & Social Theory*. Cambridge: Polity Press.

—. 2004. *Time.* Cambridge: Polity.

Adam, Shamim and Weiyi Lim. 2011. "Singapore's Election 'Watershed' May Loosen Political Hold of Lee's Party." *Bloomberg*, May 9th. Accessed Sep. 6, 2011. http://www.bloomberg.com/news/2011-05-08/singapore-s-election-watershed-may-loosen-political-hold-of-lee-s-party.html.

Barber, Benjamin R. 2003. *Strong Democracy: Participatory Politics for a New Age.* Berkeley: University of California Press.

Beck, Ulrich. 1999. *World Risk Society*. Cambridge: Polity Press.

Bokhorst-Heng, Wendy. 2002. "Newspapers in Singapore: A Mass Ceremony in the Imagining of the Nation." *Media, Culture and Society* 24 (4): 559-569.

Castells, Manuel. 1996. *The Rise of the Network Society*. The Information Age: Economy, Society and Culture, Vol. 1. Oxford: Blackwell.

Chowdhury, Neel. 2011. "Why Singapore's Election Didn't Live Up to the Hype." *Time*, May 10. Accessed June 3, 2011. http://www.time.com/time/world/article/0,8599,2070603,00.html.

Chua, Beng Huat. 1994. "Arrested Development: Democratisation in Singapore." *Third World Quarterly* 15 (4): 655-668.

—. 2004. *Communitarian Politics in Asia.* London and New York: RoutledgeCurzon.

Cottle, Simon and Mugdha Rai. 2008. "Television News in Singapore: Mediating Conflict and Consent." *Asian Journal of Social Science* 36:638–658.

Earl, Jennifer and Katrina Kimport. 2011. *Digitally Enabled Social Changer: Activism in the Internet Age.* Massachusetts: MIT Press.

Gladwell, Malcolm. 2005. *Blink: The Power of Thinking Without Thinking*. New York: Little Brown.

—. 2010. "Small Change: Why The Revolution Will Not Be Tweeted." *The New Yorker*. Oct. 4. Accessed Feb. 19th, 2012. http://www.newyorker.com/reporting/2010/10/04/101004fa_fact_gladwell.

Gomez, James. 2005. "Free Speech and Opposition Parties in Singapore." *Asiarights* 5:1-37.

Hassan, Robert. 2003. *The Chronoscopic Society: Globalization, Time and Knowledge in the Network Economy*. New York: Peter Lang Publishing, Inc.

—. 2007. "Network Time." In *24/7: Time and Temporality in the Network Society*, edited by Robert Hassan and Ronald E. Purser, 37-61. California: Stanford University Press.

Harvey, David. 1989. *The Condition of Postmodernity*. Oxford: Blackwell.

Himma, Kenneth E. 2007. "The Concept of Information Overload: A Preliminary Step in Understanding the Nature of A Harmful Information-related Condition," *Ethics and Information Technology* 9: 259–272.

Johal, Terry. 2004. "Controlling the Internet: The Use of Legislation and Its Effectiveness in Singapore". Paper presented at the *15th Biennial Conference of the Asian Studies Association of Australia* in Canberra 29 June - 2 July.

Klein, Stefan. 2008. *The Secret Pulse of Time: Making Sense of Life's Scarcest Commodity*. Translated by Shelley Frisch. Melbourne: Scribe.

Leccardi, Carmen. 2007. "New Temporal Perspective in the 'High-Speed Society.'" In *24/7: Time and Temporality in the Network Society*, edited by Robert Hassan and Ronald E. Purser, 25-36. California: Stanford University Press.

Lee, Terence and Cornelius Kan. 2009. "Blogospheric pressures in Singapore: Internet Discourses and the 2006 General Election." *Continuum: Journal of Media & Cultural Studies* 23 (6, December): 871–886.

Mazzoleni, Gianpietro and Winfried Schulz. 1999. "'Mediatization' of Politics: A Challenge for Democracy?" *Political Communication* 16:247-261.

Nair, Gopalan. 2011. "Lee Kuan Yew's Singapore, Another Egypt in the Making." *Singapore Dissident*, February 11th. Accessed Nov. 4, 2011. http://singaporedissident.blogspot.com/2011/02/lee-kuan-yews-singapore-another-egypt.html.

Parker, David. 2003. "Diaspora, Dissidence and the Dangers of Cosmopolitanism."*Asian Studies Review* 27 (June, 2): 155-179

Philemon, Jewel. 2011. "Occupy Raffles Place—We'll Do it Bigger and Better Next Time." *The Online Citizen,* October 15. Accessed Feb. 19, 2012. http://theonlinecitizen.com/2011/10/occupy-raffles-place-we-will-do-it-bigger-and-better-next-time/.

Rawls, John. 1999. *A Theory of Justice: Revised Edition.* Massachusetts: Belknap Press.

Rosa, Hartmut. 2003. "Social Acceleration: Ethical and Political Consequences of a Desynchronized High-Speed Society." *Constellations* 10 (1): 3-33.

Rosenberg, Howard and Charles S. Feldman. 2008. *No Time to Think: The Menace of Media Speed and the 24-hour News Cycle.* New York: The Continuum International Publishing Group Inc.

Rothstein, Bo. 2011. "Quality of Government and Epistemic Democracy." Paper presented at the *Workshop Epistemic Democracy in Practice* at Yale University, New Haven, USA, Oct. 20-22, 2011.

Scheuerman, William E. 2004. *Liberal Democracy and the Social Acceleration of Time.* Maryland: John Hopkins University Press.

Shirky, Clay. 2011. "The Political Power of Social Media." *Foreign Affairs* 90 (1): 28-41.

Stromer-Galley, Jennifer. 2000. "Democratizing Democracy: Strong Democracy, US Political Campaigns and the Internet." In *The Internet, Democracy, and Democratization,* edited by Peter Ferdinand, 36-58. London: Frank Cass Publishers.

Tng, Ying Hui. 2011. "Singapore: Why Occupy Raffles Failed." *The Online Citizen,* October 25. Accessed Feb. 19th, 2012. http://asiancorrespondent.com/68039/why-occupy-raffles-flopped-dismally/.

Warf, Barney. 2008. *Time-space Compression: Historical Geographies.* Oxon: Routledge.

Weng, Enqi. 2011. "Online Political Passion Vs. Political Reality in Singapore's General Election 2011." Paper presented at the *antiTHESIS Symposium* at the University of Melbourne on July 29, 2011.

Žižek, Slavoj. 2003. "Too Much Democracy?" Paper presented at the Columbia University on April 14th, 2003. Accessed Apr. 14, 2012. http://www.lacan.com/toomuch.htm.

PART IV

POLITICAL/THEOLOGICAL TIME

(DIS) ABLING TIME: THE REFUSAL OF WORK

BRADLEY KAYE

Work exists as a repressive force creating libidinal dysfunctions, but also as a means of keeping civilization intact by suppressing instinctual drives. In most Marxist literature, one finds the widely accepted axiom that work holds the key to class liberation. In the modern period, it was Marx who claimed in *Capital* ([1857] 1978) that work creates a temporal flow which becomes almost second nature to the proletarians in the factories. This presupposition seems to offer hope to able-bodied workers whose bodies are capable of hard labor. Yet, a variety of oppressed people are left out of this discourse. It is necessary to emphasize the importance of theorizing a relation to time that is outside of the flows of production such as in the case of disabled bodies, and Foucauldian mad subjects who for one reason or another cannot work within the context of capitalist labor power. As Elizabeth Grosz has claimed:

> Each of the three temporal modalities (past, present, and future, in all of their conjugative complexities) entail presumptions regarding the others that are often ill- or unconsidered: how we understand the past, and our links to it through reminiscence, melancholy or nostalgia, prefigures and contains corresponding concepts about the present and the future; the substantiality or privilege we pragmatically grant to the present has implications for the retrievability of the past and the predictability of the future; and, depending on whether we grant to the future the supervening power to rewrite the present and past, so too we must problematise the notions of identity, origin, and development. (Grosz 1999, 18)

Rather than merely accepting a hopeless position on the margins, this paper presents an active refusal as depicted in Herman Melville's novella entitled *Bartleby, the Scrivener: A Story of Wall Street* ([1853] 2010) as an opportunity to theorize unproductive time as a means to engage with Negri's concept of Love. In the story of Bartleby, Melville envisions a character going through the day-to-day grind at an office, not just any office, but significantly it is a law office. Instead of simply carrying out the office manager's orders, Bartleby repeatedly responds when given a task by saying, "I would prefer not to." The narrative is told from the perspective of the office manager who is totally baffled but also fascinated by Bartleby's stubborn refusal. He refuses to do anything the boss asks him to do, even ordering lunch. It gets to the point where at the end of the story Bartleby winds up a destitute vagabond who dies in prison. While this paper is not attempting a close reading of the text, examples such as this, where time exists outside of production, demonstrate that one can exert an active refusal which can be an act of temporary liberation. The active refusal to do anything that fits within the dominant paradigm, to merely lie seemingly dormant in a slow time of ongoing protest, is the action taken by Occupy Wall Street. It may not have occurred to #OWS members that they carry on Bartleby's tradition but they do.

The Bartleby character is Melville's technique for depicting a neo-classical-liberal set of presumptions about our illusions of liberty. Capitalism must appropriate unproductive bodies that resist its labor and either incarcerate them, throw them away, or transform them into becoming a docile-useful-body. Resisting capitalism is a zero sum game. Bartleby's refusal to work as defined by capitalism means someone else and/or Bartleby himself has to pay a price for that refusal. Within capitalism, barring a status of independent wealth, one must work so as to earn a wage in order to stay alive. The consequences of being thrown aside are usually much worse than putting up with the exploitation. Even though the outcome in Melville's story is extremely undesirable from a liberal humanistic perspective—Bartleby starves to death in prison, even refusing to take food from the guards—he negates the situation that is causing his oppression, and his refusal, while leading to his death, actually opens up a whole new temporal vista.

This mystification of Bartleby as a revolutionary subject spreads through the world of theory from Gilles Deleuze in *Essays Clinical and Critical* (1998), Giorgio Agamben in *The Coming Community* (2003), and in Slavoj Žižek, who ends his 2004 book *The Parallax View* with a scathing critique of Bartleby. Žižek claims "Bartleby couldn't even hurt a fly—that's what makes his presence so unbearable" (Žižek 2004, 385).

Although Žižek, with his rebuke of Bartleby in *The Parallax View*, cannibalizes all previous positions on Bartleby, Antonio Negri posits the best response to the critique offered by Žižek. Negri sees that the opportunity presented by Bartleby is not necessarily the refusal of work per se, but a chance to create a new temporal metrics. The refusal constitutes a non-capitalist subject stepping outside of exploitative labor-time and the axiom that "time is money." If there is to be profit, workers must produce more than they consume. This surplus labor time results in and exemplifies exploitation. Bartleby's non-action is action, *a refusal that is the affirmation of resistance.* He is in solidarity with disabled and mad subjects precisely because his refusal within a capitalist context frames him as being an abnormal body, void of the desire to work as defined by capitalism. Negri reserves the strategy of refusing work for the most severe exploitations because resistance must be irreducible to the form it resists and being in productive time includes being a subject of productive time.

Revolutionary praxis must begin to construct a new metrics of time. Time that is ontologically restless and external to the metrics of exploitation alters each moment as a differential continuum that breaks with linear modalities of temporal consistency. Unproductive time breaks with continuity and mechanical flows. In the time/place of revolution, a total re-evaluation is necessary so as to disavow time that feels natural. We must begin to think of the future as unhinged from the violence of the past and channel alienation forward towards a vision that can inspire radical praxis.

Yet, even Antonio Negri says it is through production that one exerts resistance to capital. Unproductive time provides an alternative to the dyad between disciplinary power and spontaneity. Resistance can be a disciplined intervention displayed as a spontaneous insurgency. Not wanting to go into work signifies a potential resistance. The refusal expresses the rejection of "beaten down" clock-time, re-valuing the concept of "being on time" as liberation from the ontological suppression of the differential dynamics of temporality irreducible to reification. To refuse is to come to terms with the inclination to let go and thus to lapse back into blind conformity with nature. Conceptualizing time as a differential modality without the continuity of integral social dynamics is to posit a refusal.

In taking Negri seriously we must believe his claim that in post-modernity the common presents itself as metamorphosis. It is not about the opening of Kairos time and one day waking up as an insect; rather this metamorphosis designates the creation of new forms in and by the common. Common is understood as a creative plenitude that is inexhaustible while metamorphosis is an immanent force working within the subject, not transcendentally beyond it. In this sense, Kairos involves liberating Eros (life drives, love, and creation) and Thanatos (death drives, destruction) in a safe environment at the threshold of Reason. A refusal often results in being placed in a safe haven, a prison or a mental hospital so that after a prescribed time one is expected to become productive once again.

The conception of time as active creation exacerbates the play of antagonisms constitutive of a mass strike. Refusing work constitutes the shadow cast by capitalism's inherent negation. (Dis) Abling Time involves rearticulating the death drive as resistance rather than active engagement with the exploitation of work, and by dropping out completely one disables bourgeois disciplinary time. This act is completely unexpected by the bourgeoisie who are perplexed by unemployed bodies.

A collective but individuating act of thinking constitutes the basis of generating Ideas, and Ideas are differentials of thought. Individuating ideas in capitalism more often than not culminate in a petty dogmatic conformism. Foucault and Althusser both point out that capital interpellates subjects as individuals—flexing its ideological muscles to keep our time out of joint. This means that the unconscious dimension of thought constitutes an opposition to all forms of common sense. Ideas, therefore, are reproductions of a fractured temporal "I"—a dissolved Self. Ideas result from a constitutive un-grounding as a deterritorialized line of flight. Thought constitutes an exercise culminating in a non-integral social field completely at odds with discursive homogeneity. For instance, when we glance at a page written by Slavoj Žižek, we construct lines of flight without settling in to position an entire class over the long haul. We are in a constant state of maneuvering without drawing the proletariat and lumpenproletariat into the trenches with us (materially, ideologically, politically and socially, among other ways).

The realm of the possible is not to be understood as implying a homogeneous resemblance between like-minded subjects. The "I" does not exist at all. As Deleuze and Guattari claim, we need to get to a place where it does not matter if a subject says "I" or not. With or without realizing this, we are always already immanently within a state of co-implication as beings-situated-within-the-world. The impact of the atomized interpolated individual strikes upon the Other. Each of us is uniquely enveloped in the activities of the world altogether. A-priori, the world is a common. Privation and appropriation are the illusions.

Furthermore, the world we inhabit comprises a definite quantity of spatial-temporality that operates like a force. Beings constitute a modal-center exerting a force upon other beings. This differential dynamic constitutes the metaphysical basis of reality, and it creates a co-implication in our collective understanding of space-time. Yet, this co-inhabited temporal space is also radically versatile. We can embody the space in divergent ways, manifesting the non-integral rejection of exploitation through enabling temporal space by means of constitutively rejecting severe exploitation, and (Dis) Abling the time of degradation through the refusal of work. As a communist, Negri understands the differential dynamics of time in general versus subjects relating time to each other inter-subjectively. This involves the being-alongside of time within a co-habitable space. Antonio Negri's term for this is Love. He describes Love as being eternal. This means that Love is conceptualized as transcendent of the exploitative metrics of phenomenal experience. Love is an attempt to think beyond the metaphysics of scarcity. Love is a force beyond the phenomenal space of time, and is constituted by the metaphysics of plenitude. Love can never run out, and if we patiently turn within there is always surplus of Love available to each of us that the cruelty of the world can never strip away. So, instead of basing the refusal of work on a negative affect, like Bartleby the Scrivener, one could posit this refusal on a set of positive notions, like Love. Therefore, Love constitutes the affect-based raw material necessary to stoke the flames of a new ontological mode of production based on the transcendence of linear, disciplinary, docile time.

Kairos, Love and Anti-Production

> Ecstasy is not love: love is possession to which the object is necessary, at once possessor of the subject, and possessed by it. There is no longer subject-object, but a 'yawning gap' between the one and the other and, in the gap, the subject, the object are dissolved; there is passage, communication, but not from the one to the other: the one and the other have lost distinct existence. (Bataille 2012, 59)

As the anarchist anthropologist David Graeber points out, several ideas need to be conceptualized in order for an anarchist theory to break new ground. One concept would be a viable theory of political happiness, not just why people do not experience happiness, but how they can.

Love is the co-operative understanding of political happiness and political happiness is integrally linked to a radical conception of time. For Negri, that which, in time, is generation and a formidable innovation of the eternal is spatial-operational. Co-operation is love that proceeds by forging commonalities among multiplicities. Consequently, resistance is love multiplied. Co-operation is also a common generation of space. Singularity brings productive existence out of solitude generating an unproductive productivity. Love as spatial co-operation expresses the striving of the multitude that attempts to endow being with meaning. This striving, also termed "conatus," a term deriving from Spinoza, can only exist if co-operation is an amorous force.

Love seeks the common by means of engaging with constructive space and time in an eternity. Love constitutes the generation of the "yet-to-come" of the common. Space is the projection of the accumulation of experience produced by love within the temporality rendered immeasurable between eternity and the "to-come" (Negri 2003, 213). Love is an amorous common space. Love brings the subject closer to overcoming oppression by creating cogent ties with others in meaningful ways. Opening a new temporal space can create new social bonds which could overcome the destitution of poverty by enabling affects that are irreducible to productive force. Above all, love is personal.

Many oppression theorists have noted that the first stage of being subjugated is the loss of personal identity. Amore is a gift, giving singularity back and giving an object that cares about, and nurtures the subject, and vice versa. Our identity and selfhood are returned to us thereby offering a transcendental pathway to politics. Singular time, that is personal time, the time outside of production, breaks down the walls of depersonalization. It offers a life experience that is lazy and leisurely in unrushed patience. Therefore, careful time is a time of nurturing.

The term that Negri often uses for this occurrence is "Kairos," a concept he appropriates from Christian Theology. He also claims that love happens in poverty and destitution, the first stages in philosophical discourse. Drawing on the legacy of Christ, whose famous Sermon on the Mount proclaimed "blessed are the meek, for they will inherit the earth" and "it is easier for a camel to pass through the eye of a needle than for a rich man to enter the kingdom of God," many Marxist principles demonstrate a clear Christian influence in their work. It is this sympathy for the destitute that creates a new conception of time because the destitute are outside of the capitalistic framework.

In Kairos, the future appears to be that which from here onwards is "un-finished" or what is not yet closed. Kairos is an open present, an event. It is a rupture in the continuity of time as a linear succession from one point to another point as merely one contiguous line segment. Kairos allows for the emergence of an antithetical subject as singularity. Surplus value must arise as an excess of the equivalent because it arises outside of capitalist production. In a communist transition, there will be an excess of labor-time as the worker will appropriate wealth according to basic needs and not for the sake of a surplus. Labor-time will be reduced. Liberation does not amount to a just measure, in the sense of distributive justice, but rather in a subject achieving the fullness of its expression. A radical transition must occur co-relating to a singular conception of time as restlessness.

Exploitation will be reduced as well. Superfluous labor will happen voluntarily, unforced by the capitalist labor-exploiting vampires described by Marx, feasting on the blood of the proletariat. Capital will no longer surge onwards in a hyper-manic obsession with accumulating more and more wealth. In this communist society machines will now do the work that humans used to do. Technology will allow this to happen because of the vast reserves of wealth capital reinvested into the mode of production. As more machines become available, human-labor will be replaced by mechanized labor power. Labor time will be reduced, creating a completely different metrics of time. An unproductive time will emerge that is amenable to disabled and mad subjects currently existing outside of productive time. Out of sheer inertia, when resources run out, antithetical subjects will emerge, un-subsumed by the cash-nexus of productive-consumption. A new communist society will emerge out of this rubble. At that point, the mode of production will relinquish the politics of exclusion.

Yet this is another point of concern in Antonio Negri's work. He tends to rethink time as a tumultuous set of intensifying antagonisms. Consider the end of a text that has lost currency, but that should still be studied with rigor: *Insurgencies* (1999). The end of this book rings rather hollow, because Negri smacks of realism. Rather than being inspiring Negri seems patently cynical where he writes

> Is what is awaiting us a history of freedom? It would be foolish to say so, confronted as we are by the horrid mutilations that constituted power continues to inflict on the ontological body of human freedoms and by the perpetual negation that the unbreakable series of the freedom, equality, and strength of the multitude posed in contrast. (1999, 335)

This is followed by the least inspiring statement:

> What awaits us is a history of liberation, disutopia in action, relentless and as painful as it is constructive. (335)

Where did this cynicism come from? Are things really this hopeless? Must we just forget about utopian dreams and settle for dystopian struggles in the muck? Once the mode of production shifts, the refusal of work will no longer be necessary, because the exploitation supercedes itself, and the mode of production is transformed by the progress of advanced technology—making labor power completely dependent upon machines, rather than laborers. If this transition takes place then the use of money will be outmoded, because the only reason money exists is for the bourgeoisie to appropriate it back from the working class. If the working class is completely displaced, then wage labor will no longer exist, leaving a vacuum or an opening. If this happens, then the use of money will wither away—much like the state, as Marx incorrectly predicted. As capitalism progresses, the syllogism of productive-consumption creates new products and by extension new "needs." For instance, personal computers, cell phones, plasma screen televisions, Ipods, and Ipads, unimaginable 40 years ago, yet now one cannot fathom a world without them, are luxuries sold to consumers as if they are necessities. As Marx rightly points out,

> Exchange delivers the particular products into which the individual desires to convert the portion which distribution has assigned to him; and finally, in consumption, the products become objects of gratification, of individual appropriation. (Marx [1939]1973, 89)

The refusal of work is neither utopian escapism, nor a quietist retreat into isolated consciousness, but rather a strategy to forge lasting class-composition culminating in the positive development of communist collective modes of production. The refusal of work is all about class-valorization, and the end of exploitation, but starts with theorizing about the proper time to take action, deciding when capital is most vulnerable. This vulnerability is becoming more and more prevalent as the circulation of resistances emerges in the contemporary omni-crisis of Capitalist Empire.

Time accelerates, and slows down according to productivity. Standing outside of production, in the heap of debris that contemporary man attempts to master a la Bataille, when all libidinal investments disintegrate, time reduces down to "the great Zero"—the name Lyotard gave to the instance informing a particular but insistent dispositif. The inside is then considered in terms of what is going on outside. The great Zero is an empty center, which reduces the present complexity of what happens to mere presence and absence. In this theory, all forms of signification, and all semiotic theories, are fundamentally nihilistic (Lyotard 1993).

An article in *Radical Philosophy Review* by Randall Williams pointed out that Negri's work on constituent power was instrumental in the Bolivarian Revolution in Venezuela. This is an indication that, as monumental as the theoretical insights of Negri are, his work on time has yet to be fully explored as a presage to liberation for subjects below the proletarian class.

Works Cited

Agamben, Giorgio. 2003. *The Coming Community.* Translated by Michael Hardt. Minneapolis: University of Minnesota Press.

Bataille, Georges. 2012. *Inner Experience.* Translated by Leslie A. Bolt. Albany: SUNY Press.

Deleuze, Gilles. 1998. *Essays Clinical and Critical.* London: Verso.

Grosz, Elizabeth. 1999. *Becomings: Explorations in Time, Memory, and Futures*. Ithaca: Cornell University Press.

Lyotard, Jean-Francois. 1993. *Libidinal Economy: Theories of Contemporary Culture*. Translated by Iain Hamilton Grant. Bloomington: Indiana University Press.

Marx, Karl. (1939) 1973. *Grundrisse*. Translated by Ben Fowkes. London: Penguin Books.

—. (1857) 1978. *Capital*. Translated by Ben Fowkes. New York: Vintage Books.

Melville, Herman. (1853) 2010. *Bartleby, The Scrivener: A Story of Wall Street*. Brooklyn: Melville House.

Negri, Antonio. 1999. *Insurgencies*. Minneapolis: University of Minnesota Press.

—. 2003. *Time for Revolution.* New York: Continuum.

Williams, Randall. 2011. "The Ballot and the Bullet: Anti-Juridical Praxis from Malcolm X and Nelson Mandela to the Bolivarian Revolution." *Radical Philosophy Review* 14 (1): 1-25.

Precarious Times: The Revolutionary Potentiality of the Present

Alexander Brown

The multiple struggles that broke out all over the world during the course of 2011 made it possible once more to talk about revolution. Revolutions in the Middle East toppled dictatorships and unleashed civil wars. The Occupy movement in the U.S. took to the streets and parks of New York, Oakland and cities all over the country and subsequently the world in a challenge to the might of finance capital. Movements against austerity in Europe saw squares occupied in Greece and Spain and universities and banks occupied in Britain. In Japan a powerful anti-nuclear movement emerged after the Fukushima disaster that cultural theorist Karatani Kojin argued, in a speech given at the 11 September *Genpatsu Yamero* anti-nuclear demonstration in Shinjuku, has the potential to fundamentally transform Japanese democracy through the reclamation of popular sovereignty.

Precarity has been a central question in these struggles. When a young Tunisian man, Mohamed Bouazizi, set himself alight outside a government office building on 17 December 2010 in the small town of Sidi Bouzid he drew attention to the rampant inequality in Tunisian society. The subsequent revolts in Tunisia and the unfolding of the Arab Spring were a response to the "lethal combination of poverty, unemployment and political repression" found across the Arab world as a result of neoliberal austerity and dictatorial governments (Andoni 2010). When British students took to the streets to protest cuts to education budgets they contested the precarisation of their future in order to fund bailouts for the banks. Across Europe the implementation of austerity policies in the wake of the debt crisis was seen by many as threatening the already precarious lives and livelihoods of the majority of people (Jacobson 2011).

Gill and Pratt (2008, 3) define precarious employment as "all forms of insecure, contingent, flexible work—from illegalized, casualized and temporary employment, to homeworking, piecework and freelancing."

Precarious working conditions produce an "interminable lack of certainty," and a condition of "being unable to predict one's fate or having some degree of stability on which to construct a life" (Neilson and Rossiter 2005, under "Uncertainty, Flexibility, Transformation"). Precarious forms of work have become increasingly common in labour markets in the advanced capitalist economies of Canada, the USA and Western Europe since the Oil Shock crisis of the 1970s (Waite 2009, 416) and in Japan following the economic crisis of the early 1990s (Obinger 2009). Neilson and Rossiter (2008, 58) have pointed out that the relative employment security of these advanced capitalist countries was itself always an exception to the "precarious labour [that] is the norm of capitalist production and reproduction." While precarity may not be a new phenomenon, the precarisation of not only employment but all aspects of life are accelerating. The nuclear crisis in Fukushima, for example, accentuated the growing precariousness of life in Japan by highlighting the vulnerability and risk of everyday life in advanced capitalist society.

This chapter considers precarity as a temporal phenomenon in which the entire time of social life is put to work to create value for capital. It argues that in times characterised by increasing precariousness, the potential for revolutionary social change lies in the liberation of the present moment. Drawing on Antonio Negri's essay, *The Constitution of Time*, published in English translation in the 2003 collection *Time for Revolution*, it argues that in precarious times revolution cannot be considered as a singular moment of rupture located at some point in the future but must be found in the liberation of time from capital in the present moment. This chapter highlights the prefigurative politics that animated the revolutionary movements of 2011 and argues that these are constitutive moments of liberation in which the contours of a new society are realised within the shell of the old.

Precarity and Real Subsumption

Under neoliberalism, precarity constitutes a daily experience of life,

> a juncture of material and symbolic conditions which determine an uncertainty with respect to the sustained access to the resources essential to the full development of one's life. (Precarias a la Deriva 2007, 158)

As a condition that characterises the entire time of life, precarity can be understood as a temporal phenomenon in which capitalist command is experienced all the time. Tsianos and Papadapoulos (2006) explain that

> precarity means exploiting the continuum of everyday life, not simply the workforce. In this sense, precarity is a form of exploitation which operates primarily on the level of time.

Hardt and Negri (2009, 146) concur with this view, noting that precarity "destroys the division between work time and nonwork time, requiring workers not to work all the time but to be constantly available for work." With capital accumulation no longer limited to the factory much of our leisure time is spent creating cultural and linguistic products that are in turn commodified and sold back to us. Even spending time socialising with friends on social networking sites may be exploited by marketers and made productive of cultural capital (Gill and Pratt 2008, 17). Writing about Japan's popular culture industries, for example, Anne Allison (2009, 94) notes that it is the very instability and fluidity of the precarious "life of youth at play, in their imaginations and through the intimacies they form with others" that is put to work to produce cultural commodities. Angela Mitropoulos (2005, 4) has also observed this tendency in neoliberal capitalism toward the creation of a "generalised indistinction between the labour market, self, relationships and life."

Negri understands the generalised indistinction between work and life under neoliberal capitalism in terms of Marx's notion of real subsumption. Marx developed the notion of real subsumption in order to explain the way the rule of capital slowly comes to dominate social relations (Marx 1976, 941–1084). Capital first begins to dominate the labour process by monopolising the means of production alone. During this period, which Marx called "formal" subsumption, the labour process itself is still relatively autonomous from capital. Labour is subsumed within capital through the measurement of labour-time. Marx's labour theory of value, Negri explains, is based on the reduction of the various forms of human labor to abstract labor and the measurement of discrete quantities of labor time. Negri argues that when time becomes a measure of value the diverse forms of human labor are reduced to a single, homogeneous substance. The content of labor becomes less and less important as its productivity for capital is reduced to an abstract quantity of labor time. Capital "determines [labour's] productive power . . . through the multiplication of average temporal units." Time becomes "in relation to labour . . . at once *measure* and *matter*, *form* and *substance*" (Negri 2003, 24).

However, as the capitalist mode of production becomes more fully developed it shapes the labour process itself and social relations become more and more subject to its rule. A contradiction develops between the homogenisation of labour and the complex forms of labour necessary in techno-scientific production. Negri explains that techno-scientific labour is "not reducible to simple labour, to the accumulation of units of measure" because it relies on the social reproduction of a highly educated and socialised labouring subject. Education and innovation themselves have become important sources of capitalist value and have been integrated into capitalist production. The protests of young students in Chile and London in 2011 over their precarious futures highlighted capital's reliance on a highly educated workforce and its simultaneous lack of willingness to pay for it through the welfare state. Under neoliberal capitalism, education and labour increasingly overlap. The precarity that students experience as they struggle to pay for their education is becoming a permanent condition. Once they graduate they will be expected to continually demonstrate their "work-readiness" in order to gain access to a share of social wealth whether through wage-labour or increasingly limited welfare provisions. Thus whether they are working, studying or looking for work their time is always at least potentially subject to the rule of capital.

Capitalist production relies not only on the techno-scientific development of labour but on the collective effort of workers engaged in different areas of industry, possessing different skills whose labour is coordinated across industries and regions. Capitalist value can only be produced through the socialisation of labour. Negri expresses this notion through "the concept of the socially combined worker," that, he explains, "represents the subsumption of the *social whole* by capital" (2003, 28). The socialisation of production "renders the social relations of production, reproduction and exploitation, one-dimensional" (28). However, if production takes place not in discrete periods of work time but all the time, in the daily rhythms of social life, then time can no longer serve as a measure of value. "When the entire time of life has become the time of production, who measures whom?" (29).

For Negri, real subsumption is the contradictory condition of contemporary capitalism in which time no longer serves as a measure as it did in formal subsumption but "must rather be presented as the global phenomenological fabric, as base, substance and flow of production in its entirety" (Negri 2003, 28). Negri explains that in Marx, "time is not only the instrument that constitutes the effective possibility of change—time is also the tautology of life" (33). Because Marx linked time to production he has shown how time is dynamic, productive of the new and the source of all innovation and change. In Marx,

> alongside the abstract development of social mediatisation and of the subjectification of abstract labour, time itself becomes substance, to the point that time becomes the fabric of the whole of being, because all of

> being is implicated in the web of the relations of production; *being* is equal to *product of labour; temporal being.* (33)

Negri argues that, in real subsumption, the antagonism between capitalist and proletarian time is wholly within the *Umwelt* of real subsumption. Time is the very substance of the antagonism between capital and labour.

> Being is materially given in the form of time. Every productive activity, every human action is within this *Umwelt*: a dense and strong temporal envelopment of existence is the first feature by which contemporary being reveals itself. (Negri 2003, 35)

For Negri, there is no outside to time, because time is constitutive of being in its entirety. As a consequence, the contradiction between capital and labour takes the form of a radically dissimilar, irreducible temporality. Melinda Cooper (2011, 140–41) explains that Negri's purpose in developing this notion of temporality is

> to theorize a notion of time as nondialectical difference—to conceive of time as absolute innovation, hysteresis and catastrophe—but also to construct an autoconstitutive ontology of time on the basis of this absolute difference.

In real subsumption antagonism does not disappear but is rent by a fundamental contradiction between the time of capitalist command and the times of liberation.

> In the *Umwelt* of real subsumption (of enveloping time) there emerges not only the hysteresis of the analytic of time and of multiple real times, but also and above all *the asymmetry of the time of command and of the times of liberation from exploitation.* (Negri 2003, 43)

This conflict between capitalist and communist temporality is at the heart of the contemporary experience of precarity. Precarious time is a collective experience in which work and nonwork times are practically indistinguishable and struggle requires us to liberate our time from capital in every moment of our daily lives. In these precarious times, revolutionary struggles involve "multiple subjects and forces, ideas and practices [that respond] to transformations in production and labour" (Italian Core Group 2007, 3). Precarity is not limited to precariousness, to the increasingly unstable and uncertain conditions of work and life in post-Fordist capitalism, but refers in a political sense to the emergence of struggles and the possibility of forms of resistance that can constitute new forms of social organisation (Waite 2009, 412–3; Neilson and Rossiter 2005). In precarious times nothing is fixed or absolute. Our time might be captured and turned into value for capital but the potential exists for us to liberate each and every moment and make revolution all the time.

Liberated Time

Negri challenges us to liberate our time. The time in which we refuse to produce value for capital, in which we take time for ourselves is liberated time. In precarious times we are exploited throughout the time of life and hence we need to liberate ourselves from capitalist command in every moment. For Negri, while capitalist time and the time of liberation inhabit the same immanent social field, they are antagonistic. When we stop spending our time producing value for capital we can produce something else. Liberated time is "a machine of constitution" (Negri 2003, 114) that creates a future society within the shell of the old. The basis of this revolutionary time is the collective self-activity of the multitude. Negri's understanding of liberated time is based on the idea that time is produced through doing. Therefore, when we do something different and when we do so collectively we produce a different temporality. In precarious times we are continually confronted with a capitalist timepiece that drives us to work harder and faster yet we simultaneously engage in a multitude of lines of flight and moments of liberation. When we do so, and when we organise our liberation collectively we produce the time of revolution.

In precarious times it no longer makes sense to talk of a time "after the revolution" or of a "vanguard" bringing consciousness to the class from the outside. Revolution is not a future event but an ongoing practice that pits the potentiality of living labour against the dead weight of capitalist command. In this sense, every time is a time of revolution. Negri rejects the notion of a linear transition to socialism. He sees communist time appearing "as fireworks and flares rather than [as] a secure trajectory" (Negri 2003, 45). This conception of time helps us to see the immanent revolutionary potential in our daily lives. For Negri, these singular moments of liberation, these "fireworks and flares" are not merely signs of an impending revolutionary event but are constitutive moments when communism comes into being. Negri explains that

> the idea of time . . . presents itself as the matrix of the dystopia of communism in the reality of its immensely powerful collective advance.

> Time will increasingly appear as the material from which communism is constituted. (2003, 45)

Negri's revolution of time brings about a dystopian communism because it is unfinished, incomplete and emergent. He refuses to locate communism in a utopian "no-place" and embraces the materiality of the time of struggle itself in which communism is realised in multiple, ongoing and incomplete ways. This notion of communism evokes Marx and Engels' own definition:

> Communism is for us not a *state of affairs* which is to be established, an *ideal* to which reality [will] have to adjust itself. We call communism the *real* movement which abolishes the present state of things. (Marx and Engels 1975, 49)

Japanese cultural theorist Mōri Yoshitaka describes the strategic exploration of this revolutionary potentiality of the present in recent Japanese social movements in which revolution is not thought of as a point of time located in the future but as an "ever-present possibility" located in the present moment (Mōri 2003, 167–8). Commenting on the politics of the anti-nuclear movement in Japan after Fukushima, Hirose Jun (2012) notes that the nuclear crisis necessitates an indefinite and open-ended struggle rooted in the present because the experience of the crisis is itself an ongoing part of daily life. He observes that the nuclear crisis has no discrete ending. "It is not," he explains, "that the nuclear accident *has taken place* as a solution to metastability . . . but that it *is taking place* as a duration of metastability itself." For Hirose, the ongoing nature of the crisis requires a conception of struggle that is similarly open-ended. He contrasts the notion of an ongoing insurrection against one of revolution as a moment of rupture by noting that

> the joy of revolution is discovered when it has taken place, while the joy of insurrection exists when it is taking place. All the exhaustion in revolution is requited by the joy of having shared problems, while in insurrection the exhaustion of living problems is together with its joy. (Hirose 2012)

Hirose's understanding of the simultaneity of radioactive contamination and the anti-nuclear struggle reflect the more general precariousness of the times in which we live when capitalist crisis and proletarian struggle inhabit the same immanent social field. In the midst of the horror of the unfolding nuclear crisis in Japan a multiplicity of social movements have emerged which seek to both end the nuclear industry itself and to make life liveable in the contaminated zone. In the face of an uncertain future, it is the struggle for the present that is transforming Japanese society after Fukushima.

The production of non-capitalist temporalities in the present moment can best be described as a form of prefigurative politics. As David Graeber (2011) explains with regard to the Occupy movement, "prefigurative politics involves creating a vision of the sort of society you want to have in miniature." This is a notion of revolution as a process in which the creation of a future society in the shell of the old replaces the notion of preparing for a moment of revolutionary rupture. For Negri, the purpose of prefigurative politics is not to create a perfect utopia. Their significance lies in the materiality of the multiple, imperfect acts of creation themselves. "Prefiguration", Negri argues, "is *not* in any sense *utopia*, it is concrete activity" (Negri 2003, 117).

It is the productivity of liberated time that, for Negri, holds the key to understanding its revolutionary potential. Unlike leisure time, which is time left over after work for capital has been completed, "liberated time is not the residue of exploited time, but is rather the force that breaks up and destroys all the links of capitalist society" (Negri 2003, 113). Liberated time is the time we create when we produce something other than capital. It is time stolen from capitalist work, from the production of surplus value, but it *is* productive time. "It is productive rationality torn away and isolated from the command that analysed this rationality and extorted it from the time of life" (Negri 2003, 113). The liberation of time involves the collective activity of the proletariat directly producing other values.

The revolutionary upsurge of 2011 emerged out of the ongoing experimentation with prefigurative politics in recent social movements. Carlsson and Manning (2010) describe the do-it-yourself ethos of outlaw bicyclists, community gardeners and pirate programmers as *nowtopian*. These movements embrace a strategy that closely resembles Negri's notion of the liberation of time. They establish collective projects that directly produce non-capitalist value. Nowtopian activists engage in complex techno-scientific production based on common knowledges (Carlsson and Manning 2010, 933). Carlsson and Manning (929) argue that nowtopian practices "hold moments of a post-capitalist society," moments which, for Negri, constitute liberated time. Rather than allowing their collective productive capacity to be spent producing value for capital, community gardeners, for example, engage in forms of work which "fully engage their capacities to create, to shape, to invent and to cooperate without monetary incentive" (925).

In precarious times, every moment is contested. While capital threatens to subsume every moment, struggles for liberation are necessary across the time of life. Inquiring into the anti-austerity movement in Greece, Antonis Vradis (2011, 65) asks "what happens when there is nowhere to take a day off to—or when one struggle becomes inextricably connected to another." Based on his experience of the protests in Athens he describes a temporality of protest in which struggle fills the entire time of life. This is a situation in which "the causes to fight for become a major issue in one's own everyday reality, where there is simply no escape, no way to go back home" (65–66). Comparing the gatherings in the central Athenian Syntagma Square with similar assemblies in Egypt and Spain, Vradis observes that once gathered in the square, the temporality of protest stretched out as people remained long into the night in order to take part in extensive discussions in "extended moments of struggle to take back all of their days—to take back their entire lives" (66).

The Revolutionary Potentiality of the Present

Negri's theory of time is of a material practice of human liberation. As Wolfe explains, Negri's philosophical project is a "'constitutive ontology', a discourse on transformation and liberation, rather than an explanation of the world by means of a *reductio*" (Wolfe 2007, 199). Negri's notion of the liberation of time is not meant to provide a definitive analysis of the world as it is but to challenge us to create a multitude of new worlds. Negri's constitutive temporality is a "notion of possibility as possibility of change, or better, possibility understood as *potentiality*, that is, as power" (Wolfe 2007, 209). His thought on liberated time recognises the tautology that time liberated from capital IS time liberated from capital. It is not a stepping stone towards a time "after the revolution." Through prefigurative strategies that constitute the future society we directly produce communism by liberating our time from capital. We can establish non-capitalist social relations by challenging capitalist social relations in our daily life.

Revolution is not an event but an ongoing process. While struggle necessarily involves an escape from and resistance to the time of capitalist command it is also a powerful "machine of constitution" (Negri 2003, 114) through which we create the future society within the shell of the old. The importance of Negri's understanding of liberated time lies in challenging the separation of revolutionary activity into means and ends. Revolution is not a future moment towards which we strive but a struggle to realise the potentiality of the present. We do not make revolution to realise a future communist society. We realise communism by making revolution.

We live in precarious times. The constituent struggles of the multitude are daily realising the liberated temporality of communism through prefigurative practices. However, we are threatened by the attempts of an increasingly desperate ruling class to deny us a future and reimpose capitalist command through austerity, violence and the threat of war. Negri and Hardt have argued that "we can already recognise that today time is split between a present that is already dead and a future that is already living" (Hardt and Negri 2004, 358). In order to recognise the revolutionary potentiality of the present we need to keep in mind, as Nick Southall has pointed out, that revolution is not a singular event or moment of rupture that takes us over the threshold from the capitalist present to the communist future. Negri's constitutive temporality is a powerful strategy of liberation but Negri has yet to completely let go of the notion of a revolutionary event situated in the future. With Hardt he goes on to claim in the continuation of the above paragraph that "in time, an event will thrust us like an arrow into the living future" (358). The revolution of time requires us to recognise that no single heroic event can liberate us from capitalism. Rather, revolutionary struggles take "a series of paths along which we move at various speeds, in different types of movement and direction" (Southall 2010). In precarious times we confront capitalism in every moment of our lives. We must therefore realise the revolutionary potentiality of every moment, subtract ourselves from capital and liberate our time.

It's time for revolution.

Works Cited

Andoni, Lamis. 2010. "The Rebirth of Arab Activism: How One Young Tunisian is Emerging as a Symbol of Disenfranchised and Impoverished Arab Youth." *Al Jazeera Online*, Dec. 31. Accessed Feb. 21, 2012. http://www.aljazeera.com/indepth/opinion/2010/12/2010123 1161958792947.html.

Allison, Anne. 2009. "The Cool Brand: Affective Activism and Japanese Youth." *Theory, Culture and Society* 26 (2–3): 89–111.

Carlsson, Chris and Francesca Manngin. 2010. "Nowtopia: Strategic Exodus?" *Antipode* 42 (4): 924–953.

Cooper, Melinda. 2011. "Marx Beyond Marx, Marx Before Marx: Negri's Lucretian Critique of the Hegelian Marx." In *Reading Negri: Marxism in the Age of Empire*, edited by Pierre Lamarche, Max Rosenkratz and David Sherman, 127-148. Chicago: Open Court Publishing.

Gill, Rosalind and Andy Pratt. 2008. "In the Social Factory?: Immaterial Labour, Precariousness and Cultural Work." *Theory, Culture and Society* 25 (7–8): 1–30.

Graeber, David. 2011. "You're Creating a Vision of the Sort of Society You Want to Have in Miniature." Ezra Klein's Wonkblog. *The Washington Post*. Accessed Feb. 28, 2012. http://www.washingtonpost.com/blogs/ezra-klein/post/youre-creating-a-vision-of-the-sort-of-society-you-want-to-have-in-miniature/2011/08/25/gIQAXVg7HL_blog.html.

Hardt, Michael and Antonio Negri. 2004. *Multitude: War and Democracy in the Age of Empire*. New York: Penguin Books.

—. 2009. *Commonwealth*. Cambridge: Belkin Press.

Hirose, Jun. 2012. "Reflections on Nuclear Power and Insurrection." JFissures: Cracks in the Planetary Apparatus. Accessed Feb. 16, 2012. http://www.jfissures.org/2012/01/05/reflections-on-nuclear-power-and-insurrection.

Italian Core Group. 2007. "Introduction." *Feminist Review* 87: 3–4.

Jacobson, Matthew. 2011. "Precarity and New Social Movements in Southern European Cities". Paper presented at the International Sociological Association RC21 Conference, Amsterdam, 7–9 June. Accessed Feb. 28, 2011. http://www.rc21.org/conferences/Amsterdam 2011/edocs/Session%2013/RT13-1-Jacobson.pdf.

Marx, Karl. 1976. *Capital: A Critique of Political Economy*. Vol. 1. Translated by Ben Fowkes. London: Penguin.

Marx, Karl and Frederick Engels. 1975. "The German Ideology." In Vol. 5 of *Karl Marx, Frederick Engels: Collected Works*. Moscow: Progress Publishers.

Mitropoulos, Angela. 2006. "Precari-Us?" *Mute* 29. Accessed Feb. 28, 2010. http://www.metamute.org/editorial/articles/precari-us.

Mōri, Yoshitaka. 2003. *Bunka=seiji: gurōbarizēshon jidai no kūkan no hanran* [Culture=Politics: Spatial Rebellions in the Age of Globalization]. Tokyo: Getsuyōsha.

Negri, Antonio. 2003. *Time for Revolution*. Translated by Matteo Mandarini. London: Continuum.

Neilson, Brett and Ned Rossiter. 2005. "From Precarity to Precariousness and Back Again." *Fibreculture Journal* 5. Accessed Nov. 25, 2010. http://five.fibreculturejournal.org/fcj-022-from-precarity-to-precariousness-and-back-again-labour-life-and-unstable-networks/.

—. 2008. “Precarity as a Political Concept, or, Fordism as Exception.” *Theory, Culture and Society* 25: 51–72.

Obinger, Julia. 2009. “Working on the Margins: Japan’s Precariat and Working Poor.” *Electronic Journal of Japanese Studies* February. Accessed 25. Nov, 2010. http://www.japanesestudies.org.uk/discussionpapers/2009/Obinger.html.

Precarias a la Deriva. 2007. “Adrift Through the Circuits of Feminized Precarious Work.” *Feminist Review* 77: 157–61.

Southall, Nick. 2010. “Time for Revolution and the Power of Love.” Paper presented at the conference From Empire to Commonwealth: Communist Theory and Contemporary Practice, University of Wollongong. Nov. 25–26, 2010. http://revoltsnow.wordpress.com/2011/10/03/time-for-revolution-and-the-power-of-love.

Tsianos, Vassillis and Dimitris Papadopoulos. 2006. “Precarity: A Savage Journey to the Heart of Embodied Capitalism.” *Transversal Journal* (Oct.) http://eipcp.net/transversal/1106/tsianospapadopoulos/en.

Vradis, Antonis. 2011. “A Funny Thing Happened on the Way to the Square: Thoughts in the Middle of the Athenian Autumn of 2011.” In *Occupy Everything: Reflections on Why it’s Kicking Off Everywhere*, edited by Alesio Lunghi and Seth Wheeler, 62–67. Wivenhoe: Minor Compositions.

Waite, Lousie. 2009. “A Place and Space for a Critical Geography of Precarity.” *Geography Compass* 3:412–33.

Wolfe, Charles T. 2007. “Materialism and Temporality: On Antonio Negri’s ‘Constitutive Ontology.’” In *The Philosophy of Antonio Negri*, edited by Timothy S. Murphy and Abdul-Karim Mustapha, 2:198–220. London: Pluto Press.

MESSIANIC TIME AND HISTORICAL MATERIALISM

BRAD D. HUME

Both materialism, including historical materialism, and revolution are being re-examined by scholars and those concerned about the problems and perils of the corporate-state systems and the potential (and real) damage being done to the environment (Bennett 2010; Braun and Whatmore 2010; Coole and Frost 2010). Most often these discussions are parallel even though those who are reviving the notions of revolution and communism remain materialists (though what that means differs) (Badiou 2011; Douzinas and Žižek 2010; Caputo and Alcoff 2009; Žižek 2008; Douzinas 2007; Hardt and Negri 2009, 2004, and 2000). I will bring some of these discussions together and focus on historical materialism and revolution, beginning with a basic material premise of the now: in a (human-built) world that has enough abundance for all of humanity, the existence of poverty and the denial of rights are fundamental injustices which demand change, even revolutionary change. The question, of course, is whether revolution (non- or violent) is even an imaginable project not simply because of the horrors of the past, but because our (anyone who reads or discusses these issues) relationship to history, our moment in time, has changed from the sense of History that existed, depending on your choice of dates, until 1945 or 1991.

Scholars like Paul Rabinow and Niklas Luhman have described our present as one of ambivalence toward the past and the future (Rabinow 2008; Luhman 1998). Following the revolutions of 1989-1991, after having chronicled modernity for decades, Eric Hobsbawm came to the conclusion that it may be that the past "has lost its role, in which the old maps and charts which guided human beings, singly and collectively, through life no longer represent the landscape through which we move" (Hobsbawm 1994). A key conclusion, and one that is still widely shared, is that history is not a teleological process in which humanity is destined to fulfill itself. In an attempt in part to confront the uncertainty of the present and the lack of *telos*, a number of authors have revisited the idea of messianic time as it was discussed by Walter Benjamin and in relation to the Pauline event of the proclamation of Jesus as the messiah (Caputo and Alcoff 2005; Badiou 2003; Žižek 2003; Benjamin 1968). What Benjamin actually meant by messianic time has been subject to debate, but recent authors have been more interested in the concept as an analogy to the idea of revolution or as a theistic example of a revolution which is still valid for considering revolution in general.

In what follows I take an agnostic position, if you will, on the question of messianic time and revolution. My main claim will be that from the perspective of any historical materialism, messianic time cannot be religious (supernatural) and thus requires at least a weak teleology centered on the belief that humans can in fact create a just world out of the perils of the real, that there is a greater realization of ourselves that we can in fact achieve which stems from the world and from ourselves. That potentiality would either have to be a product of evolution or a product of history (i.e., historical developments like industrial society are necessary but not sufficient conditions). I will argue that a combination of both suggests the possibility that humans could transform the currently problematic world into one that is just.

Historical Materialism, Technocratic Revolution or Apocalypse

To the evolutionary psychologist, the human subject is universal, reflecting in time the qualities all humans share (Pinker 2000; Tooby and Cosmides 1997). Evolutionary psychologists, however, flatten time, reduce the importance of history and neglect the real transformation of human subjects in time. Historians do not focus enough on this as a process. Michel Foucault and social constructivists looked at the question of the subject, providing rich descriptions of the self-articulation and production of subjects in particular historical episodes/eras as they grappled with particular knowledge-power constellations (Foucault 2001).[1] Human plasticity has been prodigious, but we must acknowledge evolution in one sense. Evolution put us on this planet with no observable purpose but with an excess that always reaches beyond the subject in history. That is, the only reason that history is possible is that wherever we can examine evidence of the past, which is real (in the form of books, speeches, monuments, buildings, whole cities, artifacts, etc), it is possible

[1] Hardt and Negri, of course, discuss the plasticity of humanity in terms of shaping a new democratic humanity, especially in *Commonwealth* (2009).

for us to communicate with that past because we recognize our common humanity, that is, within our own context.

In this sense I agree with Badiou that history is not "real" since it is always an interpretation of the (fragmentary) past, which is real but only partially recoverable and always in the realm of the symbolic, as he puts it (Badiou 2010). The excess (our humanness, the basis of a weak teleology) which makes subjects at least partially understandable allows not simply for our continued debates over the past, it also allows us to throw open the question of time (which itself has a history). In his "Theses on the Philosophy of History," Benjamin wrote that "the tradition of the oppressed teaches us that the 'state of emergency' in which we live is not the exception but the rule. We must attain to a conception of history that is in keeping with this insight" (Benjamin 1968, 257). The Cold War was a period of constant fear but also the dream time of consumerism—the camp could be ever-remembered as the price in the developed world for the glittering products. Now, however, whether one wishes to see these as "end times" (Žižek), "time that remains" (Agamben), or the crisis of empire (Hardt and Negri), it is possible to conceive of this moment not as one of ambivalence but one where "time stands still and has come to a stop" (Benjamin 1968, 261-262). Without the teleology of progressive theories it would be necessary, following Hardt and Negri, for us to already see glimmers of the redeemed world in the present one or the muddling through with all of its unnecessary misery continues. The question is to what end historical materialism can be of any service to these matters.

The loss of the acceptance of progress and *telos* in History was only partially reinvigorated, briefly in 1991. Among those who saw the end of history in 1991, as did Francis Fukuyama, the triumph of neo-liberal "democracy" and capitalism were essentially the fulfillment of the Kantian or Hegelian project (Fukuyama 1992). Fukuyama expressed his own ambivalence, suggesting that in his last man we really see Max Weber's last men, "specialists without mind, pleasure-seekers without heart," a "nothingness [that] imagines it has climbed to a level of humanity never before attained" (Weber [1920] 2002, 124). The economic disaster of 2008, resource depletion (the problem of peak oil especially), and climate change all suggest not only that history is not over, but that the current system cannot be sustained and may well result in global catastrophe (Brown 2011; Ponting 2007). Avoiding such a dismal future with a revolution that transforms the existing system is the central problem which faces us because the last crisis of capitalism—brought about by the weakness of parliamentary states—either will be one of the greatest of all human crises or we will limp along scrambling for energy in a post-petroleum world.

Apart from the historical liabilities surrounding the idea of revolution, despite the fact that many of the "liberal democracies" of today were founded based on them, a related problem is that the corporate-state system of today already claims the mantle of continuous revolution even as conditions of inequality and poverty are generated by the system itself. Within certain quarters there remain those who foresee a technocratic revolution—a future of universal robot workers, artificially intelligent (AI) machines, genetic engineering—that offer the possibility of tremendous transformation, often summarized as a "post-human" or a "transhuman" future. Whether or not such changes (which have been forecast before) will actually occur, there is very little attention among many of the scientists like Ray Kurzweil or Hans Moravec to discussing, in effect, what need there would be for humans in a world where machines have eclipsed them (unless we merge with them), a question made more serious due to the fact that ensuring that such machines would be ethical has become a serious topic among the AI community (Hall 2007; Moravec 2000 and 1990; Kurzweil 2000). Genetic engineering also promises much but would in all likelihood lead to even greater conformity to the existing system (and perhaps allow for the creation of "happy," conformist children).

In addition, and more immediately important given the uncertainties of both robotics and AI, is the very real development of military technologies which hope to securitize the future. Since the Vietnam War, the development of the military-industrial complex has been repeatedly critiqued. The military expenditure of the United States, now at over $600 billion, dwarfs the next ten states combined but exists in a world where militarism—states which expend beyond their necessity for defense in preparation for war—remains central (Martin 2007; Bacevich 2006; Hardt and Negri 2000; Felice 1998)). More importantly, all these possibilities present two immediate problems. First, because of the highly specialized societies of the developed world, this work goes on, in effect, regardless of how much controversy they entail, bypassing human emancipation in favor of machines. In this sense these activities are "colonizing" the future, potentially foreclosing our capacity to channel the energy and resources that could reduce injustice now. Secondly, because the fabled post-human future may not happen, their promise may be the shiny mask of the security corporate-state—in a society where abundance and injustice exist, what is the purpose of these advancements if they are not made in order to reduce these problems?

Historical Materialism, Messianic Time and Revolution

Amidst this looming sense of crisis, Badiou, Žižek, and Agamben, like Benjamin before them, are turning to theological analogies or allusions employing messianic time. It is just as reasonable now as it was for Benjamin's contemporaries to consider what the utility of this exercise might be (McBride 1989). As Paul demonstrates, the theistic impulse has been a revolutionary force, perhaps the only source of revolutionary force in history. It must be examined because it cannot be explained. To say that evolution has produced us as symbolic creatures is a characterization but not an explanation. The proclamation, preserved in stone, in which Hammurabi states that Marduk, slayer of the dragons of chaos and creator of the universe, has given him the authority to proclaim the law, comes after the timeless time of 100,000 years or more in which humans existed. As Blumenberg put it, in their long confrontation with the "absolutism of reality," they found(ed)/discovered/created gods (Blumenberg 1985, 1-22). But by the time of Hammurabi the process had already long been underway. It cannot be recovered; it can only be examined. It is the excess of our humanity that has made history possible. Equally important, and what is really central here, is the moment in which the early moderns decided to stop giving to God(s) all of their hope and fear and instead took a position of self-assertion toward the world. Blumenberg argued that moderns reoccupied positions which theologians had failed to resolve. He described it as a second attempt to overcome Gnosticism, by which he meant the modernist view that the bad qualities of the world could be transformed because existence was not a permanent Manichean struggle between good and evil. Both Arendt and Blumenberg reformulate Marxist alienation. By losing our cosmic home in the Scientific Revolution, they argue that humans did not become more "this-worldly" but that humanity was thrown back upon itself and is not at home in the world (the earth, universe) (Blumenberg 1983; Arendt [1958] 1988).[2]

To Marx that sense of alienation can be resolved by a classless society where humans can truly realize themselves as fully human and transform nature into a home, thus in a sense fulfilling nature as well (Casey 2005, 46-49). World alienation, whether made to appear like happiness in the empty promise of consumerism or in the real struggle and poverty of much of the world, is as strong as ever. But if the moderns reoccupied positions,

---

[2] Arendt claimed that Max Weber's discovery of World Alienation was the more important one over Marx's individual self-alienation, but of course, for Marx, alienation was not just individual to herself but also toward the rest of humanity and the world.

revolution was not one of them. Theories of resistance (from those against the breaking of divine law to the Lockean right to overthrow tyranny) were always transgressive against theology or the law. Revolution was a right moderns claimed once it passed from the hands of the apocalyptic politics of Oliver Cromwell into the secular use made by Locke (Williamson 2008).

In his "Paralipomena to 'On the Concept of History,'" Benjamin wrote that "in the idea of a classless society, Marx secularized the idea of messianic time. And that was a good thing" (2003, 401). The classless society would be the result of the transformative moment of revolution. The class struggle of the proletariat is "the course of historical development" and Benjamin blamed the Social Democrats for holding out "the notion of the 'revolutionary situation,' which, as we know, has always refused to arrive" (402). The solution was that "a genuinely messianic face must be restored to the concept of classless society and, to be sure, in the interest of furthering the revolutionary politics of the proletariat itself" (403). As I see it, Benjamin's criticism of historians was that they sought signs of the revolution in historical developments. They were misguided because it is not the revolution (the messiah) that can be found. Its moment would only come when acceptance of the truly transformative possibility of a classless society had taken root. At that moment, which is at once "every historical moment" but only that one "historical moment" where a "distinct chamber of the past, one which up to that point has been closed and locked" can open. Entrance, he wrote, "coincides in a strict sense with political action" but only when the messianic moment becomes possible (402).

In the "Theses" Benjamin compares such a moment to the historicist's empty time, "but time filled by the presence of the now [*Jetztzeit*]" (1968, 261). Robespierre and the French Revolutionaries ripped Roman republicanism out of the past but "in an arena where the ruling class gives the commands" (261). The "open air" (261) into which the proletariat leaps is the time-of-no-time but also the time that was always promised in the very existence of humanity, the moment that like a promise has accompanied us always but "only a redeemed mankind receives the fullness of its past" where it is "citable in all its moments" (254). In this sense Benjamin echoed Marx in "The Eighteenth Brumaire of Louis Bonaparte," where the communist revolutionaries "must let the dead bury their dead." Invoking Rome in the 1790s was a reference, "a phrase" that "went beyond the content; here the content goes beyond the phrase" (Marx 1968, 99).

The Past, the Real—the Problem

If we were to accept Marx's judgment, that we cannot drag "recollections of past world history in order" to "drug" ourselves concerning the content of future revolutions, then it would seem to require at the very least that we not attempt to use communism as the organizing principle (Marx 1968, 99). Hardt and Negri, who prefer the terms biopolitics and democracy, do this, however, their claims that new forms of production and social organization are making a new, democratic world possible largely repeats Marx's argument, just without "communism" or "classless society" being the crucial phrases. Hardt and Negri claim they do not want a new party, a new kind of state, and in this they are in agreement with Badiou, that the term "communist" cannot be an adjective anymore, as in modifying "party," or "state," etc. (Badiou 2010). When Hardt and Negri discuss revolutionary transformation in *Commonwealth* it sometimes seems as if they want a classless version of the world we possess today (minus the state and capitalism) rather like Benjamin's account of the *Hassidim's* description of the kingdom of the messiah:

> Just as our room is now, so it will be in the world to come; where our baby sleeps now, there too it will sleep in the other world. And the clothes we wear in this world, those too we will wear there. Everything will be as it is now, just a little different. (Quoted in Kaufman 2008)

Clearly Hardt's and Negri's description is not about a world that will be "just a little different," but if the means of production today makes a better world possible, then presumably many of those means of production will continue under new social and political circumstances. Here is where the truly "theological," if you will, problem arises: given the violence and state terror associated with communism and revolution, I am not sure even attempting to associate Paul with the messianic possibilities of revolution, to associate with it all revolutions and the one still to come, is enough. It is not that it is a difficult argument to make, which it is, but a fundamental existential problem. The Arab Spring, the Occupy movement, and the many examples that Hardt, Negri and others point to may be suggestions that more and more of the downtrodden are seeking a new order. I certainly hope so but like republicanism after the fall of Rome, revolution may require a longer historical sleep before it can be resurrected by the future's version of Renaissance Florentines who were the first to blast the Roman Republic out of the past.

If we cannot see the revolution, materialist historians should critically analyze what can be called late modernity's attempt to overcome revolution and foreclose it. We must critique the myth of politics and of economics today. As Douzinas points out, despite the revolutions that gave birth to the corporate-states of today, enshrined in founding documents but banished in politics, the world system of rights and humanitarianism blocks the economic, as liberalism always has, from appearing political (Douzinas 2007 and 2010, 98-99). The myth of the economy must be critiqued—that it is its own system which must be allowed to be run by its "invisible hand," which is belied by every economic panic-recession-depression and the creation of trusts and monopolies. The potential crises (like climate change) may make a revolution seem necessary, that is, present the opportunity for revolution, but we are faced with a paradox: historically revolutions often happen because of crises but in this historical instance we cannot wait for the crises, which may destroy the very foundations that the present system makes possible. Those who maintain these systems in the name of humanitarianism or the freedom of the market, and the impossible search for security in both, are the representatives of a wealth and a security apparatus such as the (human-built) world has never seen. It is the result of a re-doubling of efforts to ensure that (non- or violent) revolution cannot be permitted a chance because when it has ever been applied to "we the people" it has threatened the privileged and the powerful. The security corporate-state is the open face of the dream-time of consumerism and the nation state (identity politics) remains the greatest obstacle to waking up.

The threat of the potential of very real crises that I discussed above present an opportunity (by taking them seriously *before* they happen) but should they be allowed to occur the opportunity may be lost. I remain unconvinced of Hardt's and Negri's concepts of biopolitics and "immaterial labor" that can produce democracy. In our networked world, which could easily be extended, the capacity for the people to embrace the political in the form of a new kind of democracy is undoubtedly real. But here we are confronted again with the messianic kingdom where "everything will be as it is now, just a little different" and it is there that, as a historian, I see problems. During the early years of the revolution in Russia, Alexandra Kollantai saw the same problems. If we are to have the means of production that we have today then will it require the kind of specialization and all the training required for it now? For every moment of wireless time I need not just my laptop but also my router, the fiber-optic cables that carry the information to me, and all the infrastructure they require and the people who make it all possible. Pick a hospital, pick a power plant (solar or coal) and imagine the nodes of infrastructure and the people required to make them run. "Immaterial labor" floats on a majoritarian sea of material laborers, many of them specialists, engineers, and technicians. In 1921 Kollantai lamented that the bourgeois managers and specialists were dominating the workers' collectives and that the party was taking their side (Kollantai 1921). Lenin embraced the Taylor System of scientific management in order to teach the Soviet people "how to work" (Hughes 2004, 255-257). According to Kollantai, with their bourgeois skills and knowledge the specialists also brought *bourgeois consciousness*. In 1846, Marx could imagine that

> in communist society, where nobody has one exclusive sphere of activity but each can become accomplished in any branch he wishes, society regulates the general production and thus makes it possible for me to do one thing today and another tomorrow, to hunt in the morning, fish in the afternoon, rear cattle in the evening, criticize after dinner, just as I have a mind. (Marx 1932)

I can be a sewer worker today, a subway driver tomorrow (with some training), but neither a neuro-scientist nor a petroleum refiner. The material production of the world and its social and economic relations need real critique and a real attempt at understanding how a re-organization of the "economy" and "politics" could take place.

The corporate-state system is producing alienated, oppressed subjects or subjects who need to be awakened but I see little evidence of democratic subjects who want to govern everything. Even if it can rouse the people, I am left with what one can call the material labor or infrastructure problem I identified. In terms of the very complexity of real production, training, and specialization we are confronted not simply with a system which produces abundance but is unjust, but also the question of what a democratic oil production system looks like, or whether this new democracy will want biotechnology. Even so, I am agnostic because the theistic impulse, the evolutionary excess of our humanness (my weak teleology), and history itself, demonstrate a potential in us the limits of which we do not know. The democratic, classless future is truly analogous to a messianic time, but the moment of Paul shows us that we can change the world.[3]

[3] Some of the thinking and research on these topics benefited from my time as a research fellow at the Max Planck Institute for the History of Science.

Works Cited

Agamben, Giorgio. 2005. *The Time That Remains: A Commentary on the Letter to the Romans*. Translated by Patricia Dailey. Stanford: Stanford University Press.

Arendt, Hannah. (1958) 1998. *The Human Condition*. Chicago: University of Chicago Press.

Bacevich, Andrew J. 2006. *The New American Militarism: How Americans Are Seduced by War*. New York: Oxford University Press.

Badiou, Alain. 1982. *Theory of the Subject*. Translated by Bruno Bosteels. New York: Continuum International Publishing Group.

—. 2003. *Saint Paul: The Foundation of Universalism*. Translated by Ray Brassier. Stanford: Stanford University Press.

—. 2010. "The Idea of Communism." In *The Idea of Communism*, edited by Costas Douzinas and Slavoj Žižek, 1-14. New York: Verso.

—. 2011. *Second Manifesto for Philosophy*. Translated by Louise Burchill. Malden: Polity Press.

Benjamin, Walter. 1968. *Illuminations: Essays and Reflections*. Translated by Harry Zohn. New York: Schocken Books.

—. 2003. *Selected Writings, Volume 4: 1938-1940*. Edited by Michael W. Jennings. Cambridge, MA: Harvard University Press.

Bennett, Jane. 2010. *Vibrant Matter: A Political Ecology of Things*. Durham: Duke University Press.

Blumenberg, Hans. 1983. *The Legitimacy of the Modern Age*. Translated by Robert M. Wallace. Cambridge: MIT Press.

—. 1985. *Work on Myth*. Translated by Robert M. Wallace. Cambridge, MA: MIT Press.

Braun, Bruce and Sarah J. Whatmore, eds. 2010. *Political Matter: Technoscience, Democracy, and Public Life*. St. Paul: University of Minnesota Press.

Brown, Lester R. 2011. *World on the Edge: How to Prevent Environmental and Economic Collapse*. New York: W. W. Norton and Company.

Caputo, John D. and Linda Martin Alcoff, eds. 2009. *St. Paul Among the Philosophers.* Bloomington: Indiana University Press.

Casey, Timothy. 2005. "Nature, Technology, and the Emergence of Cybernetic Humanity." In *Is Human Nature Obsolete: Genetics, Bioengineering, and the Future of the Human Condition*, edited by Harold W. Baillie and Timothy K. Casey, 35-65. Cambridge: MIT Press.

Coole, Diana and Samantha Frost, eds. 2010. *New Materialisms: Ontology, Agency, and Politics*. Durham: Duke University Press.

Douzinas, Costas and Slavoj Žižek, eds. 2010. *The Idea of Communism*. New York: Verso.

Douzinas, Costas. 2007. *Human Rights and Empire: The Political Philosophy of Cosmopolitanism*. New York: Routledge-Cavendish.

Felice, William F. 1998. "Militarism and Human Rights," *International Affairs* 74 (1, January): 25-40.

Foucault, Michel. 2001. *The Hermeneutics of the Subject: Lectures at the College de France, 1981-1982*. Translated by Graham Burchell. New York: Palgrave.

Fukuyama, Francis. 1992. *The End of History and the Last Man* New York: Free Press.

Hall, J. Storrs. 2007. *Beyond AI: Creating the Conscience of the Machine*. New York: Prometheus Books.

Hardt, Michael and Antonio Negri. 2000. *Empire*. Cambridge: Harvard University Press.

—. 2004. *Multitude: War and Democracy in the Age of Empire*. New York: Penguin Books, 2004.

—. 2009. *Commonwealth*. Cambridge: Harvard University Press.

Hobsbawm, Eric. 1994. *The Age of Extremes: A History of the World, 1914-1991*. New York: Vintage Books.

Hughes, Thomas P. 2004. *American Genesis: A Century of Invention and Technological Enthusiasm, 1870-1970*. Chicago: University of Chicago Press.

Kaufman, Eleanor. 2008. "The Saturday of Messianic Time (Agamben and Badiou on the Apostle Paul)," *South Atlantic Quarterly* 107 (1, Winter): 37-54.

Kollantai, Alexandre. 1921. "Individual or Collective Management," *Pravda*, Jan. 25. Accessed Apr. 26, 2012. http://www.marxists.org/archive/kollonta/1921/workers-opposition/ch01.htm.

Kurzweil, Ray. 2000. *The Age of Spiritual Machines: When Computers Exceed Human Intelligence*. New York: Penguin Books.

Luhman, Niklas. 1998. *Observations on Modernity*. Translated by William Whobrey. Stanford: Stanford University Press.

Martin, Randy. 2007. *An Empire of Indifference: American War and the Financial Logic of Risk Management*. Durham: Duke University Press.

Marx, Karl. 1932. "Idealism and Materialism." *Marxists.org.* Accessed Apr. 26. 2012. *http://www.marxists.org/archive/marx/works/1845/german-ideology/ch01a.htm.*

—. 1968. *Karl Marx and Frederick Engels: Selected Works*. New York: International Publishers.

McBride, James. 1989. “Marooned in the Realm of the Profane: Walter Benjamin's Synthesis of Kabbalah and Communism.” *Journal of the American Academy of Religion*, 57 (2, Summer): 241-266.

Moravec, Hans. 1990. *Mind Children: The Future of Robot and Human Intelligence*. Cambridge: Harvard University Press.

—. 2000. *Robot: Mere Machine to Transcendent Mind*. New York: Oxford University Press.

Pinker, Steven. 2000. *The Blank Slate: The Modern Denial of Human Nature*. New York: Viking Press.

Ponting, Clive. 2007. *A New Green History of the World: The Environment and the Collapse of Great Civilizations*. New York: Penguin.

Rabinow, Paul. 2008. *Marking Time: On the Anthropology of the Contemporary*. Princeton: Princeton University Press.

Tooby, John and Leda Cosmides. 1997. “Evolutionary Psychology: A Primer.” Accessed Feb. 22, 2012. http://www.psych.ucsb.edu/research /cep/primer.html.

Weber, Max. (1920) 2002. *The Protestant Ethic and the Spirit of Capitalism*. Translated by Stephen Kalberg. Los Angeles: Roxbury Publishing Company.

Williamson, Arthur H. 2008. *Apocalypse Then: Prophecy and the Making of the Modern World*. Santa Barbara: Praeger.

Žižek, Slavoj. 2003. *The Puppet and the Dwarf: The Perverse Core of Christianity*. Cambridge: MIT Press.

—. 2008. *In Defense of Lost Causes*. New York: Verso.

Glorious Instants, Perpetual Recurrences: Georges Sorel, Vilfredo Pareto, and the Dismissal of Progress as a Political Category

M.F.N. Giglioli

Introduction: Dissidents in the Age of Progress

A rather peculiar coincidence of opposites forms the topic of the present chapter. Georges Sorel and Vilfredo Pareto, the two social thinkers of the *Belle Epoque* to be discussed, were acquainted, corresponded, and respected each other from an intellectual point of view[1]—not an insignificant detail for two men known for their irascible intractability (Busino 1967, 131n)—but were, nonetheless, on opposite ends of most spectrums and on opposite sides of most issues. Sorel was a committed proponent of working class emancipation, Pareto was a detached observer of social stratification and the dominance of elites; Sorel shared many intellectual affinities with Bergsonian vitalism (Julliard 1985), while Pareto was a staunch positivist; Sorel was a master of conversation and episodic, polemical prose, while Pareto was an eccentric, aristocratic recluse who shrouded his scientific knowledge in hypertrophic systematizing (Aron 1967, Bobbio 1996). Despite the distance that separates their worldviews, an element of commonality is apparent: neither theorist's conception of politics is much concerned with the notion of historical change.

To understand how idiosyncratic this shared position was, it is helpful to place Pareto and Sorel in their context, both intellectual and social. Intellectually, the leading doctrines of the 19th century were dynamic:

[1] For general biographical information on Sorel, see Andreu 1982; Berlin 1980. On Pareto, see Bousquet 1960; Bobbio 1973; Busino 1967.

liberalism, socialism, evolutionism all focused on processes of change in society, on stages, shifts, mutations, progress. All conceived of the present as an essentially new age with respect to the past, all expected the future to be profoundly and importantly different from the present. Even those intellectuals (such as De Maistre or Carlyle, for instance) whose normative preferences led them to decry and oppose the forward march of history were obliged to concur with this basic descriptive characterization (Sternhell 2006). The major social thinkers, from Hegel to Comte, John Stuart Mill to Marx, and Tocqueville to Spencer, perceived fundamental breaks in history, and took it to be the goal of theorizing to analyze the workings of social change. Hence, even the cultural interest in concepts of decline and decadence that spread through Europe at the close of the 19th century, a lineage running *inter alios* from Gobineau to Taine to Nietzsche to Spengler, could appear in this context as nothing but a mirror-image reaction to progressivism, preserving a dynamic view of history while reversing its direction (Forth 1993; Weber 1982; Mazgaj 1982).

If such widely differing voices were in unison with regard to the importance of historical change, the cause can be found in no small part in social life itself. For a citizen of the Victorian or the Bismarkian era, the notion that the present was an age marked by momentous transformations must have appeared as the most self-evident of platitudes. The economic and technological developments of the second industrial revolution, the social and demographic explosions of mass urbanism, transcontinental migration, and disease control, the political and organizational leaps of universal manhood suffrage, national conscript armies, State-supported literacy and the first forms of welfarism all contributed to afford contemporaries the impression that they were witnessing the birth of a new world, the age of the masses.[2]

Confronted with this environment, however, both Georges Sorel and Vilfredo Pareto developed political theories that, in different ways, ignored or minimized the notion of historical change in general, and the significance of modernity in particular, while attacking the belief in progress as an ideological mystification. This chapter will first of all attempt to establish this claim by discussing the relevant aspects of the political thought of Sorel and Pareto, in turn. Subsequently, it will offer a few hypotheses about the driving forces behind the aversion for the theorization of change in these authors as part of a broader cultural dynamic. The chapter will conclude by suggesting some reasons for the

---

[2] For an overview, see for instance Hinsley 1967; Hobsbawm 1987; Prochasson 1991; Blom 2008; Kalifa 2001; Ozouf 1963.

relevance of the issue in our contemporary predicament, at the close of the age of ideologies.

Sorel: The Moment of Revolutionary Mobilization

Georges Sorel is generally remembered as the theorist of revolutionary syndicalism, an unmediated politics of direct action and violence on the extreme Left of the socialist movement in pre-war France, aimed at achieving the collapse of the capitalist order by means of the general strike (Sorel 1999, 118). He is further remembered for his anti-intellectualism, and for the reliance of his political thought on the ideational construct of the social myth, a slogan, a rallying cry, a synthetic image that permits collective action without exhibiting any of the conceptually schematic and utopian characteristics of Jacobinism, social reformism, or academic socialism (Kolakowski 1975; Paquot 1982; Portis 1989).

For the purposes of the present discussion, however, this standard picture must be amended, or rather deepened, in two main directions. On the one hand, it is important to underline the relative lack of importance for Sorel's revolutionary theory of any evaluation of the chances for success in the struggle, and the absence of any palingenetic quality of the proposed post-revolutionary State. The life of the producers would in fact continue in much the same fashion in material terms and with regard to the organization of labor, according to Sorel, once the dominance of the parasitic classes has been shaken off.[3] Therefore, paradoxically the material grievances of the workers are not the fulcrum of motivation for revolutionary action. What really counts about class struggle and the general strike, rather, is the calling forth, in the charged atmosphere of mass mobilization, of the spirit of altruism, self-sacrifice, tenacity, and heroism that already resides in the working class, constitutes its social character, and grounds its claim to pre-eminence (Vernon 1978, 33). These traits are the foundation of ethical life for Sorel (as stated explicitly in the letter to Benedetto Croce of 6 May 1907, quoted in Hughes 1958, 170), and their social deployment is the goal and only worthwhile characteristic of politics as direct action.

The second revision of the received image of his thought has to do with the wider historical applicability of this dynamic: the mechanism leading to productive social tension by means of the social myth is in no way limited to revolutionary syndicalism— the general strike is simply its

[3] It is not difficult to discern the influence of Proudhon on this matter (Vernon 1978, 15-21).

most recent instantiation. The past offers many analogous cases, from the early Christian communities to the first German followers of Martin Luther to the soldiers in the French revolutionary armies of 1792 (Sorel 1999, 13 and 240-242). The same dynamic reappears any time a marginalized group possessing strong internal cohesion and altruistic mores happens upon the mobilizing images capable of igniting collective action.

This particular view of politics as a cycle of conflict and crisis is both condensed in the moment of the clash of worldviews and indefinitely iterated in historical examples from all ages—irregular, unpredictable, but recurrent. Such an understanding of the essence of politics finds further expression in Sorel's attack on the notion of progress as a typically bourgeois ideology, to which the working class, as the bearer of a new spirit for the future organization of society, opposes a messianic pessimism (Sorel 1908). I will refrain for present purposes from engaging with the thorny interpretative problem of moral progress in Sorel's thought, which merits more detailed reconstruction. Nonetheless, if Sorel could interpret any lessening of the passion for social opposition as a sign of decadence, the historical process as a whole could then be understood by him only as an irregular, random recurrence of cycles of social tension, from which any notion of Marxian teleology (despite the lip-service that Sorel for a time paid to the idea) had been essentially expunged (Horowitz 1961; Humphrey 1951; Meisel 1950; Roth 1980).

In short, it is in the glorious instants of collective action that Sorel's politics lives: not only, then, are the specificities of modern society of no particular strategic relevance to a politics of direct action, but the fundamental normative value of duration as a mark of success (and the attendant structuring of revolutionary politics as a competition for the founding of a stable political order) is radically called into question.

Pareto: The Immutability of Social Hierarchy

Vilfredo Pareto is generally associated with the Italian school of sociology and with the *Realpolitik* critique of democracy that goes by the name of the theory of elites. In order to understand the implications of the architectural structure of Pareto's view of society, it is useful to recall his intellectual trajectory, and in particular to note the homologies that exist with his seminal contributions to the field of economics.[4] Accordingly,

[4] The best introduction to the argumentative structure of the *Traité de sociologie générale*, Pareto's main sociological work, is Bousquet 1925.

social science consists for Pareto in the analysis of invariance and general equilibrium conditions: actual empirical states can either manifest themselves as convergent and stable, or as divergent and unstable, but all crises prefigure the appearance of a new static equilibrium, indeed can only be conceived in terms of it. Therefore, the dynamic aspect of the system is limited to the description of a cycle, all shocks being ultimately subsumed by the recomposition of the *status quo*.

What this *status quo* amounts to is well known: the fundamental fact of politics, according to Pareto, is social heterogeneity—a few rule, everyone else obeys. And as it is scarcely possible for the few to oppress the many by force alone, the great edifice of the *Traité* is devoted to the study of ideological obfuscation (La Ferla 1954, 44). In this sense, Paretian sociology decisively parts ways with economics (*homo oeconomicus* is no longer the object of study: Busino 1967, 130), and, ultimately, with the *Raison d'État/Realpolitik* tradition, as well, especially in terms of its pragmatic intent and its ambiguous relation with the notion of publicity. The main building blocks of the analysis are the positing of original *residues* (i.e. psychological ideal-types), the elaborate taxonomy of *derivations* (i.e. ideological formations), and the linkage between changes in the composition of residues within the ruling class and elite circulation. Pareto's discursive strategy, which depends on extensive discussion of historical examples, and relies heavily on the Ancient world for this purpose, is certainly motivated by his inductive method (since controlled experiments are typically unavailable to the social sciences, the need arises for as much raw historical data as possible), but is also an implicit claim about the epistemological unity of social facts across time, and a refusal of historicization (Busino 1967, 132).

From this point of view, it is clear that progress, and, for that matter, decadence,[5] has no true scientific significance within the bounds of Paretian theory: the trans-historical truths about society concern heterogeneity and ideological discourse, and a cult of progress (or a demonization of decadence) are precisely nothing more than derivations—instances of politically tainted and thus intellectually bankrupt reasoning (Pareto 1964, sections 301-4).

While the theory thus commits itself to conceive of history as an endless replaying of the same script (a view that naturally affords no special status to the present, or modern society as such), it also renders change essentially parametric: inasmuch as the circulation of elites depends on unobservable variations in residues, it cannot meaningfully be

[5] *Contra*, see Valade 1990.

predicted (even if warning signs can be perceived in the variations of ideological discourse). By a different route, Pareto seems to be rejoining Sorel's conclusions on this count as well: not only is history a series of equivalent iterations, but these also recur irregularly, and are therefore fundamentally unpredictable.

Very different conclusions ensue, however, with regard to personal engagement in real-life political struggles. Pareto maintains that a necessary condition of intellectually defensible investigation in social matters is a radical separation between theory and praxis: a view from nowhere, or a retreat to Galapagos (Finer 1968), from which to contemplate the perpetual recurrences of different configurations in political inequality and social stratification. This ascetic position is linked to the status of rationality and the opposition between knowledge and action, a position that is brought to the fore in Pareto's extensive critique of the language of politics and in his scientistic attachment to definition and neologism.[6]

The Opacity of Mass Society and the Crisis in the Understanding of Collective Agency

If the brief interpretative reconstructions attempted above appear persuasive, it would seem that Sorel and Pareto's views of politics, by focusing either on the unique moment of mobilization or on the long-term recurrence of hierarchical arrangements, seriously discount the notion of historical change. In the specific context, this stance amounts to a denial of, or at least a significant challenge to, the view that the rise of the masses and the "Social Question" were world-historical events, irreversible transformations of the configuration of society—a view held by most European social thinkers, at least from the time of Tocqueville's meditations on the first wave of democratization. Sorel and Pareto were not traditionalists, or untimely in a Nietzschean sense:[7] they were both conflict theorists who directed their analyses to the heart of the new industrial mass society, with its technical and disciplinarian conception of

---

[6] La Ferla 1954, 9-14; Busino 1967, 39. Gramsci's rebuttal of Pareto's many tirades on the imprecision of current political vocabulary can be considered emblematic of the passing of an era, and the dawn of the "linguistic turn" in political analysis (Gramsci 1975, Notebook 11, section 24).

[7] *Pace* Sheldon Wolin (cf. Vernon 1973 and Vernon 1981).

time.[8] However, their perspective transcended the struggles of the *fin de siècle*: they believed in the trans-historical nature of conflict; they recognized historical multiplicity, but they did not locate their theory of politics at the level of historical change. Thus, they differed even from an earlier alternative to the linear, progressive notion of time, namely, the cyclical view of history held by thinkers such as Polybius or Machiavelli. What we find in Pareto and Sorel, rather, is something much more radical: a de-valuation of the historical process *per se*. This negation of teleology in Pareto's case prefigures the impartial stance claimed by the modern social sciences. It is hardly possible, however, to miss the formal similarity between the conclusions to be drawn from his stance and from Sorel's, which stems from a completely different source, an essentially aesthetic revaluation of the (ephemeral) moment as the only possible locus of revolutionary enthusiasm. Consequently, we may deduce by implication their shared belief that what, if anything, is distinctive about modernity has nothing to do with politics.

The question then arises: what is the determinant, the driving impetus, the pre-analytical appeal of this conception of time? The issue is complex, and allows for no straightforward solutions. The direction I propose to follow in order to offer a hypothesis leads towards the shared generational perception of a failure of politics as a dialogic or deliberative experience. Otherwise stated: an irreducible disconnect between public discourse and political decisions and actions characterizes the entry into the political age of the masses. The issues of what counts as a persuasive argument in the political realm, how consent is to be acquired, maintained, and even measured, and what the springs of collective agency are, all confronted the turn-of-the-century generation in fundamentally novel guises. The problem of the conditions under which one can talk about politics in modernity became central.

One analytically available answer to these dilemmas of modernity was an epistemology of suspicion, allied to a rhetoric of unmasking and denunciation, such as is to be found in Marx, in Nietzsche, and to a certain extent in Freud. I believe that both Sorel and Pareto can be shown to operate within this paradigm; in the historical circumstances of the *fin de siècle*, however, this amounted to a more general loss of faith in the possibility for a rational, justifiable organization of society, and thus of

[8] Frederic Jameson (personal communication, February 2012) has pointed out that much of the reflection on different regimes of time in high modernism, both in the realm of philosophy (Bergson) and in that of literature (Proust, Mann), was driven by the perception of the effects of a still *incomplete* process of modernization, most visible in the opposition between the rhythms of town and country.

applying political doctrines in the world. This crisis of rational agency at the level of large aggregates in turn implied a crisis of principled political engagement: the interplay between the two can be observed for instance in the period's intellectual fascination with the irrationality of crowd behavior (Nye 1973; Nye 1975; Barrows 1981). Consequently, the problem of the social role of the intellectual was also posed for both in an essentially unsolvable form.[9]

While political voluntarism was under attack in the closing years of the 19th century, comparable problems had surfaced with regards to social theories based on non-agentic, structural determinants of social change. The paradigmatic real-world case was the issue of revolution and its timing. Observers on both sides of the socialist-bourgeois cleavage were mystified by the apparent decoupling of economic developments and working-class organization. On the one side, the empirical falsification of Marx's general immiseration thesis did not seem to have any effect on the growth of explicitly Marxist socialist parties. On the other, the growing strength of working class mobilization seemed unable to reverse the receding of the revolutionary horizon. The peculiar political opacity of mass society seemed to lie precisely in this limbo, where the reasons for the subsistence of the *status quo* had been refuted, yet revolutionary change did not erupt. Different ambits of society seemed to be moving in a wholly uncoordinated manner: quantitative development in material conditions found no translation in qualitative change in the field of politics.

The views, then, of Sorel and Pareto can be read as two symptomatic attempts to account for phenomena of revolution and order in an intellectual context in which previous theories of social action had been discredited, yet the concrete political situation imposed the issue as one of immediate practical concern. Their respective normative preferences concerning revolution account for the polarity of the viewpoints adopted. Nonetheless, the solidarity between their dismissals of progress as a key political category can be connected with a broader crisis in the reflexive understanding of modern European societies, which took the form of a reaction against Victorian rationalisms. Inasmuch as political action could not be understood (either on its own discursive merits or in terms of deeper structural forces) except as prevarication or blind mythical enthusiasm, progress was not, and could not be, a meaningful interpretive or normative category.

---

9 Of the vast secondary literature on this topic, see especially Charle 1990.

This theoretical stalemate can best be understood when compared with the value orientation and specific personality requirements that Max Weber attributed to political and intellectual pursuits, the stoicism of vocation when confronted with a similarly disenchanted view of history and the potential meaninglessness of human fate. The Weberian problem of the ultimate ends of politics escapes the bounds of this position, which at the same time has abandoned any commitment to a liberal or Marxian humanism, and thus any teleology of deliverance or emancipation.

Conclusions for Contemporary Victorians

In a political environment such as has prevailed in most Western countries since the fall of the Berlin Wall, an environment pervaded by temptations to declare itself the end of history, and indeed to look for its civilizational antecedents in late Victorian globalization, it may prove fruitful to revisit the circumstances that caused the demise of the previous liberal end of history. If we take seriously the historical interpretation of the first half of the 20th century as a protracted European civil war, the theories of Pareto and Sorel will figure as non-marginal elements in the narrative of its cultural and intellectual origins. Most importantly, perhaps, focusing on their thought, and on their peculiar refusal to theorize change can alert us to the fundamental differences between the liberal-democratic polities they were discussing and our own. Without this preliminary move of estrangement, I would claim, the antinomies of our present institutions and political practices will remain largely unintelligible.[10]

Works Cited

Andreu, Pierre. 1982. *Georges Sorel, entre le noir et le rouge*. Paris: Syros.

Aron, Raymond. 1967. *Les étapes de la pensée sociologique*. Paris: Gallimard.

Barrows, Susanna. 1981. *Distorting Mirrors: Visions of the Crowd in Late Nineteenth-Century France*. New Haven, CT: Yale University Press.

Berlin, Isaiah. 1980. *Against the Current*. New York: Viking.

Blom, Philipp. 2008. *The Vertigo Years: Change and Culture in the West, 1900-1914*. London: Weidenfeld and Nicolson.

[10] An early version of this chapter was presented at the conference "The Idea of Politics: History and Theory," held at King's College, Cambridge, in March 2009.

Bobbio, Norberto. 1973. *Pareto e il sistema sociale*. Firenze: Sansoni.
—. 1996. *Saggi sulla scienza politica in Italia*. Bari: Laterza.
Bousquet, G. H. 1925. *Précis de sociologie d'après Vilfredo Pareto*. Paris: Payot.
—. 1960. *Pareto (1848-1923): Le savant et l'homme*. Lausanne: Payot.
Busino, Giovanni. 1967. *Introduction à une histoire de la sociologie de Pareto*. Genève: Droz.
Charle, Christophe. 1990. *Naissance des «intellectuels», 1880-1900*. Paris: Editions de Minuit.
Finer, S. E. 1968. "Pareto and Pluto-Democracy: The Retreat to Galapagos." *American Political Science Review* 62:440-450.
Forth, Christopher E. 1993. "Nietzsche, Decadence, and Regeneration in France, 1891-1895." *Journal of the History of Ideas* 54:97-117.
Gramsci, Antonio. 1975. *Quaderni del carcere*. Edited by Valentino Gerratana. Torino: Einaudi.
Hinsley, Francis, ed. 1967. *Material Progress and World-wide Problems, 1870-1898*. Volume XI of *The New Cambridge Modern History*. Cambridge: Cambridge University Press.
Hobsbawm, Eric. 1987. *The Age of Empire, 1875-1914*. New York: Vintage.
Horowitz, Irving L. 1961. *Radicalism and the Revolt Against Reason: The Social Theories of Georges Sorel*. London: Routledge and Kegan Paul.
Hughes, H. Stuart. 1958. *Consciousness and Society: The Reorientation of European Social Thought, 1890-1930*. New York: Knopf.
Humphrey, Richard. 1951. *Georges Sorel, Prophet Without Honor. A Study in Anti-Intellectualism*. Cambridge, MA: Harvard University Press.
Julliard, Jacques, ed. 1985. *Georges Sorel en son temps*. Paris: Seuil.
Kalifa, Dominique. 2001. *La culture de masse en France*. Paris: Découverte.
Kolakowski, L. 1975. "Georges Sorel: Jansenist Marxist." *Dissent* 22:63-80.
La Ferla, Giuseppe. 1954. *Vilfredo Pareto filosofo volteriano*. Firenze: La Nuova Italia.
Mazgaj, Paul. 1982. "The Young Sorelians and Decadence." *Journal of Contemporary History* 17:179-199.
Meisel, James H. 1950. "Georges Sorel's Last Myth." *Journal of Politics* 12:52-65.
Nye, Robert. 1973. "Two Paths to a Psychology of Social Action: Gustave LeBon and Georges Sorel." *Journal of Modern History* 45:411-438.

—. 1975. *The Origins of Crowd Psychology: Gustave LeBon and the Crisis of Mass Democracy in the Third Republic*. London: Sage.

Ozouf, Mona. 1963. *L'Ecole, l'Eglise et la République 1871-1914*. Paris: Armand Colin.

Paquot, Thierry. 1982. Introduction to *La décomposition du Marxisme et autres essais*, by Georges Sorel. Paris: PUF.

Pareto, Vilfredo. 1964. *Traité de sociologie générale*. Volume 12 of *Œuvres complètes*, edited by Giovanni Busino. Genève: Droz.

Portis, Larry. 1989. *Georges Sorel: Présentation et textes choisis*. Montreuil: La Brèche-PEC.

Prochasson, Christophe. 1991. *Les années électriques 1880-1910*. Paris: La Découverte.

Roth, Jack. 1980. *The Cult of Violence: Sorel and the Sorelians*. Berkeley, CA: University of California Press.

Sorel, Georges. 1908. *Les illusions du progrès*. Paris: Marcel Rivière.

—. 1999. *Reflections on Violence*. Translated by Jeremy Jennings. Cambridge, Cambridge University Press.

Sternhell, Zeev. 2006. *Les anti-Lumières: du XVIII*e *siècle à la guerre froide*. Paris: Fayard.

Valade, Bernard. 1990. *Pareto: La naissance d'une autre sociologie*. Paris, PUF.

Vernon, Richard. 1973. "Rationalism and Commitment in Sorel." *Journal of the History of Ideas* 34:405-420.

—. 1978. *Commitment and Change: Georges Sorel and the Idea of Revolution: Essay and Translations*. Toronto: Toronto University Press.

—. 1981. "'Citizenship' in 'Industry': The Case of Georges Sorel." *American Political Science Review* 75:17-28.

Weber, Eugen. 1982. "Introduction: Decadence on a Private Income." *Journal of Contemporary History* 17:1-20.

The Final Revolt, The True Life: Perfecting Time and Timing Perfection in Ecumenicity

Christos Retoulas

> Be you therefore perfect as your heavenly Father is Perfect.
> —The LORD Christ

> Perfection has but one limit, that of having none.
> —St. Gregory of Nyssa

> Don't say "I have read, I have learned," Don't say "I have worshipped a lot." If you do not accept the Perfect Human Being, all other works are futile.
> —Yunus Emre/İmre (c. 1240 – 1329)

> Die before you die so you do not die when you die.
> —A dictum abundantly found on the walls of the Mount Athos Monasteries

> That is what our Master, the Prophet of Allah (peace and blessings be upon him) meant when he said: "Die before dying," meaning, "Know yourself before you die."
> —Ibn Arabi

The true source of all revolutions of imperfection is divorce and allergy. Divorce signifies διάστημα, or distance (spatial or otherwise), dimensional opening and apartness. While on the one hand allergy presupposes some kind of contacting closeness between the allergen and the healthy part, on the other, it concurrently points to an inner distance between the two not by way of non-contactability but in terms of an incompatible state, i.e. an uneasy state of being. As such, in general, interested distance and co-incidence, meaning relation and participation, although different in nature are organically and differentiatedly united, i.e. they inter-encapsulate. The notion of inter-encapsulation (αλληλοπεριχώρησις) denotes a constant tripartite linear, cyclical and helix-like unifying inter-movement without violation of the common essence and the incommunicable individualities of the unified elements (see Retoulas 2011). Yet in revolutions of imperfection this particular unification is unfulfilled and problematic in the sense that it is based on διάστασις (separation through stasis (emanated or hidden)). In other words, their inter-encapsulation has not been perfected in actuality. And the arche (beginning, source, first principle) and telos (end, purpose, aim) from which the διάστασις (apartness) of the current (μέσον, the middle) is concurrently measured is the true Self itself. Therefore, a true and full revolution can only be founded upon, evolve towards and aim at self-knowledge in actualization. And this is not merely a purely personal-subjective matter, but at the same time social, objective and even cosmological, i.e. Being-relating, since the self might not be everything but everything is through the self. For that matter, deprive the human being of both memory and hope, and time becomes unrecognizable—if existent at all.

Understanding revolutions in general and their relation to time in particular require an inter-encapsulating approach. That is, they need to be assessed not only from within an "indigenous" standpoint but, above all, from incommensurables ones. And because inter-encapsulation provides for the correctness of things yet without denying anything as such, this standpoint needs to be theophilosophical—instead of it being merely philosophical. Only then are we truly able to speak about Revolution as well as about revolutions. And only then can we really know whether revolutionary paths that present themselves as constituting different alternatives are truly so.

Let us look for example at two apparently antithetical revolutions: Liberal Capitalism and Socialist Communism. In their apparent antithesis lies veiled a fundamental unity because the presence and identification of this antithesis presupposes the acceptance of the key precepts of the Enlightenment framework, i.e. the individualization of the person and the instrumentalisation of reason—with the economic accepted as the main objective factor of life itself. As Caroll states, "in term of first principles," socialism and Marxism were "to prove no more than a subspecies of liberalism" (2004, 143). Papapetrou describes Western socialism as constituting a kind of communitarian individualism where individualism is provided with a wider social base (1973, 32). Gyömörey argues, "the most important defect is the one-sided application of Cartesian thought, which Marxism fully shares with all the forms of bourgeois rationalism" (1977, 233). In this light, Carlton is right when he states: "progressivism does not offer a real alternative. Capitalism is a modernist economic system and progressivism is a modernist palliative—not an alternative" (Carlton 2009). Let us now see an incommensurable revolutionary path, that of Ecumenicity and its historical actualisation, the ecumenical Romanity.

Roman Christianity refers to the ecumenical Christianity of the One, Holy, Catholic, and Apostolic Ecclesia of the Symbolum Nicaenum, the Seven Ecumenical Councils and the Five Patriarchates of (Elder) Rome, New Rome (Constantinople), Jerusalem, Antioch and Alexandria,[1] which after the fall of the Roman West to the Carolingian Franks survived only in the remaining free part of the Roman Empire, which in Frankish historiography has distortedly been called the "Greek" or "Byzantine Empire"—despite the fact that "no state titled Byzantium by either its population or its government ever existed" (Kazhdan 1990, 1). The "Byzantine-Ottoman Continuum" of Roman Ecumenicity is generated out of the historically enacted theophilosophical commensurability between Roman Christianity (Orthodoxy) and the central Ottoman religiosity of Islamic vahdet-i vücud tasavvuf ("the unity-of-Being" Sufism)[2]—including their organic connection to Judaism (canonical and/or evolved such as Sabataism).[3]

In general, after the fall of the Roman West to the Carolingian Franks from about 800 AD onwards, the tremendous power of Augustinian thought (having being violently separated from the overall Roman Patristic tradition which it belonged to) created in the Frankish-occupied ex-Roman milieus an overall Manichaean-derived dualist religio-moral order based on the concept of, on the one hand, the religious universal, contemplative upper reason, sainthood-related upper will, and on the conflictual other, the worldly human singular, instrumental lower reason, survival/power-related lower will. The emancipation of the notion of singularity and individuality in such an essence-dominated scheme transformed (via conflict and bloody strife) the pre-existing theocentric conception of Being into the modern secular one centered fundamentally on mere individuality, instrumental reason and, finally, when the quest for universal reason failed, on purely individualized power. Historically, this is the actualized movement from the Medieval Augustinian tradition, to the Encyclopaedia

---

[1] For this understanding of "Roman" I am indebted primarily to Fr. John Romanides (1927–2001), regarded by many as the most important dogmatic Orthodox theologian of our times, or as Sopko calls him, the "Prophet of Roman Orthodoxy" (2003).

[2] For an expository treatment of vahdet-i vücud tasavvuf, see, for example, Chittick 1989.

[3] For a full and detailed discussion of the relevant notions and terms, see Retoulas 2011. On the issue of Sabataism, see also Şişman 2008 and Baer 2009.

Enlightenment and, in the end, to the "Counter-Enlightenment" of Nietzschean Genealogy (see MacIntyre 1990; Caroll 2004). It is in this framework that the fourteenth-century merely cyclical conception of time as "revolution and turning" (originally of celestial bodies) was transformed into an abruption-centered linearity of time and social progress. The "revolutions of our times" take place within this later framework of the Nietzschean genealogical notion of Being as "nothing more than 'the recurrence of difference'" (MacIntyre 1990, 208) which feeds individualistic power-centric Globalization. As such, it is not accidental—attesting for that matter to the inherent failure of the Enlightenment's quest for the holy grail of the "neutral rational standpoint"—that the intrinsically Enlightenment-derived categories of Capitalists and Socialists, Right and Left remain in a state of deep fusion, murkiness and a search for meaning and relevance. It is in this framework that the revolutions of today and their relations to time should be understood.

In Roman reality, the most comprehensive form of Time within the Built Creation is the aeon (αιών, the "age," aevum, al-darh). A composite of αεί (ever) and ών (being), aeon pertains to true and real Being. It is the highest form of time which, in the post-Fall cosmos, can and is experienced only by those who have attained theosis ("deification," or otherwise, "befriending God" via their experiencing-participation in the divine uncreated energies). Chronos (zaman, time as we ordinarily know it) constitutes the lowest part of the aeon. Chronos is born out of the reality known as "the First Aeon" (al-dahr al-awwal) or the "Aeon of Aeons" (dahr al-duhur) (Yousef 2008, 30). The maximum limit of chronos is the aeon (al-dahr) which is infinite, does not repeat itself and has no daytime and night (Yousef 2008, 59). As every simple nature found in existence which is either noetic (heart (nous)[4] and/or intellect-related) or sense-related, "the aeons are the measures of the noetic [entities] while chronos

[4] It is important to note that in the Roman tradition the heart does not pertain to emotions but it is essentially considered the psychosomatic centre of the human being and the locus of manifestation of the divine uncreated energies by way of which human deification (theosis) is achieved. As Dounetas writes, "the Orthodox tradition does not attribute emotions to the heart . . . Contrary to the followers of [Western] scholasticism, the Orthodox ascetics warn that the confusion of emotions with the presence of God and of the saints constitutes a deception by the Devil" (Dounetas 2010, 15).

is of the sensible ones."[5] While aeon is the subject of the category of "when," chronos pertains to that of "number" (a countable existent (hours, days, years, centuries, etc)). In the Roman Patristic tradition,

> chronos is a dimension of a specific part of the built universe. This is so because the first built thing was the aeons, not chronos. Chronos was created by God [only] afterwards [, i.e. together with the cosmos—yet after the noetic realm of the angels]. Thus the main differentiation between aeons and chronos is that in the latter we have events in succession, while in the aeons we do not necessarily have a sequence of events, but a co-existence of events in a coexistence of things, without necessarily co-relate events with sequentiality. The human being is restricted within the sequentiality of the states [of being] because he/she exists within the chronos. The human being does not [normally] experience a being which is based on non-sequentiality. Yet he/she does that only within the experience of theosis. (Romanides 2004, 229)

The aeon (al-darh) is the time dimension of the whole Cosmos through all places and chronos (times) (Yousef 2008, 98). Although in rather spatial terms and focused on the present, Monk Nikolaos gives us a passing taste of the aeon in the discussion of his concept of the unified optical field: "Then attention rests on a wider area of our optical field and our optical relations to all things remain undisturbed; each of the particulars is connected with the other parts of the total, it is integrated into the unified field naturally and unforcibly, as naturally and unforcibly little children make friendships" (Nikolaos 2003, 65). Unite this momentary unified image with all moments in past and future (both noetically and sensibly) and you get a fuller taste of the aeon: "the Day of the Age is all a 'daytime' [i.e. utterly bathed in uncreated divine energies]: it does not have a night because there is no orb above it; it is the truly all encompassing orb that includes all material and spiritual orbs" (Yousef 2008, 70).

As a created existent, Time (aeon and chronos) itself is subject to arche (origin, beginning), middle (sustenance, movement) and telos (end). In this setting, time inter-encapsulates with διάστημα (spatial distance and dimension) and motion. Within the divine economy (i.e. God's salvific plan for the Creation), the natural movement of time within Creation is from its arche (genesis) towards its telos, i.e. the return to God in an attempt to cover the in-between διάστημα and put an end to διάστασις

[5] Maximus the Confessor, Του Αυτού Μαξίμου Σχόλια εις το Περί Θείων Ονομάτων (Edjustem S. Maximi, In Librum de Divinis Nominibus [of Dionysus the Aeropagete] Scholia), *Patrologia Graeca* 4, 181AB.

("separation" and "divorce" through stasis). And, this return to God passes through the theosis of the Perfect Human Being due to his/her microcosmic constitution:

> While chronos exists within the aeons, not all created beings are delimited within chronos. This is so because there are some creatures (the angelic hosts, the demons and the souls of those [the saints] who have past away) who are independent of chronos because they live in the aeons and not in chronos. For this reason they move at [such] enormous speeds . . . so it seems that they are at many places at the same time. In other words, that they are not delimited by space and chronos. (Romanides 2004, 280-1)

—something that partially echoes memories of modern physics string theory.[6] In a certain sense, as Maximus states, "aeon is the chronos when it ceases its motion, and chronos is the aeon when it is counted in motion; so I can call aeon the chronos without motion, and chronos the aeon counted in motion" (Maximus the Confessor, Περί Απoριών (Quaestiones). *Patrologia Graeca* 91, 1164C).[7] Yet exactly because of tashbīh (similarity between God's Self-manifestations and the Creation),[8] this stasis in the aeon is actually not immobility as such but the perfecting στάσις ἀεικίνητος καί κίνησις στάσιμος (ever-moving stillness and ever-resting motion) (Mpetsakos 2006, 231). It is in this light that the Frankish teaching on the staticity of Paradise-related immutability bears, from the point of view of Roman religiosity, more connotations of Hell-related corruption than of ever-living Paradise.[9]

---

[6] On the issue of the relation of Sufi Cosmology and modern, see Hämeen-Anttila 2005 and Yousef 2008.

[7] On the general subject of the theology of one of the pre-eminent Roman Church Fathers, Maximus the Confessor, see, for example, Thunberg 1995.

[8] In Roman reality, the tashbīh principle unceasingly inter-encapsulates with that of the tanzīh (incomparability between God and the created cosmos in terms of essence).

[9] Contrary to the Roman one, "in the Augustinian tradition exists the idea that God is in ever-present, i.e. . . . for God both future and past are a constant present. Consequently, Augustine and his followers, and the Papist tradition, were given the opportunity to think that if the human being was able to get rid himself/herself of time, he/she would be able to intellectually arrest the continuous present, the eternal present. And also that when he/she attained that intelligible state, he/she will be able to comprehend the uncreated, even the essence of God Himself [or at least part of the essence (something totally heretical in the Roman Christian tradition!)]. This Augustinian line of thought on time was crucially formative for the line of thought of the Scholastic tradition" (Romanides 2004, 280).

In the Creation before the Fall, the Human Being lived in a state of theosis within perfected time (aeon and its chronos). After the Fall, which in the true Roman Christian tradition has nothing to do with legalistic punishment due to the Original Sin (sin is identified merely with an error, a mistake exactly because his/her nous was cut-off from the divine uncreated energies and "darkened," gradually losing the "memory" of God (i.e. death)), the human being was able to reside only within chronos. Because of that, although the Prophets, the saints, etc. could reach theosis, it could not last beyond physical time (chronos). So when they died they did so both noetically and corporeally. With the incarnation and resurrection of the LORD Christ, noetic death is lifted, opening access to the ever-life in the aeon. Within this framework, John 11:25-26 properly reads (following the original Greek text): "And Jesus said to her, 'I am the resurrection and the life. He who believes in me will live, even though he dies; and whoever lives and believes in me will not die in the aeon [οὐ μὴ ἀποθάνῃ εἰς τὸν αἰῶνα]'" (our translation).[10] After death, all human beings see God's Uncreated Energies, each according to his/her state. For those found in the virtuous state of theosis, these Energies are experienced as a pleasant Light ("Paradise"), while for those in a debased state these same Energies will feel like a consuming Fire ("Hell"). That is why in Roman religiosity, there is no Paradise and Hell from the point of view of God, only from the point of view of the Human Being. Yet this is not "full life" but only the noetic aspect of the "Kingdom of God" which will be fully restored by the inclusion of the corporeal with the second Coming. And this overall framework is the true Revolution in Time and Being: neither its late fourteenth-century conception as a mere cyclical turning nor its replacement by the mere linearity of progress that Enlightenment modernity has imposed—both finally concluding in the nihilistic power-centric Nietzschean predicament humanity finds itself in today.

[10] From the point of view of ecumenical Romanity, irrespectively of whether or not its Islamic parts accept the incarnation as such (notional and/or actual), the accessibility of the perfect human being to the aeon via "Muhammadan Reality-Logos" (*Hakîkat-ı Muhammediyye*) is objectively a post-Incarnation possibility which is opened by Christ as the Logos Incarnate who is Himself not only "the Door" (John 10:9) but the Alpha (the Arche) and the Omega (the Telos): "I am the Way/Path and the Truth and the Life. No one comes to the Father except through Me" (John 14:6).

Our dictum, *Revolt to Perfection*, or else, "the Last True Rebellion,"[11] is simply the return of the uncreated energies of God to the centre of the everyday life of Humanity with One Goal: "Know Yourself," or else, the Return to the True and Full Self. In fact, Roman religiosity has less to do with religion-ideology as it is usually defined and more with personality therapeutics around the axis: Catharsis (Praxis) (cleansing of the passions); Enlightenment (Knowledge as Seeing) (via the intellect too); and Theosis (Deification as Becoming) (via the nous (the noetic heart)) (Romanides 2004, 46–49).

> Be gracious[12] to me, O GOD, according to Thy steadfast love: according to the multitude of Thy tender mercies blot out my transgressions. . . . Wash me thoroughly from my iniquity, and cleanse me from my sin. . . . Behold, Thou desirest truth in the inward parts: therefore teach me wisdom in the innermost heart. ... Create in me a clean heart, O GOD, and renew a steadfast spirit within me. Cast me not away from Thy presence; and take not Thy holy spirit from me. Restore to me the joy of Thy salvation. (Psalms, 49: 2-15)

It is in this framework that Romanity considers "religion" (in its common understanding) as a kind of sickness which begs for a cure. And unless the problem is first tackled at the personal level, no Revolution can be of any use. A Revolution that believes merely in structures and forms cannot but contain a lobotomized conception of humanity as such—missing some of the human constituting characteristics, e.g. individuality, sociality, autexousiotis, etc. Structures are fine and inescapable, yet it is content that gives them their meaning. Fill a water pipe system with acid and see what happens. In the Frankish context (both religious and secular), a prerequisite for human perfection is the restoration of αυτεξουσιότης (autexousiotis, the moral and actual freedom, not simply to freely will, but

[11] To partly quote a renowned Roman Orthodox site, "Death to the World: The Last True Rebellion" (named according to the fashion of the punk youth) where "world" here is defined spiritually vis-à-vis esoteric imperfection ("Death to the World" 2012; "Desert Wisdom" 2012).

[12] In Roman reality, God's grace/eleos (usually yet inadequately translated as "mercy") signify the divine uncreated energies. Contrary to the Frankish Christian legalistic term "mercy," the word ἔλεος (eleos) noetically bears the material properties of its relative word, ἔλαιον (eleon, olive oil) such as light proving, as food for life sustenance, as therapeutic for healing, as a perfumed cosmetic for cleansing and beautification—to mention the most important of them. It should not be forgotten that Christ means the "anointed one, the one covered in olive-perfumed oil."

above all to govern oneself) to the human being which in the common Christian tradition before the "Frankish" Augustine was "virtually synonymous with 'the gospel'" (Elaine Pagels quoted in Webb 1992, 4). Without an initial "implosion" to a new conception of human personhood that transforms Globalization to Ecumenicity, any "revolutionary explosion" might be not only futile but in reality a facilitator towards the meeting of humanity with that which awaits her at the next corner: nihilism as non-being. And nihilism as non-being can be reached from two directions: via either essentialism or individualism. How ironic is the fact that the revolt to reinstitute Perfection to Being requires again a Creation ex-nihilo yet this time not through the original ontological nihil but its human-"made" syzygy...

And if you think this is somehow a Platonizing—and for that matter intellectually an Augustinian—exercise, you are in for a big surprise. For a mere glance at the lives and works of—let us say for example—the Three Hierarchs (Basil the Great (329/330–379), John Chrysostom (c. 347–407) and Gregory of Nazianzus (c. 329–389/390)) proves otherwise:[13]

> The Three Hierarchs support the poor, the persecuted and the defenseless of their time in every way possible. They consider it natural that they have to sacrifice their own lives for each and any of those people. The marginalization of the disadvantaged is not consistent with the Orthodox ethos. Each human being is an unrepeated personality, an image of God. "By what right," asks Chrysostom "one can despise those whom God honors so much as to give them the Body and Blood of His own Son?" His persistence to build a leprosarium—not in a deprived area of Constantinople but in the richest district of the suburbs where large landowners resided who as a result saw the economic value of their luxury properties being reduced due to the proximity to that building—constituted the pretext for his final persecution, which would lead him into exile and excruciating death. During the great famine that struck his region, Basil deplores the action of profiteers who strive to get rich at the expense of their starving compatriots and organizes soup kitchens for all the people offering help without any discrimination to Christians, pagans, Jews and heretics, saving thousands from certain death. Sometimes woefully and sometimes forcefully, he asks the rulers to exempt the poor from taxation, while he does not fail to intervene in the interests of the workers in the mines of the Taurus Mountains. (Argyropoulos 2009b)

And Gregory cries out:

---

[13] For a detailed examination of the subject, see Dounetas 2010.

> Do not open your hands towards the heavens but towards the hands of the poor. When you open your hands towards the hands of the poor, then you have touched the top of the heavens. (Argyropoulos 2009b)

One thing must be clear here though: all these are done for a one and single telos for all persons involved, i.e. the participation in the silence of the aeon of True Being both in this life and the after-life through theosis—and for no other reason at all. That is why ecumenical Romanity does not divide the human beings into pious vs. impious, good vs. evil ones:

> There is no such thing in the Orthodox Church [which is the community of the friends of God, i,e the human beings in theosis (in addition to its common understanding)]. In the Orthodox Church there are ranks/stages of perfection. Not moral or immoral, good or evil [human beings]. (Romanides in Hierotheos 2010, 153)

At the level of pure Divinity which had yet to be fully revealed,

> The LORD preserves the simple: I was brought low, and He saved me. Return to thy rest [/stillness/silence (menuchah)], O my soul; for the LORD has dealt bountifully with thee. For thou hast delivered my soul from death, my eyes from tears, and my feet from falling. I will walk before the LORD in the land of the living. (Psalms, 116:6-9)

At the Level of the Divine Humanity of the "ancient mystics and magicians" of Roman Hellenity (Hellenes-Latins-Romans) which had yet to be perfected,

> the journey into a greater reality was a journey made through silence, in silence and into silence. The noise of the syrinx is the ultimate password. It's the sound of silence. (Kingsley 1999, 130)

And the Apollonian syrinx (later transformed into the Sufi ney) was travelling back and forth from the Mediterranean to Mongolia (see Kingsley 2010). And as for Aeon and Perfected Telos:

> Then the angel showed me the river of the water of life, as clear as crystal, flowing from the throne of God and of the Lamb down the middle of the great street of the city. On each side of the river stood the wood of life, bearing twelve crops of fruit, yielding its fruit every month. And the leaves of the wood are for the healing of the nations. No longer will there be any curse. The throne of God and of the Lamb will be in the city, and His servants will serve Him. They will see His face, and His name will be on

> their foreheads. There will be no more night. They will not need the light of a lamp or the light of the sun, for the Lord God will give them light. And they will reign in the aeons of aeons. (John, Revelation, 22: 1-5)

Or transliteratedly,

> Alif Laam Mim
> Certainly, the Romans have been defeated.
> In the nearest land. Yet it is they who, notwithstanding their defeat, shall be victorious.
> Within a few years. With Allah rests all power of Decision, first and Last: on that Day shall the Believers rejoice.
> In God's victory. He grants victory to whomever He wills.
> He is the Almighty, Most Merciful.
> It is the promise of Allah. Never does Allah depart from His promise— but most men understand not.
> They know but the exoteric in the life of this world: but of the End of Things/Hereafter they are heedless. (The Holy Quran, The Romans (ar-Rum) 30:1-7)

Works Cited

Argyropoulos, Andreas. 2009. *Το Επαναστατικό Μήνυμα των Τριών Ιεραρχών και η Εποχή μας* [The Revolutionary Message of the Three Hierarchs and our Era]. Σειρά: Έξοδος στην κοινωνία και τη ζωή. Athens.

———. 2009b. Το Επαναστατικό Μήνυμα των Τριών Ιεραρχών και η Εποχή μας [The Revolutionary Message of the Three Hierarchs and our Era]. [Speech delivered at the Municipality Theatre of the island of Lesvos]. Accessed Jan. 30, 2012. http://religiousnet.blogspot.com/ 2009/02/to.html.

Baer, Marc David. 2009. *The Dönme: Jewish Converts, Muslim Revolutionaries, and Secular Turks.* Stanford: Stanford University Press.

Caroll, John. 2004. *The Wreck of Western Culture: Humanism Revisited.* Melbourne: Scribe Publications.

Carlton, Clark. 2009. "My Two Cents on Capitalism." Ancient Faith Radio, November. Accessed Mar. 12, 2012. http://ancientfaith.com/ podcasts/carlton/my_two_cents_on_capitalism.

Chittick, William. 1989. *The Sufi Path of Knowledge: Ibn al-'Arabī's Metaphysics of Imagination.* Albany: SUNY Press.

"Death to the World: The Last True Rebellion." 2012. Accessed Feb. 23. http://www.deathtotheworld.com.

"Desert Wisdom." 2012. Accessed Feb. 23. http://www.desertwisdom.org/dttw/index.html.

Dounetas, Kyriakos. 2010. *Beyond Wealth: Orthodoxy, Capitalism, and the Gospel of Wealth*. Etna: Center for Traditionalist Orthodox Studies.

—. 2010. *Ορθόδοξη Κοινωνιολογία* [Orthodox Sociology]. Athens: Εκδόσεις «Ίδρυμα Δημήτρη Κιτσίκη» [Διάθεση: Εκδόσεις Αθαν. Σταμούλης].

Fountoulakis, Georgios. 2010. *Η Έννοια του Χρόνου στον Αρχαίο Φιλοσοφικό Στοχασμό και σε Μεγάλους Πατέρες της Εκκλησίας* [The Concept of Time in Ancient Philosophic Thought and in Great Church Fathers]. Athens: Ιδιωτική Έκδοση.

Gyömörey, Lorentz. 1977. *Η δύση της Δύσης: Η Απομυθοποίηση της Ευρώπης και ο Ελληνισμός.* [The Sunset of the West: The De-mythologisation of Europe and Hellenism]. Athens: Εκδόσεις Παπαζήσης.

Hämeen-Anttila, Jaakko. 2005. "The Immutable Entities and Time." *Journal of the Muhyiddin Ibn 'Arabi Society* XXXIX. Accessed Apr. 18, 2012. http://www.ibnarabisociety.org/articles/ immutable_entities.html.

Hierotheos (Vlahos) (Metropolitan of Nafpaktos and St Vlasios). 2010 and 2011. *Εμπειρική Δογματική της Ορθοδόξου Καθολικής Εκκλησίας (σύμφωνα με τις προφορικές παραδόσεις του π. Ιωάννου Ρωμανίδη) (Τόμοι Α' κ' Β').* [Empirical Dogmatics of the Orthodox Catholic Church (according to the oral lectures of Father Ioannis (John) Romanides) Vols. I & II]. Livadia: Ιερά Μονή Γενεθλίου της Θεοτόκου (Πελαγίας).

Ibn Arabi. 1997. *Divine Governance of the Human Kingdom.* Interpreted by Shaykh Tosun Bayrak. Louisville: Fons Vitae.

Kazhdan, Alexander and Ann Wharton Epstein. 1990. *Change in Byzantine Culture in the Eleventh and Twelfth Centuries.* Berkley: University of California Press.

Kingsley, Peter. 1999. *In the Dark Places of Wisdom*. Inverness, CA: The Golden Sufi Center.

—. 2010. *A Story Waiting to Pierce You: Mongolia, Tibet and the Destiny of the Western World*. Inverness, CA: The Golden Sufi Center.

MacIntyre, Alasdair. 1990. Three Rival Versions of Moral Inquiry: Encyclopedia, Genealogy and Tradition. Notre Dame: University of Notre Dame Press.

Mpetsakos, Vasileios. 2006. *Στάσις ἀεικίνητος: Η Ανακαίνιση της Αριστοτελικής Κινήσεως στον Μάξιμο Ομολογητή.* [Ever-Moving Stasis: The Renewal of Aristotelian Movement by Maximus the Confessor.] Athens: Εκδόσεις Αρμός.

Nikolaos. 2003. *Θέαση* [Vision]. Athens: Εκδόσεις Αρμός.

Papapetrou, Constantinos. 1973. *Η Ιδιοτροπία ως Πρόβλημα Οντολογικής Ηθικής: Συμβολή εις την μελέτην της σχέσεως μεταξύ της καθολικότητος του προσώπου και των εξ αυτής ατομικών αποκλίσεων.* [Peculiarity as a Problem of Ontological Ethics: Contribution to the Study of the Relation between the Universality of the Person and the Individual Diversions stemming from it]. Athens.

Patrologia Greaca. 1857. Edited by Jacques Paul Migne. Paris.

Retoulas, Christos. 2011. *Religion and Secularisation in the 'Byzantine-Ottoman' Continuum: The Case of Turkish Secularism*. DPhil Thesis, Oxford: University of Oxford, St Cross College/St Antony's College.

Romanides, John S. 1975. *Franks, Romans, Feudalism and Doctrine (Part III): The Filioque.* Accessed Feb. 7, 2012. http://www.romanity.org/htm/rom.03.en.franks_romans_feudalism_and_doctrine.03.htm.

—. 1975b. *Franks, Romans, Feudalism and Doctrine (Part II): Empirical Theology Versus Speculative Theology.* Accessed Feb. 7, 2012. http://www.romanity.org/htm/rom.03.en.franks_romans_ feudalism_ and_doctrine.02.htm.

—. 2002. *Ρωμηοσύνη*. [Romanity.] Thessaloniki: Εκδόσεις Πουρνάρα.

—. 2002b. *The Ancestral Sin.* Translated by George S. Ridgwood. New Jersey: Zephyr Publications.

—. 2004. *Πατερική Θεολογία.* [Patristic Theology.] Thessaloniki: Εκδόσεις Παρακαταθήκη.

Sopko, Andrew J. 2003. *Prophet of Roman Orthodoxy: The Theology of John Romanides*. Dewdney: Synaxis Press.

Şişman, Cengiz. 2008. *Sabatay Sevi ve Sabataycılar: Mitler ve Gerçekler*. [Sabatay Sevi and the Sabataists: Myths and Realities]. Ankara: Aşina Kitaplar.

The Jerusalem Bible. 1989. Jerusalem: Koren Publishers.

Thunberg, Lars. 1995. *Microcosm and the Mediator: The Theological Anthropology of St Maximus the Confessor*. Chicago: Open Court.

Webb, Eugene. 1992. "Augustine's New Trinity: The Anxious Circle of Metaphor." In *Religious Innovation: Essays in the Interpretation of Religious Change*, edited by Michael A. Williams, Collett Cox and Martin S. Jaffee, 191-214. Berlin: Mouton de Gruyter. http://faculty.washington.edu/ewebb/Gus.pdf.

Yousef, Mohamed Haj. 2008. *Ibn 'Arabi: Time and Cosmology*. Abington (Oxon): Routledge.

Yunus Emre. 1989. *The Drop That Became the Sea: Lyric Poems of Yunus Emre*. Translated by Kabir Helminski and Algan Refik. Putney: Threshold Books.

PART V

CREATING/MAKING TIME

COSMETIC PHARMACOLOGY IN (CON)TEMPORARY SOCIETIES

ŁUKASZ KAMIEŃSKI

> Let every man be master of his time . . .
> —Shakespeare, Macbeth

> They literally make productivity enjoyable. Even too enjoyable . . . I can accomplish more in one hour taking 20 mg [of Ritalin], than I could normally accomplish in a full workday.
> —Project manager at a CAD design firm for military contractors

The Pharmacologization of Society

Economic forces of globalization, new modes of work, and shifting lifestyles have transformed the cultural construction of time in contemporary postindustrial societies. One of the effects of this change has been the growing consumption of performance enhancing drugs. This has been a truly revolutionary transformation of vital consequences for the future.

Sociologists investigating the changing social role of pharmaceuticals have for some time been writing about "the medicalization of society" (Conrad 2007), "the pharmaceuticalization of everyday life" (Williams et al. 2008), and "the domestication of pharmaceutics" (Fox and Ward 2008). Undoubtedly, a "new emphasis on lifestyle in the production, marketing, and consumption of pharmaceuticals" has been on the rise (Fox and Ward 2008, 856). Until recently, most people would consider taking medication primarily to lessen the symptoms of a disease but this is no longer the case as drugs are increasingly expected to also improve the quality of life.

On the one hand, new medications have been developed for those who would like to enhance their physical appearance or the performance of their bodies. For sexual potency men take Viagra or Cialis, while women can improve their libido with Lovegra. Those who want to lose weight faster can consume such supplements as Xenical, Alli or Adipex. For a bad mood or depression one can take Prozac, the famous "happy pill," or new-generation antidepressants such as Xanax. Although various types of lifestyle drugs are already numerous, their numbers have grown very fast in response to social expectations as well as to artificial "needs" created by the pharmaceutical industry through high-pressure advertising (in the United States also through extremely powerful direct-to-consumer advertising). After all, drug companies are transformers of social norms who (de)construct new pathological conditions for profit. Additionally, the pharmaceuticals already employed for the treatment of certain diseases are finding new applications. Paradoxically, we live in times of not only "a pill for every ill" or "an ill for every pill" (Mintzes 2002, 909), but increasingly also of "more ills for an old pill." These drugs, when discovered to overcome the natural limits of a healthy body, begin their "second lives." For example, the drugs used for the treatment of attention deficit hyperactivity disorder, or ADHD (Adderall, Concerta, Focalin, Ritalin), Alzheimer's (Namenda, Aricept), narcolepsy and sleep apnea (modafinil sold in the United States under the brand name Provigil) have become a popular means of increasing alertness and concentration "on demand" in otherwise healthy people. The consumption of "smart drugs" for non-medical reasons has risen sharply, as exemplified by the sales growth of Provigil: from \$5 million in 2005 to \$1 billion in 2009 and projections of up to \$10 billion by 2018 (Armstrong 2010, 58). Psychopharmacological uppers, by enhancing cognition and eliminating sleepiness, allow for lengthier and more productive activity. Another medicine, propranolol (Inderal), used for the treatment of high blood pressure and arrhythmia, has been found to reduce anxiety and block emotional memory, thus probably preventing post-traumatic stress disorder.

The term "Viagra phenomenon" captures the process of blurring distinctions between the curative and enhancive effect of drugs, between healing the sick body and granting the fit one capabilities that surpass its natural limits. The practice of healthy individuals using enhancement neuropharmacology was famously referred to by Peter Kramer as "cosmetic psychopharmacology" (Kramer 1993, 246-49) as it gives the psyche what cosmetic surgery gives the body: artificial improvement. Cosmetic pharmacology, or the pharmaceuticalization of life in general, which promotes socially desirable behavior changes, is developing in the broader context of social, economic, technological, and mental transformations of late modernity. One of the most crucial dimensions of these changes is the new sociocultural deconstruction of time which, like an out of control metronome, clicks faster and faster.

Speeding up Fast Capitalism, Speeding up Fast Time, Speeding up Everything…

In *A Geography of Time* Robert Levine, an American psychologist, presented the results of his research on various approaches to time depending on the speed of life in diverse societies (Levine 1997). In 31 cities around the world he measured three indicators: the average walking speed, how long it took postal clerks to sell a stamp, and the accuracy of bank clocks. Levine concluded that "People are prone to move faster in places with vital economies, a high degree of industrialization, larger populations, cooler climates, and a cultural orientation toward individualism" (Levine 1997, 9). Thus, the more industrialized and individualized a society, the faster the social time it constructs. In 2007 British psychologists headed by Richard Wiseman repeated Levine's research by measuring the walking speed in 32 countries. They discovered that over 10 years it had increased by 10 percent on average (and in Singapore by 30 percent, while in New York by 12 percent) (Henderson 2007, 5). Two main catalysts of the acceleration of time and lifestyle in post-industrial societies have been rapid high-tech innovations and transformations in the global economy. The two are so interdependent that they represent a powerful techno-economic symbiosis: technological developments generate economic changes, which in turn stimulate further waves of innovation.

In 1989 Ben Agger published a book with the expressive title of *Fast Capitalism* (1989). Fifteen years later, when he returned to the topic, he wrote a postscript study under the even more telling title *Speeding Up Fast Capitalism*. He confessed: "I was correct, in 1989, to notice that capitalism had sped up since Marx's time, and even since the post-World War II period . . . But I didn't foresee the extent of acceleration and instantaneity we have come to know today" (Agger 2004, 3). Agger was definitely right. Goods and services are produced faster. For example, in 2005 Boeing was able to reduce the manufacture time of its "737" plane from 28 to 11 days, and it estimates that in the future it will produce planes in just 3 days (Friedman 2007, 227). On another front, it takes less than 90 seconds to purchase a Big Mac at a McDonald's Drive-Thru in the United States. Faster production is inescapably followed by faster consumption, which as such leads to changing lifestyles. The contemporary global market depends on real-time, high-speed communications. Owing to IT technologies and software, money transfers, trade transactions, and stock exchange can flow non-stop. Thus, international markets and global trade have become a real 24-hour "casino capitalism," bringing into life what Susan Strange foresaw in the mid-1980s (Strange 1997). Great cities no longer go to sleep as they have developed into always-open 24-hour metropolises. E-commerce, e-services, e-banking, e-pharmacies, e-government and so on are increasingly shaping what Ede Moore described as the "24/7 society" (Moore 1993). Therefore, the current stage of capitalism, which itself enforces the utmost effective use of time, is marked by the unprecedented dynamics of acceleration. To paraphrase Lenin, today, "the highest stage of capitalism" is the imperialism of time.

Individuals may experience shock when confronted with too many changes in too short a time. In 1968 Warren Bennis and Philip Slater claimed that the advancing speed of technological, economic, and social change was giving rise to the "temporary society"—highly mobile, flexible, adaptable, with liquid moral norms, but also alienating and less secure. Such a society becomes a society of events, which, living in the now and paying its main attention to superficiality, is inherently unstable. In the late 1960s their vision of American society was probably premature, but at the turn of the 21st century it has become a reality. Therefore, when in 1998 they republished *The Temporary Society*, their description could no longer be contested (Bennis and Slater 1998). Western postindustrial temporary societies have eagerly preached what Naomi Klein calls a "shock doctrine" (Klein 2007). The West, by speeding up fast capitalism (in the United States in particular), promotes rapid and radical changes around the world, thus "exporting" its own tempos of life and socioeconomic transformations, even in waging war. In this manner, in 2003 the Americans implemented the "Shock and Awe" doctrine during Operation Iraqi Freedom.

Because ever-accelerating fast capitalism compels people to work and live quicker, the members of "(con)temporary societies" are almost always in a constant hurry. Time turns out to be our most precious resource. New technologies, which aim at making life easier and save time, regularly blow back by creating greater time-pressure, consuming more leisure time, and making fast time even more tyrannical. Paradoxically, the time-saving technologies save time but only for more things to be done than before, thus leaving people with less time than before (Eriksen 2001).

Nowadays, the market is governed by a new Darwinian rule of the survival of the fastest and most efficient. Taylorism, which in the early 20th century introduced a golden rule of HR management (that only an efficient employer is a good employer), has been taken to the extreme. People must synchronize with the "technopulse" of their faster societies by working more intensively from one deadline to another and multitasking. Paradoxically, the more flexible work-time becomes (*flexitime*), the more increasingly oppressive it is perceived to be. This "chronos syndrome," to use Hervé Fischer's term (2006), combined with the all encompassing advanced technologies, intensifies "technostress" (Brod 1984). Humans are becoming more stressed because the gap between their psychophysical capabilities and those of technological devices increases nearly exponentially. Writing in 2008, Philip Zimbardo and John Boyd noted that "a typical human has a processing speed of about four hertz. In comparison, modern desktop computers have central processing unit (the brain of a computer) clock speeds of over three gigahertz; they are roughly 750,000,000 times faster than we are" (Zimbardo and Boyd 2005, 30-31). Since they finished their book in 2005 the speed of computer processors has accelerated, at the time of this writing, by a factor of almost five, confirming the validity of Moore's law, which states that the capacity of the microchip processor (i.e. its speed) doubles every 18 months (Moore 1965). The human being is frequently regarded as the weakest element of the system that needs to be improved. Many find the ultimate solution for such a "timelock" (Keynes 1991) in cosmetic pharmacology, which gives humans hope for full control over time. Such chemical time-management can be seen as a postmodern form of Taylorism. As people increasingly suffer from what Linda Stone calls "continuous partial attention," their minds are absorbed by multitasking technologies and hyperlinked information. Would not the medications for hyperactive children be the best solution for such "a self-induced state of ADHD" (Armstrong 2010, 64)? The motto coined by Corneliu Girugea, a Romanian chemist who in 1962 developed the first brain enhancer, becomes increasingly true: "Man is not going to wait passively for millions of years before evolution offers him a better brain" (quoted in Armstrong 2010, 64). He has no more time to wait.

Viagras for the Brain

It has been said that Jean Paul Sartre "habitually wore moccasins to save the time he would otherwise have used tying his shoelaces" (Eriksen 2001, 21). Psychopharmacology, being the postmodern equivalent of Sartrean moccasins, has become trendy in various walks of life, from corporate business to universities, from the military to night clubs. Also, since the late 1970s classical musicians have often used propranolol to manage stage fright and improve performance. A brief look at some cases exemplifies the phenomenon.

Work in business corporations requires high productivity. To meet the great expectations of employers and face up to severe competition, some office workers consume "little helpers." It is an open secret that self-prescribed stimulation is exceptionally popular in the IT sector, in Silicon Valley in particular (Arrington 2008). As long as employees meet their goals, employers do not care. What is more staggering, however, is that some bosses unofficially suggest and/or supply their subordinates with "steroids for the brain." This has been the case in big foreign corporations, notably in emerging economies like Poland. It is not easy to refuse such an offer, especially in times of economic turmoil and high unemployment. Of course, this is not as coercive as, say, in wartime Japan, when everyone working for the defense industry was forced to take a methamphetamine pill in the morning to ensure more efficient work (Vaughan, Huang, and Ramirez 1995, 497). However, the psychological pressure might be equally compelling and stressful. Businessmen are also prone to pill-popping in order to enhance their mental prowess on the job.

Brain-boosters have also become eagerly used in the university world. Provigil, Ritalin (known as "Vitamin R" on the campuses), and Adderall are the most popular "cognitive enhancers" consumed by students in the United States and Great Britain. In writing essays and preparing for exams they assist themselves with "study drugs" that improve short-term memory, increase cognition, and boost alertness. Stimulants also often enable them to reconcile studies with work. In a word, "Viagra for the brain," by extending the day and strengthening mental ability without the side effects of traditional stimulants such as amphetamines or caffeine, allow for more productive studies and the achievement of better grades. In 2009 Adderall was the most popular upper at the London School of Economics and Political Science; nonetheless, students experimented with everything that could help them during the exam period. According to a study, conducted also in 2009 at the University of Michigan, 8 percent of American undergraduates used "study drugs" at one time or another to boost their brains (Tahir 2009, 12). It is estimated that 5 to 15 percent of U.S. students take stimulants but, because of the difficulty in assessing this off-label administration of prescription drugs, some experts claim it to be as high as 25 percent (Editorial 2008; Armstrong 2010, 58).

The use of performance-enhancers is not limited to students only, as it has grown fashionable among professors as well. Academics take such stimulants in order to improve sustained hard thinking, be more prolific, complete the ever piling up tasks, meet the endless deadlines, fight jet lag and boost performance during presentations. Barbara Sahakian, professor of neuropsychology at Cambridge University, researched the use of modafinil in academia, inspired by her own experience. Will Iredale quotes Sahakian as having said: "I flew over to a conference in Florida and one of my colleagues offered me some modafinil. I found at least four colleagues took it and that was without me even asking them. One was at Oxford, two were from The Hague and one was in Washington" (quoted in Iredale 2007, 7).

In January 2008, over 1,400 readers from 60 countries, mainly from academia, responded to an informal online survey on the use of brain enhancers conducted by *Nature*. The result was that 20 percent admitted using drugs to "stimulate their focus, concentration or memory." The most popular drugs were: Ritalin (62 percent), Provigil (44 percent), and betablockers such as Inderal (15 percent) (Maher 2008). One of the responses in another poll, carried out by *Wired* magazine, came from a post-doc: "I wrote my high-intensity Ph.D. while on Adderall, wrote a 600-page monograph while on Namenda. I would recommend cognitive enhancement to everyone" (quoted in Madrigal 2008).

Aristotle remarked that "the essential object of any social system must be to organize the military institution *like all the others*" (quoted in Virilio 2006, 88). Thus, changes in the military reflect transformations in the society at large. The American way of war has dramatically sped up due to the developments in conventional high-tech weapon systems and the breakthroughs in IT that enable real-time military communications. The initial air campaign and its first several dozen hours have become a critical phase of combat operations. Therefore, it is decisive for pilots to be capable of conducting the opening sorties for 48 hours at a stretch. Although the U.S. Air Force has used wakefulness-promoting agents, especially amphetamine pep-pills, since World War II (Rasmussen 2008), it has been searching for a stimulant with fewer side-effects. In 1989 Michel Jouvet, one of the modafinil researchers, noticed that it "could keep an army on its feet and fighting for three days and nights with no side effects" (quoted in Westcott 2005, 334). Two years later, during the Gulf War, the French army experimented with modafinil on the Foreign Legion. In 2000 the U.S. Air Force conducted initial tests on helicopter pilots and in 2001 also on pilots in Afghanistan flying 40-hour missions during Operation Enduring Freedom. Although the U.S. military assumed Provigil to be an extremely effective "super medicine," highly promising research is underway to invent a "magic pill" that would enable soldiers to function for up to five days without sleep or with very limited rest.

More and more people of various occupations are using neuroenhancers to gain a competitive edge, like the American professional poker player, Paul Phillips, who revealed that Provigil and Adderall helped him earn more than $2.3 million (Haggo and Haggo 2008, D12). Furthermore, cosmetic pharmacology spills over from public to private life, as exemplified by a 31-year-old software developer from Seattle who regularly took Provigil, not having time for a full night's sleep. He said: "I find I can be very productive at work. I'm more organized and more motivated. And it means I can go out partying on a Friday night and still go skiing early on Saturday morning" (quoted in Lawton 2006). Those taking Provigil can spend the whole night clubbing and, instead of wasting time in the morning sleeping it off, can continue enjoying themselves.

Conclusion

A new generation of performance-enhancing drugs will become common in the near future. Globalization already enables credit card holders, who have access to the Internet, to purchase prescription drugs without a prescription. In a couple of decades cognitive enhancers may come to be as popular as coffee, but more powerful and cheaper (in e-pharmacies modafinil costs about $1.50 per pill). One can argue that since both caffeine and modafinil stimulate the central nervous system there's no need to differentiate between prescription and non-prescription enhancers. The further acceleration of life will only make brain-boosting more desirable and socially acceptable. After all, initially cosmetic surgery was highly controversial, but today it is more widely accepted. Cosmetic pharmacology will likely follow the same path because, as Anjan Chatterjee observes, "although technical and scientific developments were a necessary prerequisite" for the popularization of plastic surgery, "cultural, sociological, and economic forces paved the way. Analogous forces seem poised to pave the way for cosmetic neurology" (Chatterjee 2007, 132).

Should we, however, treat "smart drugs" like a cup of coffee? Should we see cosmetic pharmacology as similar to plastic surgery? Not really, because neuroscience and the development of brain enhancers are not only a part of the big scientific revolution, but also a catalyst for significant social change. In fact, we face the Huxleyan dystopian vision of a happy society of individuals fed with drugs that prevent maladjustment without reducing efficiency. This, however, raises some large ethical questions. First, what about authenticity, honesty and social trust? Is taking uppers a form of cheating, like doping in sport? Second, can brain-boosting lead to the corrosion of character? Who is the real "me"—the one with limited psychophysical capabilities or the one locked in the "stimulant-sedative loop"? Third, will not freedom be endangered if market competition and corporate expectations force employees to psychostimulate themselves? Fourth, what about distributive justice? Will the pharmacologization of society produce a new type of discrimination and a new underclass of those who do not want cosmetic neuropharmacology or cannot afford it? Fifth, what about global justice? Neurotechnology may accelerate economic growth for the countries which encourage the cognitive enhancement of workers. Will cosmetic pharmacology expand the global divide between the rich and the poor nations? And finally, although it is claimed that these drugs have minimal side effects, little is known about the effects of their longer-term use by healthy subjects. They might, at least, be habit-forming. Taking psychostimulants may be like using credit cards—you buy now, but will have to pay later, most likely with interest. Despite all these dilemmas, which deserve separate analysis, cognitive enhancers are with us and here to stay. We have already said "hello" to the 24-hour society and there is no turning back.

However, this does not mean that there are no signs or prospects for resistance. Every revolution produces a counterrevolution. Certainly, this will also be the case with the cosmetic pharmacologization of society. Human beings are psycho-physically limited and cannot be turned into chemically powered robots. They are also rational actors often capable of recognizing the possible consequences of uncontrolled pharmacologization of their lives. Today, no one knows long-term effects of the regular use of "brain botox" by healthy subjects. In the times of the fashionable healthy lifestyle obsessively promoted by media and pop-culture, of spreading environmentalism and growing spiritualism many would ask themselves the question brought up by Anjan Chatterjee: "Are any risks tolerable when the alternative is normality?" (Chatterjee 2006, 111). Finally, Western postindustrial societies prize freedom as the fundamental moral and social value. Thus, drawing on Isaiah Berlin's famous dichotomy of positive and negative liberty, it can be argued that we not only have and should have the freedom "to" use cognition enhancers but also equally have the freedom "from" coercion and social pressure to take them (Berlin 1969).

As for now, the most vocal resistance against manipulating our brain chemistry comes from some prominent bioconservative and antienhancement ethicists (Fukuyama 2002; McKibben 2003; Sandel 2007) and was aptly expressed by president George W. Bush's Council on Bioethics (*Beyond Therapy* 2003). Even some of the most zealous proponents of neuropharmacology recognize the risks, especially of over-medicalization, which "gives pharmaceutical companies the ability to capitalize on human suffering and exploit insecurities and unhappiness in order to increase drug sales" (Henry, Fishman and Youngner 2007, 17–18).

Pharmacology is not the only way that people try to deal with the growing tempo of social change. In March 2012, the Italian Association *L'arte del Vivere Con Lentezza* (The Art of Slow Living) celebrated the 6th "slowing day". By organizing Global Days of Slow Living in more and more places around the world, such as Milan, London, New York, and Tokyo, the movement has been getting internationalized. The idea is to encourage people to slow down and change their attitude toward fast and flying time. It is one of the examples of civic resistance against speeding up our lives.

Cosmetic neurology and a changing cultural conception of time (which defines the character of a society) are strictly connected. On the one hand, "mind hacking" is the outgrowth of the social revolution driven particularly by the acceleration of time. The cult of extreme individualism in postindustrial societies has also brought into public discourse the issue of body management. Additionally, social pressure and administrative practices, which promote socially desired models of behavior, encourage what Michel Foucault would refer to as "disciplining individual bodies" (Foucault 1971). Cosmetic pharmacology can thus be seen as a Foucauldian disciplinary technology of social power which squeezes bodies into a space of specific social and power-political constructs. Throughout history societies have disciplined the body in the name of economy and efficiency. Today, individuals are neuropharmacologically *self*-disciplining their bodies to calibrate themselves to the evolving time-profile of postmodern societies. In the time-obsessed (con)temporary societies, in which people have so much to do with so little time, "brain botox" helps us to stay awake long enough to function better and obey the iron law of boosting productivity.

On the other hand, not only does cosmetic pharmacology reflect social changes and the shifting cultural time-profile of the postindustrial societies, but it also remarkably transforms patterns of social life. The innovative classes of sleeping pills, now under development, will soon enable the matching of sleep to lifestyles. Experts in sleep research, like John Caldwell from the U.S. Air Force Research Lab in San Antonio, Texas, predict that novel drugs will deliver deeper and more refreshing sleep, thus allowing us to reduce by half the time "wasted" on resting (Lawton 2006, 37–38). Will neuropharmacology turn people into machine-like automatons, able to be switched on and off depending on the demands of productivity? The culture directed at the management of the sleepy and exhausted body will inevitably spill over into the (bio)political sphere of governance and become a new disciplinary technology of power.

Brain-boosters have such "magic" properties for enhancing our natural capabilities that, for his premodern ancestors, turbo-stimulated contemporary man would be a demigod-like being of superhuman strength. Those preindustrial people, however, lived much more in harmony with nature and themselves. Today, neuropharmacology, by changing the natural rhythms of life and by threatening to destroy our respect for human weakness (and thus for our humanity), seems to be doing with the human sleep-wake cycle what contraception has done with sex: release it from nature. This time, however, it might turn out to be a Faustian bargain. Although the Huxleyan vision of the "societies with eyes wide open" is disturbing, hopefully, cosmetic pharmacology will also trigger a firm counterrevolution in the name of getting back to nature. Our limited capabilities and our weaknesses are nothing to be ashamed of. They are signs of our humanity that have to be respected if we want to remain human without turning ourselves into chemical cyborgs.[1]

Works Cited

"About Us." 2010. *Go Slow Be Happy*. Accessed Mar. 12, 2010. http://www.goslowbehappy.com/about_us.

Agger, Ben. 1989. *Fast Capitalism: A Critical Theory of Significance*. Urbana: University of Illinois Press.

—. 2004. *Speeding Up Fast Capitalism: Cultures, Jobs, Families, Schools, Bodies*. Boulder: Paradigm Publishers.

Armstrong, Walter. 2010. "Brave Neuro World." *Pharmaceutical Executive* 30 (4): 58–60, 62, 64.

Arrington, Michael. 2008. "How Many Silicon Valley Startup Executives Are Hopped Up On Provigil?" TechCrunch Blog, July 15. Accessed Feb. 16, 2012. http://techcrunch.com/2008/07/15/how-many-of-our-startup-executives-are-hopped-up-on-provigil.

Bennis, Warren, and Philip Slater. 1998. *The Temporary Society*. San Francisco: Jossey-Bass Publishers.

Berlin, Isaiah. 1969. "Two Concepts of Liberty." In *Four Essays On Liberty*, Isaiah Berlin, 118–72. Oxford: Oxford University Press.

Beyond Therapy. Biotechnology and the Pursuit of Happiness. 2003. Washington, DC: A Report of The President's Council on Bioethics.

[1] This essay is part of the research project: "New Technologies and the Changing Face of War. Science Fiction or the Future of the Western Way of War?" supported by the Polish National Budget Funds 2010-2011 for science under the grant N N116 389539.

Brod, Craig. 1984. *Technostress: The Human Cost of the Computer Revolution.* Reading, Mass.: Addison Wesley.

Chatterjee, Anjan. 2006. "The Promise and Predicament of Cosmetic Neurology." *Journal of Medical Ethics* 32 (2): 110–13.

—. 2007. "Cosmetic Neurology and Cosmetic Surgery: Parallels, Predictions, and Challenges." *Cambridge Quarterly of Healthcare Ethics* 16 (2): 129–37.

Conrad, Peter. 2007. *The Medicalisation of Society*. Baltimore: Johns Hopkins University Press.

Editorial. 2007. "Enhancing, Not Cheating." *Nature* 455 (November 15): 320.

Eriksen, Thomas Hylland. 2001. *Tyranny of the Moment. Fast and Slow Time in the Information Age.* London: Pluto Press.

Fischer, Hervé. 2006. *Digital Shock. Confronting the New Reality.* Montreal: McGill-Queen's University Press.

Foucault, Michel. 1971. *Discipline and Punish: The Birth of the Prison.* London: Allen Lane, Penguin Press.

Fox, Nick J., and Katie J. Ward. 2008. "Pharma in the Bedroom…and the Kitchen…The Pharmaceuticalisation of Daily Life." *Sociology of Health & Illness* 30 (6): 856–68.

Friedman, Thomas L. 2007. *The World Is Flat. A Brief History of the Twenty-First Century.* New York: Picador.

Fukuyama, Francis. 2002. *Our Posthuman Future: Consequences of the Biotechnology Revolution.* London: Profile Books.

Haggo, Douglas, and Regina Haggo. 2008. "One Pill Makes You Larger, One Pill Makes You Smart." *The Hamilton Spectator*, January 12: D12.

Henderson, Mark. 2007. "Modern Life: It's One Step At a Time, Only Much Quicker." *The Times*, May 2: 5.

Henry, Michael, Jennifer R. Fishman, and Stuart J. Youngner. 2007. "Propranolol and the Prevention of Post-Traumatic Stress Disorder: Is it Wrong to Erase the 'Sting' of Bad Memories?" *The American Journal of Bioethics* 7 (9): 12–20.

Iredale, Will. 2007. "Tired Professors Take 'Smart' Pills To Perk Up Lectures." *Sunday Times*, July 15: 7.

Kantrowitz, Barbara. 2010. "The White House Mystery Drug." *The Daily Beast*, March 4. Accessed Feb. 14, 2012. http://www.thedailybeast.com/articles/2010/03/04/the-white-house-mystery-drug.html.

Keynes, Ralph. 1991. *Timelock: How Life Got So Hectic and What You Can Do About It.* New York: HarperCollins.

Klein, Naomi. 2007. *The Shock Doctrine: The Rise of Disaster Capitalism*. New York: Metropolitan Books.

Kramer, Peter D. 1993. *Listening To Prozac: A Psychiatrist Explores Antidepressant Drugs and the Remaking of the Self*. New York: Viking Press.

Lawton, Graham. 2006. "Get Up and Go." *New Scientist* 189 (2539): 34, 36–38.

Levine. Robert. 1978. *A Geography of Time: The Temporal Misadventures of A Social Psychologist, or How Every Culture Keeps Time Just A Little Bit Differently*. New York: Basic Books.

Madrigal, Alexis. 2008. "Wired.com Readers' Brain Enhancing Drug Regimens." *Wired*, Apr. 24. Accessed Feb. 13, 2012. http://www.wired.com/print/medtech/drugs/news/2008/04/smart_drugs

Maher, Brendan. 2008. "Poll Results: Look Who's Doping." *Nature* 452, April 10: 674–75.

McKibben, Bill. 2003. *Enough. Genetic Engineering and the End of Human Nature*. London: Bloomsbury.

Mintzes, Barbara. 2002. "Direct To Consumer Advertising is Medicalising Normal Human Experience." *British Medical Journal* 324 (7342): 908–9.

Moore, Ede. 1993. *The 24/7 Society: The Risks, Costs and Consequences of A World That Never Stops*. London: Piatkus.

Moore, Gordon E. 1965. "Cramming More Components onto Integrated Circuits." *Electronics* 38 (8, Apr. 19): 114–17.

Rasmussen, Nicholas. 2008. *On Speed. The Many Lives of Amphetamine*. New York: New York University Press.

Sahakian, Barbara, and Sharon Morein-Zamir. 2007. "Professor's Little Helper." *Nature* 450 (December 20-27): 1157–59.

Sandel, Michael J. 2007. *The Case Against Perfection. Ethics In the Age of Genetic Engineering*. Cambridge Mass: The Belknap Press of Harvard University Press.

Strange, Susan. 1997. *Casino Capitalism*. Manchester: Manchester University Press.

Tahir, Tariq. 2009. "Drugs To Write Essays." *Sunday Times*, May 3: 12.

Vaughn, Michael S., Frank F.Y. Huang, Christine Rose Ramirez. 1995. "Drug Abuse and Anti-Drug Policy in Japan. Past History and Future Directions." *The British Journal of Criminology* 35 (4): 491–524.

Virilio, Paul. 2006. *Speed and Politics*. Los Angeles: Semiotext(e).

Westcott, Kelli J. 2005. "Modafinil, Sleep Deprivation, and Cognitive Function in Military and Medical Settings." *Military Medicine* 170 (4): 333–35.

Williams, Simon J., Clive Seale, Sharon Boden, Pam Lowe, and Deborah Lynn Steinberg. 2008. "Waking Up To Sleepiness: Modafinil, the Media and the Pharmaceuticalisation of Everyday/night Life." *Sociology of Health & Illness* 30 (6): 839–55.

Zimbardo, Philip, and John Boyd. 2008. *The Time Paradox: The New Psychology of Time That Will Change Your Life*. New York: Free Press.

THE NON-LINEAR FUTURE OF ALTERNATIVE PRESENTS: TEMPORAL UTOPIANISM AND IMMANENT REVOLUTION

RHIANNON FIRTH AND ANDREW ROBINSON

This paper will examine a number of theories of time and revolution from the standpoint of "temporal utopianism," which we have developed in another article (Firth and Robinson, *in press*). Temporal utopianism applies the ideas of critical utopianism (Moylan 1986; Sargisson 1996; Firth 2011) to theories and views of time. Contemporary utopianism rejects the view that utopias are necessarily absent or future blueprints, instead suggesting that utopia can be materialised in present spaces. Utopia is theorised in terms of a utopian function, in which practices or relations are utopian if they actualise affects of hope and desire. This field is expanded by the theory of temporal utopianism that we develop in our forthcoming article in theorising the utopian affects of particular theories of time. Theories of temporal utopianism suggest that utopia can be temporal as well as spatial, expressing conceptions of time which transgress the linearity of what Benjamin calls the "homogeneous empty time" of capitalist modernity. They advance various heterogeneous theories in which time is viewed as synchronous, reversible, multiple, and open to networked connections. In this chapter we develop our theory of temporal utopianism by considering its implications for social change and revolution.

Temporal utopianism has extensive implications for the relationship between time, utopia and revolution, and for the modalities and goals of revolutionary practice. In traditional revolutionary politics, revolution is located temporally in the future. This is achieved by means of a historical teleology which views revolution as the end-point of an evolutionary progression. Such visions are criticised for internalising within revolutionary theory the temporal framework of the very capitalist system they aim to overcome (Dirlik 1994, 11). Scholars of capitalist time frequently typify it as "linear" or "homogeneous empty time," characterised by a sequential and irreversible progression and by the equivalence of moments, and the subordination of experienced time to mechanical time (al-Saji 2004, 204; Bloch 1970, 125; Negri 2005, 23; Thompson 1967; Adam 1998, 14). One difficulty with teleological views of revolution is that they reproduce this view of time by locating revolution in a sequential future. This type of narrative locates revolutionary theory within homogeneous empty time. It fundamentally repeats capitalist temporality in its overcoming.

But if teleology is rejected, how can a temporal space be created for alternative possibilities? The current world is beset by the problem of temporal closure, the experience of the future as indistinct from the present and the resultant experience of an eternal present which repeats itself indefinitely. The future is seen as indistinct from the present because it is a structural continuation of the same social arrangement; the possibility of a radical break is foreclosed, through devices constituting an "eternal present" (such as naturalisation and the happy consciousness). This temporal experience makes it difficult to imagine other worlds as possible (Neilson and Rossiter 2012). This chapter will suggest that the temporal stance specific to revolution is the occupation of a zone of affect within the present, by means of an event. It will examine the temporal theories of Walter Benjamin, Gilles Deleuze, Antonio Negri, Giorgio Agamben and Alfredo Bonanno. The five theories explored here apply various temporal utopian positions to the possibility of revolution. While they differ widely, the common focus of such theories is on the location of the experience of revolution in the present, in the occupation of different affective or temporal zones, rather than in an aspirational future. The implication is that revolution consists in the escape from homogeneous empty time into other temporal zones. A revolutionary process creates experiential and material abundance in the present, as a starting-point for revolutionary lines of flight from capitalist modernity. These lines of flight are emergent in the present, and are neither linear teleologies nor eternal presents.

Gilles Deleuze: The Event as Crystal of Time

Deleuze adopts a theory of time based on the philosophy of Henri Bergson. Time is here understood in relation to affect, as the field of inner experience or "auto-affection" (Deleuze [1982] 1998, 114-15), of rhythms and durations specific to each form of life. According to Deleuze and Bergson, each person's usual temporal stance is limited by its containment within their own "attention to life," limited to their own resonances. This is characterised, among other things, by treating the present as a set of problems and therefore as tolerable (Deleuze 1989, 19-20). Such a problem-solving orientation is a major barrier to revolutionary praxis, since it creates a stance of adapting to the system rather than transforming or overcoming it. Deleuze is effectively discussing the blockage in revolutionary praxis we face today: people are too caught-up in present life-problems to feel revolution as a possibility (Neilson and Rossiter 2012).

To overcome this blockage, Deleuze looks to a theory of multiple temporalities. Deleuze's theory suggests that the past and present are simultaneous, rather than sequential. The past consists of a range of affective "zones," sometimes called zones of affect or names of history. Each zone can be related to different layers of reality (Deleuze 1989, 199). Usually, this multiplicity is invisible as one is trapped within one's own duration. The different zones of the past become apparent to people in the time-image or virtual image, which appears when people are able to see different durations or sensory-motor zones (33, 77). This image only arises when one makes an intense effort to break with the closure of one's own perspective and duration (al-Saji 2004, 227). It happens when people are able to transcend their confinement within a particular sensory-motor present, defined by their own duration, to also apperceive other durations.

This, in turn, creates a possibility of revolutionary change. Revolution in Deleuzian theory occurs through the "event" (or "intensity"). This is a movement of differentiation which escapes the present arrangement of zones of the past (Deleuze [1964] 2006, 155). Revolutionary events occur due to an opening to the temporal vistas of the various zones of the past, so as to occupy different zones distinct from those of current social assemblages. The future arises from the gesture of searching the zones of the past through desire (Lampert 2006, 51). It occurs when one reads time as *aion* instead of *chronos*, as a zone in which connections can occur between distant points (Deleuze, [1969] 1990, 5), or as *kairos*, the condensation of points in a particular moment (Deleuze [1964] 2006, 190). It occurs at the level of absolute deterritorialisation, which is a multi-levelled temporal standpoint (Lampert 2006, 139-40). Events are present in the field of the past or the virtual, a field of possible zones from which one can select.

Events are incarnated in a "crystal of time" in which two distinct flows are combined without touching (Deleuze 1989, 79). Through the crystal, one passes between durations, actualising a different zone of the past and creating a new reality (1989, 85). It is this dimension of transformation which prevents closure from occurring, allowing becomings to occur (Deleuze [1968] 2004, 145) and calling forth a new people and a new earth. This temporal zone is also the zone of utopia, which is "the becoming of unforeseeable futures" (Parr 2008, 48). Deleuzian becomings recreate zones of affect through a temporal move (Deleuze 1989, 119). The event itself makes present and past phenomena seem to be its precursors (Lampert 2006, 66). Events can constantly be reactualised by returning to the crystal of time. A revolutionary stance towards the past does not seek to copy or duplicate it, but rather, to repeat it as virtuality, as zone of affect (Lampert 2006, 93-4). Events are temporal changes, but not in the sense of a linear passage. For Deleuze, revolution is not the creation of something new. Rather, it is something which is tucked in-between the layers of history, pairing points in the past and future (Lampert 2006, 139). It enacts a rupture in the linear progression of capitalist time and in the sequence of being (Virtanen and Vahamaki 2006). It creates a situation in which one can select which zone of affect to "territorialise" or occupy.

Walter Benjamin: Now-time and Revolution as Redemption

Benjamin's theory of messianic time emphasises the experience of moments of "now-time" (*jetztzeit*) or immediacy, outside the framework of homogeneous empty time. Now-time creates connections between past and present revolutions, "redeeming" past revolutions in the present (Benjamin [1940] 2005, Thesis VI). Past moments become active and present in contemporary revolutions, creating connections which bypass homogeneous empty time. Time is connected directly, between particular points, rather than through a line (Gibbs 2005, 214). This experience appears to be achieved through allegory, or the slippage of signification. By creating a situation where anything can mean anything else (Handelman 1991, 127-8), the process of remembering is altered, bringing the past into contact with the present (1991: 164). It consists in "tak[ing] in the time and render[ing] it up in altered form—that of expectation" (Benjamin [1930-5] 1999, 107). Remembrance is radicalised through the immanence of ritual which connects to prehistoric pasts and unprecedented futures (Benjamin [1955] 1970, 182; Handelman 1991, 152-3; Gibbs 2005, 209). It creates a "standstill" or "cessation of happening," rupturing the smooth process of history (Benjamin [1940] 2005, Thesis XVI).

It is because of the ruptural character of messianic time, because it is excessive over homogeneous empty time, that it is able to put to an end the otherwise inescapable sequence of complicities. Without such a rupture, an iron continuity exists between past victories of oppressors and future developments. It is only by brushing history against the grain (Benjamin [1940] 2005, Thesis VII) that one can avoid a future built on the treasures of past domination. Instead, it reconstructs history from the standpoint of the oppressed (Leslie 2000, 202). In effect, Benjamin theorises revolution as a rupture with path-dependencies which are constructed through the continuousness of homogeneous empty time. It is the irruptive nature of messianic time which provides its revolutionary force. It blasts an epoch out of homogeneous empty time, creating instead a monad or "zero-hour" (Benjamin 1940, Thesis XVII). Messianic time is not the continuation or end of history. Rather, it terminates history and initiates a leap into another era (Wolin 1994, 57-8).

For Benjamin, revolution arises in messianic now-time, not in homogeneous empty time (Benjamin [1940] 2005, Thesis XIV). Or rather, it is a kind of transubstantiation, in which messianic time enters and reconfigures a social space previously constructed by homogeneous empty time. Benjamin is deeply critical of the influence of teleological views within the socialist movement: "Nothing has corrupted the German working class so much as the notion that it was moving with the current" (Benjamin [1940] 2005, Thesis XI). Revolution occurs in the reconciled life which is possible in the monadic sites in which now-time erupts (Wolin 1994, 58). It occurs by making things leap out of their context (Nägele 1986, 22). This escape is known in Benjamin's theory as revolution or redemption. It is the moment when messianic time overcomes homogeneous empty time. This process of redemption is not simply a change in perceptions. It leads to different actions in the present (Leslie 2000, 200), unsticking the present from the continuous future which seems necessary (Gibbs 2005, 214). It is because it consists in rupture that revolution is unique to the excluded or revolutionary classes (Benjamin [1940] 2005, Thesis XV). Since the present generation was "expected on earth" by those which went before, in the form of their messianic desires, each generation is endowed with a weak messianic power, and the past has a claim on this power (Benjamin 2006, 390).

Agamben: Transforming the Experience of Time

Agamben, like Benjamin, develops a messianic theory of time. His theory of temporal play stresses immediacy and the rejection of linear historical narratives as aspects of revolutionary temporality. It focuses on a conflict between diachrony and synchrony. Play is taken to destroy linear time, replacing synchrony with diachrony and initiating historical movement, whereas ritual fixes linear time in place, replacing diachrony with synchrony (Agamben [1978] 2007, 77-8). Western views of time are criticised for focusing on the precise moment and succession of points, which are confused with "the real time of experience" (110). The effect of such confusion is to transform experiential time into "waiting and deferral," instead of "something full and singular" (111). Today's social system is so fixated on synchrony that it prefers to keep the past alive as phantoms than to bury it, and is terrified by the signifiers of the present (95).

Against this view, Agamben argues for a non-teleological conception of time. Agamben criticises teleological views of history for establishing ideals which are absent from experience (Agamben 1999a, 56). Instead he seeks an alternative conception of time "among the folds and shadows of the Western cultural tradition" (Agamben [1978] 2007, 110). Agamben's approach stresses the creation of a "coming" community or politics of a utopian kind, but counterposed to transcendent, fixed and deferred utopias. Events in Agamben involve the removal of elements from the economic or practical field (Agamben [1978] 2007, 82). Redemption has the effect of redeeming the whole of the past and of history. It is not, therefore, an end of history (Agamben 1999b, 152). Crucially, it is something which is present and past, not future: it "has already taken place" (Agamben [1978] 2007, 110). As in gnosticism, Agamben suggests that utopia can be seized in the present by transforming experience, replacing homogeneous empty time with an experience of "an incoherent and unhomogeneous time, whose truth is in the moment of abrupt interruption" (111). Time is here re-grounded in the experience of pleasure, which is unquantifiable (114).

Agamben thus radicalises Benjamin's view, suggesting that revolution consists in a shift in perspective which overcomes the effects of capitalist reason and time (ranking, segmentation, sovereignty and bare life) and uses a different temporality to create a singular experience of "whatever-singularity," in which life is self-valorising without qualification. It is not simply a rewriting of history from the standpoint of the oppressed. Rather, it is formally distinct from linear histories (Agamben 1999b, 152-3). Such an experience is taken to be revolutionary, transforming the values taken from the past. A revolutionary stance is a stance which has re-remembered history from the *epoche* of pleasure, carrying this memory to each instant as a promise (Agamben [1978] 2007, 115). The effect in terms of practical revolutionary activity is an exodus: the slave who abandons work without seeking recognition from the master. By refusing the struggle for recognition, the slave also refuses historical teleology (Prozorov 2009, 527).

Negri: The Time of Liberation

Negri's theory of the "time of liberation" provides a collectivist variant on the approaches taken by Benjamin and Deleuze. Time is here associated with becoming and with constitutive power (Negri 2005, 45). This type of time is the core of utopia (62). Negri calls for the reconnection of temporal experiences to social processes of creativity and life, to the "global phenomenological fabric" (29). Autonomism, like orthodox Marxism, maintains that class struggle is recurrent and periodic, evolving through a series of stages. Negri's is not, however, a teleological theory; it is more like a series of strategic manoeuvres against an adversary who constantly changes. There are basically two modes of time in Negri's theory, those of capital and the proletariat. Negri sees it as a "collective constitution of time" through which being is created (59, 115). It is through cooperation that a specifically proletarian temporal duration is created (74). Proletarian time is itself pluralised, however. It consists of a multiplicity of experiential times, in contrast to the unified time of capitalist command (55). These multiple times tend "towards explosion and diffusion" (40-1) and rest on "the irreducibility of the many to the one" (43). It is unclear how Negri reconciles the irreducible plurality and the cooperative universality of proletarian time; his temporal theory repeats the paradoxes of his wider discussions of the "common" in this way.

Negri focuses on the incommensurability of capitalist and proletarian constructions of time. Capitalism spatializes time, rendering it synchronic and structural. In contrast, revolution involves temporalizing space, seeing space as well as time as being located in a process of becoming (35). Revolution is necessarily directed against equivalential time (41). It realises a "contemporaneity of the future and the present" (97). The self-transformation of workers in struggle imposes a metamorphosis on capitalism from outside its own temporal frame. Class struggle appears as a long-term temporality through which counter-power is accumulated (Negri 2009). It is processual, and does not acquire a determinate form until liberation is complete (Negri 2006, 103). Negri differs from the other theorists discussed here in viewing the alternative temporality in terms of social production and substance, and dismissing the idea of a leap into a different experience except on the basis of a new collectivity. Whereas for the other authors a new way of seeing is followed by or creates a new collectivity, for Negri the collectivity precedes the new temporality.

Bonanno: Revolutionary Immediacy

A rather different instance of activist appropriation of alternative temporalities arises in the field of post-left anarchy, a body of activist theory emerging from networked anarchist critiques of socialist transformative models from the 1970s onwards (Robinson and Tormey 2009). The Italian insurrectionary anarchist Alfredo Bonanno is one of its most influential theorists. His work, in particular the essay "Propulsive Utopia," is relevant to temporal utopianism because of its location outside future-oriented politics. Post-left anarchy rejects deferral as well as teleology, believing that desires must be actualised in the present, and emphasising the propulsive force of active desire.

For Bonanno (1988, 20), the "craving desire for action" contradicts historicist teleologies. Utopia, or rather, its "propulsive" subtype, is here seen as the inspiring energy of real movements, as the "thunderbolt of a propelling idea." Utopia is theorised as a means for breaking through the boundaries set by those in power, seeing social problems in different terms. A utopian concept is necessarily irreducible to the social conditions in which it emerges. Propulsive utopia exists in the field of becoming and agency, in contrast to existing structures. It is only when utopia is present and immanent that it is a living part of movements. When it is abstracted into a desire for the future, it becomes a Stirnerian "spook" (Stirner [1844] 1993, 39). Crucially, it is the affective reality of a utopian idea, as distinct from the flattened objectivity of spook-like concepts, which makes an idea propulsively utopian (Bonanno 1996). Utopian energies are constantly at risk of being rendered ineffectual and passive by the disconnection of ideas from active force.

Plural Times of Revolution

Comparing these five distinct theories, what characteristics might a non-teleological view of revolutionary time be said to have? A first similarity is that expressive experiences of time are privileged over instrumental theories of time. Deleuze's time-image is counterposed to a sensory-motor present, which is defined by the self's attention to life, and hence is somewhat instrumental (Deleuze [1966] 1990, 70). The time-image provides an expressive, dialogical alternative. Benjamin's now-time is lived as an intense experience of presence, akin to intoxication (Benjamin [1930-5] 1999, 462-3). Agamben stresses the "real time of experience" against externalised conceptions of time (Agamben [1978] 2007, 110), Negri the lived "fullness and difference" of a time without measure (Negri 2005, 45), and Bonanno the affective reality of propulsive energies. Revolutionary time can thus be seen as a subjective time of energies, flows and fluxes, as distinct from an "objective" time based on measurement.

A second similarity is the replacement of linear connections with rhizomatic or transversal connections within the field of time. Deleuze's event folds the past and the future together. Benjamin's theory suggests a possibility of temporal montage, bringing dissimilar elements together (Taussig 1984, 89) into "constellations" of present and future elements (Handelman 1991, 148), an approach largely imitated by Agamben. In Negri, it is the process of forming a "common," rather than an external frame, which connects different temporalities. A non-teleological revolutionary time is thus non-linear, involving connections between distinct temporal points or zones of affect which are not reducible to a sequential series.

A third similarity is that all the theories reject the idea of a unified, single temporal experience, instead insisting that temporalities are multiple, and encouraging polyphony among them. Time can be separated into "zones" of the past or of affect, as in Deleuze's theory. Benjamin's idea of "practical memory" involves actualising several different durations at once (Benjamin [1930-5] 1999, 205). Agamben emphasises the "incoherent and unhomogeneous" nature of liberatory time (Agamben [1978] 2007, 111). Negri's theory insists on the Manichean antagonism of two counterposed temporal frames (Negri 2005: 93).

Finally, each of these theories maintains that revolution is in some sense immediate: as now-time (Benjamin), event (Deleuze), propulsive force (Bonanno), lived social substance (Negri) or present transformative experience (Agamben). None of the theories deny that revolution is necessarily transformative, and discontinuous with the field of relations constituted by capitalism. Rather, they suggest that the overcoming of this particular temporal zone is theorised in terms of an immanent, present practice which realises alternative zones of affect in the present. It is thus distinct from teleological views in which different zones of affect are displaced to the future, and are tended or aspired towards, often while remaining concretely within the temporal experience of capitalism. The sense of revolution as a rupture between present and future is weakened in these theories. Rather, revolution is treated as immanent. The rupture occurs between zones of affect which are simultaneously past, present and future. Utopia is created within temporal zones which are actualised in the present.

What does temporal utopianism mean for revolutionary practice? We do not wish to lay out blueprints for movements, but can suggest various practical examples which are clearly instances of temporal subversion of the kind discussed here. Immediacy finds practical political application in theories such as Hakim Bey's immediatism. According to Bey, to find time outside the temporal colonisation by capitalism is already a victory. To meet with a social group other than family, work-group or consumption-group is already to achieve immediatism's goals—but such a meeting is usually precluded by not having the time (Bey 1994). The immediate is also visible in experiences of direct action. Accounts of activism often emphasise the immediate intensity arising from the performance of ritualised actions (Peterson 2001; Juris 2008). One might also discuss in this context the restoration of ideas of natural cycles and variant times in New Age and "rewilding" theories. There is also the phenomenon of revolutionary rehearsals in the form of activist commemorations, such as the annual 17th November demonstrations commemorating a student uprising in Greece. Activist commemorations are temporal zones which resist the fading into memory of the experience of presence in past revolutionary upsurges, while also projecting immanent futures. A further instance of alternative temporality would be processes of consensus deliberation which are by their nature endless. Accounts of negotiations between corporations and states and indigenous groups and social movements suggest that the former are often frustrated with the temporally extensive deliberations of the latter (Smith 2005).

To conclude, therefore, revolution necessarily has a utopian temporal dimension, if it is to overcome the entrapment of modern societies within the homogeneous empty time of capitalism. So intertwined is this temporal model with capitalist modernity that it is hard to imagine a revolution which does not transform it, yet revolutionary theory has until recently neglected the temporal problem, repeating linear teleologies which reproduce homogeneous empty time. Against this approach, revolutionary temporal utopians such as Agamben, Benjamin, Negri, Deleuze and various post-left anarchists have argued for temporal challenges in the present, and/or through the non-linear connection of past, present and future temporalities. Time is restored to a subjective, expressive meaning, against the measuring and homogenising functions of capitalist time. The recreation of a world of multiple temporal zones is a necessary part of a revolutionary overcoming of capitalism, and for this reason, a revolutionary perspective without a temporal dimension is necessarily incomplete.

Works Cited

Adam, Barbara. 1998. *Timescapes of Modernity*. London: Routledge.

Agamben, Giorgio. 1999a. "Language and History: Linguistic and Historical Categories in Benjamin's Thought." In *Potentialities*, by Giorgio Agamben, 48-61. Stanford: Stanford University Press.

—. 1999b. "Benjamin and the Demonic: Happiness and Historical Redemption" In *Potentialities*, by Giorgio Agamben, 138-159, Stanford: Stanford University Press.

—. (1978) 2007. *Infancy and History: On the Destruction of Experience*. Translated by Liz Heron. London: Verso.

al-Saji, Alia. 2004. "The Memory of Another Past: Bergson, Deleuze and a New Theory of Time." *Continental Philosophy Review* 37:203-39. http://www.marxists.org/reference/archive/benjamin/1940/history.htm.

Benjamin, Walter. (1955) 1970. *Illuminations*. London: Jonathan Cape.

—. (1930-5) 1999. *The Arcades Project*. Cambridge, MA: Belknap.

—. (1940) 2005. "On the Concept of History." Translated by Dennis Redmond. Accessed 20 Jan. 2012. http://www.marxists.org/reference/archive/benjamin/1940/history.htm.

—. 2006. *Walter Benjamin: Selected Writings, Volume 4: 1938-1940*. Edited by Howard Eiland and Michael W. Jennings. New York: Cornell University Press.

Bey, Hakim. 1994. "Immediatism vs. Capitalism." In *Immediatism: Essays by Hakim Bey*, 19-23. Edinburgh: AK Press.

Bloch, Ernst. 1970. *A Philosophy of the Future*. New York: Herder and Herder.

Bonanno, Alfredo M. 1996. *The Anarchist Tension*. Translated by Jean Weir. London: Elephant Editions.

—. 1988. *Propulsive Utopia*. London: Elephant Editions and Work in Progress.

Deleuze, Gilles. 1989. *Cinema 2: The Time-Image*. Minneapolis: University of Minnesota Press.

—. (1966) 1990. *Bergsonism*. Translated by Hugh Tomlinson. New York: Zone.

—. (1969) 1990. *The Logic of Sense*. Translated by Mark Lester and Charles Stivale. New York: Columbia University Press.

—. (1988) 1992. *The Fold: Leibnitz and the Baroque*. Translated by Tom Conley. Minneapolis: University of Minnesota Press.

—. (1968) 2004. *Difference and Repetition*. Translated by Paul Patton. London: Continuum.

Dirlik, Arif. 1992. *After the Revolution: Waking to Global Capitalism*. Hanover and London: Wesleyan University Press.

Firth, Rhiannon. 2011. *Utopian Politics: Citizenship and Practice*. London: Routledge.

Firth, Rhiannon and Andrew Robinson. In press. "For the Past Yet to Come: Utopian Conceptions of Time and Becoming."

Gibbs, Robert. 2005. "Messianic Epistemology: Thesis XV." In *Walter Benjamin and History*, edited by Andrew Benjamin, 192-214. London: Continuum.

Handelman, Susan A. 1991. *Fragments of Redemption: Jewish Thought and Literary Theory in Benjamin, Scholem, and Levinas*. Bloomington: Indiana University Press.

Juris, Jeffrey. 2008. "Performing Politics: Image, Embodiment, and Affective Solidarity during Anti-Corporate Globalization Protests." *Ethnography* 9 (1): 61-97.

Lampert, Jay. 2006. *Deleuze and Guattari's Philosophy of History*. London: Continuum.

Leslie, Esther. 2000. *Overpowering Conformism*. London: Pluto.

Moylan, Tom. 1987. *Demand the Impossible: Science Fiction and the Utopian Imagination*. New York: Methuen.

Nägele, Rainier. 1986. "Benjamin's Ground." In *Benjamin's Ground: New Readings of Walter Benjamin*, edited by Rainier Nägele, 19-38. Detroit: Wayne State University Press.

Negri, Antonio. 2005. *Time for Revolution*. Translated by Matteo Mandarini. London: Continuum.

—. 2009. *Communism: Some Thoughts on the Concept and Practice*. Translated by Arianna Bove. Accessed 20 January 2012. http://www.generation-online.org/p/fp_negri21.htm.

Neilson, Brett and Ned Rossiter. 2012. "From Precarity to Precariousness and Back Again: Labour, Life and Unstable Networks." Accessed January 20. http://publication.nodel.org/From-Precarity-to-Precariousness.

Parr, Adrian. 2008. *Deleuze and Memorial Culture: Desire, Singular Memory and the Politics of Trauma*. Edinburgh: Edinburgh University Press.

Peterson, Abby. 2001. *Contemporary Political Protest: Essays on Political Militancy*. Aldershot: Ashgate.

Prozorov, Sergei. 2009. "Giorgio Agamben and the End of History: Inoperative Praxis and the Interruption of the Dialectic." *European Journal of Social Theory* 12 (4): 523-542.

Robinson, Andrew and Simon Tormey. 2009. "Utopias Without Transcendence? Post-Left Anarchy, Immediacy and Utopian Energy." In *Globalization and Utopia: Critical Essays*, edited by Patrick Hayden and Chamsy el-Ojeili, 156-175. Basingstoke: Palgrave Macmillan.

Sargisson, Lucy. 1996. *Contemporary Feminist Utopianism*. London: Routledge.

Smith, Richard C. 2005. "Can David and Goliath have a Happy Marriage? The Machiguena People and the Camisea Gas Project in the Peruvian Amazon." In *Communities and Conservation*, edited by J. Peter Brosius, Anna L. Tsing and Charles Zerner, 231-56. Lanham, MA: AltaMira.

Stirner, Max. (1844) 1993. *The Ego and Its Own*. Translated by Steven T. Byington. London: Rebel Press.

Taussig, Michael. 1984. "History as Sorcery." *Representations* 7:87-109.

Thompson, Edward P. 1967. "Time, Work-Discipline, and Industrial Capitalism." *Past and Present* 38 (1): 56-97.

Vähämäki, Jussi and Askeli Virtanen. 2006. "Deleuze, Change, History." In *Deleuze and the Social*, edited by Martin Fuglsang and Meier Sørensen, 207-228. Edinburgh: Edinburgh University Press.

Wolin, Richard. (1982) 1994. *Walter Benjamin: An Aesthetic of Redemption*. Berkeley and Los Angeles: University of California Press.

THE LONG NOW: SYMBOLIC AND MONUMENTAL, OR LIVED REALITY?

LINDA LEVITT

Temporality is deeply implicated in determining the cultural shifts that mark the onset of modernity. A break with the past, and the opening of the future, were fundamental to defining the modern era. The dramatic cultural and social transformations that accompanied the industrial revolution and burgeoning urban life led to a schism between the past and the present: it was difficult to imagine life without electricity, mass transportation, and standardized time. One could hardly imagine the wonders yet to come. Because of the rapid pace of change in modern life, however, the future no longer holds the promise it once did. In many circles, the future extends no further than the lifespan of one's children or grandchildren, rather than an endless vista of expansive possibility. In the absence of teleology, belief in progress inherent in the modern vision has disintegrated. This lack of faith in the future on a conceptual level translates into a desire to live "in the moment," in a present in which a vision for the future has been extremely foreshortened. If the future is unimaginable, we are left with an extended present in which our concerns are personal and local, rather than communal and global. The danger of the extended present is not only that it leads to the kind of short-term thinking that puts the fate of humanity in peril; it also shifts the focus in the present away from humanity and onto innovation and technology.

This concern is articulated by inventor, designer and author Stewart Brand who, in his book *The Clock of the Long Now: Time and Responsibility*, proposes the need for long-term thinking as automatic and common instead of difficult and rare (Brand 1999, 2). Brand and his fellow board members of the Long Now Foundation are constructing a 10,000 year clock and adjoining library oriented toward establishing an icon that will "give permission to think long term" (Brand 1999, 48). In a 2011 article praising recent inventions, *Time* magazine notes that the first Clock "will be 61 m [200 feet] tall and housed in a remote West Texas cave. Built primarily out of steel, titanium and ceramic ball bearings the clock will play a unique melody once each day and when prompted by visitors to the site" (Brock-Abraham 2011, 80). The magnificent clock face features six concentrically arranged dials. The position of the sun is marked on a golden ring, and the moon appears on a band that makes its rotation around the center black-dome face that shows the stars in the night sky. The Foundation has also purchased land in an ancient pine tree forest in Nevada, implying that thinking of ten millennia in the midst of an ancient forest will create a resonance between nature and technology, as well as bringing light to the conflict between natural time and mechanical time, inspiring visitors to think long term.

This essay argues that a significant gap exists between theoretical ideas of the future and the lived experience of time in contemporary urban cultures in which the accelerated pace of life renders the long now increasingly difficult to imagine. Addressing the impact of the Long Now Foundation's efforts to transform thinking about the future, the essay considers whether this revolutionary approach can have the impact to which it aspires.

Technology: Mired in the Past, Intensifying the Present, Foreshortening the Future

Stewart Brand's initial public prominence came in 1966 when he launched a campaign calling for NASA to release to the public the first photographs of earth taken from space aboard Apollo 8. While it may be difficult for those born after the heyday of the space program to imagine never having seen a photograph of the entire planet, Brand believed that a view of the whole earth could lead to a paradigm shift. It could help reframe thinking about the planet as a whole rather than as disparate, disconnected nations. His thinking about the long now has a similar perspective: "Just as the Earth photographs gave us a sense of *the big here*," Brand says, "we need things that give people a sense of *the long now*" (Brand 1999, 133, emphasis original). Where technology—namely aeronautics and photography—previously enabled a new worldview, now digital technology renders the present so saturated with experience that the future is difficult to consider. Global connectivity brought about through the internet, smartphones, and other advances in telecommunications opens broad vistas of experience and access. Amid the abundance of information and data, many individuals feel a strong compulsion to remain connected, informed and "always available." In a crush of interpersonal correspondence, social media networking, and streaming media content, cyberculture overwhelms the present.

Simultaneously, the digital world allows unprecedented access to the personal and cultural past, enabling a deep descent into memory. Regarding the nostalgia made possible by the vast online cultural archives, Carl Wilson wrote in an August 4, 2011 *New York Times* article that nostalgia "raises exception to the great leveling effect of the Internet, the perpetual digital now that tears cultural artifacts out of context to make them into objects of curiosity or pastiche. And it's a reminder that it matters not only that an idea or an image was created, but when." Cyberculture and its digital archives entrench us in the past, in a nostalgia where longing and memory can be barriers to attending to the present or focusing on the future. In a culture already preoccupied with the busyness of the present, it takes little effort to be enamored of the past, putting more psychical distance between the now and the future.

Indulgence in the past may also provide relief from a culture in which the pace of life continues to accelerate. Brand argues that because of the constant quickening pace, the future is no longer anticipated in the manner the modernists once looked forward to the "new" that the future would bring with it. "Now that we have progress so rapid that it can be observed from year to year, no one calls it progress," Brand says. "People call it change, and rather than yearn for it, they brace themselves against its force" (Brand 1999, 15-16). Business plans, career goals, marriages: across various dimensions of culture, continuity is not to be expected, and subsequently, people find themselves in a state of constant planning and contingency. When the concerns of the immediate future fully occupy one's time, even ten years seems like a lifetime away. Despite the accelerated pace of life, time itself does not change. As Brand points out, an hour is an hour. The Clock of the Long Now, oriented toward 10,000 years, ticks once a day, and bongs once a year. If time does not change, then perceptions of it must change in order to experience time differently.

Personal Time, Social Time and the Present

Crucial to Brand's argument for the need for the long now is an awareness of and potential reconciliation of *chronos*, our routinized, socialized clock time, and *kairos*, a kind of deep "time outside of time" in which the ticking of the clock of *chronos* falls away. In *kairos*, one might experience what psychologist Mihaly Csikszentmihalyi refers to as flow, a highly focused, deeply satisfying state of consciousness (see Csikszentmihalyi 2008).

By combining the immediacy of *kairos* with an awareness of *chronos*, Brand proposes that it is possible to live in a present that is imbued with a sense of the long now. Offering a global perspective, he suggests that the *kairos-chronos* balance can be achieved by integrating characteristics of American and European culture. "While Europe specializes in deep continuity with occasional equally deep discontinuity, America specializes in perpetual petty turmoil. America provides the stimulation in the arrangement, Europe provides the wisdom" (Brand 1999, 79). While these are broad generalizations that aspire to localize cultural differences, Brand does have a point that much can be gained by integrating different perspectives on the future. Such congruity should be increasingly possible as cyberculture enables us to interact with one another and our sense of real people living in another part of the world is more commonplace than ever before. Ideally, ongoing synchronous and asynchronous communication across cultures should encourage us to learn about different ways of life and bring the best to bear on our own perspectives toward time. Yet inherent in this cultural leap is an understanding and acceptance of other cultural norms and traditions. Balancing *chronos* and *kairos* requires time, and therein lies the great challenge for the Long Now: if one is already overwhelmed by the pace of daily life, how can time be set aside to contemplate how to live differently?

Today's extreme shortage of time can be largely attributed to the effects of worldwide simultaneity. The mere idea that one can be connected nearly instantly with any part of the world, that the economic markets continue churning around the clock, and that significant events are occurring elsewhere *while you sleep* has created a new sense of urgency, driven largely by the technological and economic spheres. Austrian sociologist Helga Nowotny notes that as a result, "people again feel a peculiar and hitherto unheard-of desire: *they want to have more time for themselves*" (Nowotny 1994, 18), pointing to the awareness of the split between objective time and subjective time and a longing for the experience of *kairos*. Can the long now provide deep time?

Sociologist Michael G. Flaherty also has studied the differences between objective and subjective time, or what he terms "lived duration." Flaherty identifies and explores three forms of lived duration: protracted duration, in which time seems to move more slowly than it does; temporal compression, in which time is perceived as moving more quickly than it does, and synchronicity, in which the subjective sense of temporality matches objective temporality (Flaherty 1999, 34). In synchronicity, in other words, ten minutes feel like ten minutes. This is the most frequently occurring sense of temporality because it is "part of the underpinning for

interpersonal coordination and social order" (35). Objective time and subjective time are synchronized necessarily because although time is experienced idiosyncratically, objective time becomes internalized even when it is contrary to subjective time. Flaherty notes that temporality is not necessarily viewed as an important aspect of one's experience (37). This may be, as Erving Goffman points out, because lived duration is a "taken-for-granted feature of 'the organization of experience'" (40). Time is so essential to being-in-the-world and dictates the structure of being that it is difficult to isolate temporality as an element of experience.

In *Modernism and Time*, Ronald Schleifer proposes that abundance—material abundance, manifold interpretations of social and personal circumstance, and specifically a duality in the sense of temporality—are key elements in the experience of modernity. At issue with regard to temporality is the struggle to reconcile the past and the present in an effort to understand the now. Abundance, Schleifer posits, led to significant transformations in modes of explaining and understanding experience, seen most clearly in "changing phenomenologies of temporality governing explanation, conditioning experience, and apprehending value" (2000, 109). The freedoms afforded by modernity, especially newly-founded leisure time, combined with an increasingly secular worldview that erased the idea of religious eternity (Schleifer 2000, 3), resulted in a panic rooted in discontinuous temporality (58).

Musician Brian Eno, a member of the Long Now Foundation board, relates the experience of discovering a "conceptual poverty" (quoted in Brand 1999, 28) within his social circle. Noting that his acquaintances conceived of a sense of "here" that did not extend beyond their front doors, Eno realized people were "just as local in their sense of 'now'" (quoted in Brand 1999, 28). The future is thought of in immediate and personal terms, rendering nearly impossible the notion of contemplating one's actions with 10,000 years of impact in mind. While Eno's Baby Boomer friends may have a conceptual poverty about time, today's digital natives are critiqued for their impoverished sense of self-reflection. With unprecedented access to screened media in the form of smartphones, computers, and televisions, one need never truly be alone. Those coming of age in a digital world are also more likely to equate a sense of being busy with a feeling of accomplishment: the more items one can check off a list, the more successful one's day has been. Taken together, these qualities are another means of rendering the future as very short term. If making plans for ten years from the present seems exasperating, how can we make plans for ten thousand years from now?

Time Out

Computer designer Danny Hillis, co-founder with Stuart Brand of the Long Now Foundation, had the original vision for a 10,000-year clock. In a November 1998 article in *Forbes* magazine (1998), Hillis writes about his experience with the Foucault Pendulum at the Smithsonian Museum of American History. Observing museum visitors, he notes that the pendulum "inspires silence." Families that otherwise rush through the museum linger before the pendulum, captivated by its slow, steady movement. "I suspect that the real reason the pendulum is so well loved is because it is slow," Hillis says. "Its leisurely swing, just slower than the human breath, calms audiences and entrances them with a slower beat." The Long Now Foundation aims to achieve a similar response with the clock which, Hillis says, "provides what the pendulum only promises: the calibration of the rhythm of life, the definition of Now." He aims to suspend the "temporal disorientation [that] is an unwanted side effect of modern life." Modernity welcomes the future in such a manner that we are "dazzled by progress, rushed by events" and subsequently disconnected from the natural, steady unfolding of time.

At root, the long now encourages thinking beyond a five-year economic forecast or "long-term" planning for one's retirement by reframing our daily lives in such a way that we neither rush toward the future nor allow it to rush at us; rather, that our day-to-day actions are considered with a view toward their impact on the world in 10,000 years. The proliferation of messages from the environmental movement makes it difficult not to think of the long now in terms of environmental impact, but Hillis and his cohorts are as interested in the well-being of humanity as they are in the well-being of the planet. In an interview with Peter Leyden, Brand cites fellow Long Now Foundation board member Paul Saffo in pointing out that "we overestimate how rapidly change is coming in the short term and underestimate how much change is coming in the long term" (Leyden 2003). The sense of change Brand refers to here is technological at its core. Science fiction writers like Bruce Sterling and Bill Gibson, Brand says, understood "that there would be a noir version of high-tech that was not going to be glossy—it was going to be ugly. It was not going to be freeing; it was going to be enslaving. It wasn't going to make everybody rich; it was going to make some people rich." It is the ugly, enslaving future that the Long Now Foundation wants to avoid by, as Brand says, "reframing human endeavor" (Brand 1999, 48).

In their writing, interviews, and public lectures, both Hillis and Brand point out that one must *make* time to think about time. Running counter to

frenetic everyday life, with its impossibility to disconnect, is an awareness of the value of time for oneself. The desire for retreat encourages meditative space in which one can give pause in the way Hillis describes museum-goers contemplating the Foucault Pendulum. This meditative space is often rendered timeless, a secular use of the religious "non-time" that Brand describes:

> Religious time is time out. Time out from personal striving or suffering, time out from the chaos of history. In the sacred place set apart, in the sacred ritual changeless and timeless, in the sacred communion with a higher order, we step out of ordinary time and thereby make life meaningful, or at least bearable. (1999, 43)

Moving from the religious sense of eternity to a secularized temporality is a marker of modernity. The possibility that individuals may seek a slower temporality that allows for contemplation is an encouraging gesture for those who advocate the long now. Secular non-time remains, however, problematic. In a culture focused on time management and the tools for organizing both leisure time and work time, individuals may use their time out to determine how to more effectively manage their daily lives with a focus on maximizing productivity rather than allowing for reflection.

Reframing Time

The desire to "reframe human endeavor" (Brand 1999, 48) is an extraordinarily ambitious goal for the Long Now Foundation. That significant social change has been initiated via social media raises interesting questions about the possibilities for circulating the idea of the long now through Facebook, Twitter, and other instances of online communication. The project benefits from the involvement of Amazon.com founder Jeff Bezos, who owns the land on which the first clock, called The Clock in the Mountain, is currently being constructed. Bezos's financial and cultural investment can benefit the foundation broadly, as his cultural capital can be used to disseminate information about the long now.

Initially envisioning the project in 1999, Brand thought first about building an iconic object that would give materiality to the ideas about time embraced by the Long Now Foundation. Yet in an era moving away from materiality, does the existence of the object matter? Choosing mp3files over CDs and streaming video over DVDs is not merely a matter of convenience and storage; in the long term, choosing digital media

begins to obscure the material object and change relationships to the built environment. Do we indeed need *things*, or will representations of things suffice? The photograph, of course, is merely a representation of the earth, and yet it provided a view that reshaped perceptions. The Clock of the Long Now can be an iconic object even without being visited, yet Brand offers an apt description of how the Clock would function as an icon:

> For the Clock to be such a point of reference it must deliver an experience. After an encounter with the Clock a visitor should be able to declare with feeling, 'Whew. *Time*! And me in it.' It is not so much a conversion experience as a deep pause, like coming upon the Grand Canyon by surprise, where you simply want to sit and watch for a while and let your life adjust to two million years visible in one glance. (Brand 1999, 49)

In order to have this transformative experience Brand describes, people must travel to the Clock. The location selected for the first Clocks in Texas and Nevada are not easily accessible, and visitors would need to make a pilgrimage to encounter the Clock. While encouraging such travel requires a good deal of promotion, there are positive examples of timeless landscapes that draw visitors from all over the world. Stonehenge sees nearly one million visitors each year, and the National Park Service reports annual visitation of nearly five million at the Grand Canyon. Neither may, as Brand points out, offer a conversion experience, but travelers would certainly seek some kind of transformation from the pilgrimage in which they have invested their time and resources.

Fundamentally, the possibility of revolutionary thinking about time rests in the requisite investment of time to think about time. Yet this kind of mindfulness is a choice, and nurturing a sense of the Long Now can become habitual if deemed useful. Novelist and essayist Michael Chabon, writing in the January 2006 issue of *Details* magazine, notes that the Clock need not even be built in order for it bring about the social change the Long Now Foundation intends. "The point of the Clock is to revive and restore the whole idea of the Future, to get us thinking about the Future again, to the degree if not in quite the way same way that we used to do" (91). We cannot take responsibility for that which we no longer contemplate. Through digital culture, through its iconic Clock, through its series of seminars about long-term thinking that are also made available as podcasts, the Long Now Foundation hopes to encourage a paradigm shift that is hopeful, rather than dystopian, about the future. In 1966, Stewart Brand printed buttons with the phrase "Why haven't we seen a photograph of the whole earth yet?" and sold them for a quarter at the University of California at Berkeley, then at Stanford. The Long Now Foundation has

far more sophisticated means to ask whether an iconic clock can inspire us to think differently about time, yet it is only a matter of the means of dissemination being more complex than they were a generation ago. And while an hour is always an hour, it only takes a minute to think about 10,000 years.

Works Cited

Brand, Stewart. 1999. *The Clock of the Long Now: Time and Responsibility*. New York: Basic Books.

Brock-Abraham, Cleo, Nick Carbone, Eric Dodds, Jeffrey Kluger, Alice Park, Nate Rawlings, Claire Suddath, Feifei Sun, Mark Thompson, Bryan Walsh and Kayla Webley. 2011. "The 50 Best Inventions." *Time*, Nov. 28. 55-82.

Bronson, Po. 1998. "The Long Now: Time-traveling with Danny Hillis." *Wired*, May. Accessed Feb. 10, 2012. http://www.wired.com/wired/archive/6.05/hillis.html.

Chabon, Michael. 2006. "The Future Will Have to Wait." *Details*, January/February. Accessed Feb. 10, 2012. http://longnow.org/static/ djlongnow_media/press/pdf/0200602-Chabon-TheFutureWillHaveto Wait.pdf.

Csikszentmihalyi, Mihaly. 2008. *Flow: The Psychology of Optimal Experience*. New York: Harper Perennial.

Flaherty, Michael G. 1999. *A Watched Pot: How We Experience Time*. New York: New York University Press.

Hillis, Danny. 1998. "Impatient Pendulum." *Forbes ASAP*, Nov. 30. Accessed Mar. 10, 2003. http://www.forbes.com/asap/1998/1130/049.html.

Leyden, Peter. 2003. "Taking the Really Long View." Long Now Foundation, Jan. 21. Accessed Mar. 10, 2003. http://www.longnow. com/ about/articles/ArtTakingLongView.htm

Nowotny, Helga. 1994. *Time: The Modern and Postmodern Experience*. Translated by Neville Plaice. Cambridge: Polity Press.

Schleifer, Ronald. 2000. *Modernism and Time: The Logic of Abundance in Literature, Science and Culture, 1880-1930*. Cambridge: Cambridge University Press.

Wilson, Carl. "My So-Called Adulthood." 2011. *New York Times*, Aug. 4. Accessed February 16, 2012. http://www.nytimes.com/2011/08/07/ magazine/the-gen-x-nostalgia-boom.html?pagewanted=all.

An Ecology of Times: Modern Knowledge, Non-modern Temporalities

Martin Savransky

Introduction: "Been there; done that. We used to believe; now we know."

Among the set of practices that are central to understanding the role that time plays in the configuration of worlds, in our various modes of existence, and which allows us to establish differences and distinctions between an "us" and a "them," however ontologically and politically variable those pronouns might become, it is knowledge-practices, particularly as they have been constituted in the so-called "modern West," that are arguably of utmost importance to this task (Latour 1993; Sousa Santos 2004; Stengers 2000), and that will constitute the central concern of this chapter.

What is the relationship between time and modern knowledge? How has modern knowledge contributed to the shaping of a certain temporality that in its turn not only affects our particularly modern mode of existence but that also allows "us," as moderns, to judge and disqualify the co-presence of other practices and collectives in the process? What is the relationship between the temporality of the moderns and the possibility of something we might call a "revolution"? How might we escape modern temporality? These are some of the questions that I will explore in what follows.

As Barry Hindess (2008) argues, the modern, Western concern with putting knowledge production at the service of the demarcation of hierarchical differences between an "us," endowed with "rational" disciplines and practices authorised to speak in the name of knowledge, of nature and, indeed, of universal truth, and a "them," haunted by irrational beliefs and fetishes, is one which not only makes itself visible through a spatialisation of difference that would divide the rational "West" from the fetishistic non-West (on this issue see also Latour 2010), but it is also one that is historically constituted around a developmental, or progressive, understanding of time. Such a demarcation also involves a *temporalising* of difference:

> we should first note the implication that, if some of our contemporaries belong to the European or Western past, then their own pasts and their present must be conditions that can already be found in *this* past. It suggests, in other words, that the basic features of their past are already known to us, if only in principle, and that we [moderns] in the West have already 'been there. Done that.' (Hindess 2008, 202; emphasis added)

"Been there; Done that. We used to believe; Now we know." This has been the basic leitmotif of modern knowledge to assert itself against and above other knowledge-practices, to construct a form of spatial, epistemological, and ontological demarcation of and judgment upon what we deem part of a dark past: a confusion of fact and value, subject and object, discourse and matter, essence and appearance, nature and society, etc. The modern prerogative of progress inscribed in the singularity of modern science is that of sharpening distinctions, of purifying enquiries, in order to be able to separate, once and for all, knowledge from belief, science from politics (Latour 2004).[1] Moreover, the modern experience of progress is one brought about through a series of ruptures with the past. As Latour (1993, 68) argues:

> The moderns have a peculiar propensity for understanding time that passes as if it were really abolishing the past behind it . . . They do not feel that they are removed from the Middle Ages by a certain number of centuries, but that they are separated by Copernican revolutions, epistemological breaks, epistemic ruptures so radical that nothing of that past survives in them—nothing of that past ought to survive in them.

Thus, according to such a view, "we" moderns inhabit a world whereby time becomes a simple matter of fact, a container of events with the form of a universal, linear arrow of only one unambiguous direction. The knowledge produced by modern science is the one responsible for leading us towards an always brighter and epistemically purified future

---

[1] It should also be noted that this historicist rhetoric has not only allowed for a demarcation whereby only "us" moderns inhabit the present while "others" inhabit the past, but that insofar as their present is our past, the modern West has imperialistically authorised itself to narrate their history. On this issue see Seth 2004.

where distinctions are sharper and more incise, and in the process, we break with a past of confusion, of conspicuous and multiple associations of entities that ought to be clearly separated. Moreover, if those others engaged in knowledge practices outside the epistemic framework of modern scientific rationality are to belong to our world in any way, they have to become "archaic;" placed at the rear end of time's arrow, they become "pre-modern."

In this sense, one of the problematic implications of modern temporality is that the notion of "revolution," both in its political and scientific senses (Cohen 1985), already situates us, for scientific, political and social revolutionary changes presuppose a radical rupture with what precedes them and is, in many ways, a condition for them. Thus, "revolution" has become such common currency in e.g. modern physics, a science the vocation of which has been constituted by the continuous search for a "beyond," that, as Stengers (2011, 19) asserts:

> In today's speculative physics, the effects of a possible revolution, far from being unexpected, are often already taken into account and exploited. . . . 'Revolution' is no longer an event, it is a component of the physicist's management of future perspectives.

Even the so-called "post-moderns," supposedly enemies of modernity, who claim to have destabilised the modern territory in a variety of ways, who claim, moreover, to have stopped the "machine of modern time" (Latour 2004, 189), are still situated within such a temporality. Obviously, the very prefix "post" gives them away, and their critical, ironic denunciations of modern knowledge ultimately assert the implicit ambition of a last revolution that would stop temporality altogether (see for instance Stengers 2000).

Are we irredeemably trapped then? If the attempt to revolutionise the time of modernity brings us back to our starting point, is there anything else to be done? I would argue that there might indeed be something to be done. However, before we jump too rapidly into addressing potential solutions to the problem, this needs, as Stengers (2011) would suggest, to be 'slowed down' and considered carefully. For, after all, Time, in the modern sense just presented, and thus also its entanglement with knowledge-production, is but one particular form of narrating history, one particular interpretation of the passage of time which has nothing necessary about it. This begs the question: How can we rethink temporality in modern science?

The Hybrid Times of Practice: Modern (Medical) Knowledge and the Persistence of The Past

To claim, therefore, that our modern, temporal mode of existence is a form of narration implies, on the one hand, that there may be other forms of narrating the passage of time. On the other hand, it suggests that in order to understand how knowledge is produced in modern science, we need to learn a different way of narrating that production, a way that stresses less emphatically the sudden "discoveries" of science, and that draws a stronger emphasis on the practical elements of such an activity, that is, on the "repeatable sequences of activities on which scientists rely in their daily work" (Pickering 1995, 4).

The focus needs to be displaced from a conception that equates science with knowledge and therefore narrates retrospectively how these achievements where brought about, to one that attends to science as an inventive practice. This is what the heterogeneous field now known as "science studies" has been attempting to develop in a variety of ways, and with interesting, even surprising, results (see for instance Knorr-Cetina et al. 2000; Latour 1988; Pickering 1995; Stengers 1997).

In this sense, one of the advantages of attending to practices when attempting to understand the activity of scientists is that it allows for the possibility of doing away with the fixed "subject-object," "nature-culture," distinctions that characterise conventional historical narratives of scientific "discoveries." As Latour (1993, 70) illustrates as regards the "discoveries" of 17th century scientist Robert Boyle:

> If you suppress Boyle and Hobbes and their disputes, if you eliminate the work of constructing the pump, the domestication of colleagues, the invention of a crossed-out God, the restoration of English Royalty, how are you going to account for Boyle's discovery? The air's spring comes from nowhere.

Thus, according to Latour (1993), it is the clear-cut separation between subject and object that gives place to the miraculous discoveries and unexpected achievements of human rationality that emerge *ex nihilo*, thereby allowing an understanding that privileges the modern conception of time as an arrow always leading us to an ever more rational future while putting the past behind us.

Conversely, when one does attend to knowledge-practices in order to account for how such scientific inventions are achieved, it is not uncommon to find oneself engaged in a narration that is necessarily less able to draw clear-cut distinctions and separations. The actualities of practice thus make the study of science a messy, heterogeneous and multiple process of constructing precarious and changing associations between scientists, technicians, machines, bodies, spaces, and of course, temporalities. Furthermore, because we are forced to abandon clear-cut distinctions between subject and object, nature and culture, discourse and matter, etc., we are immediately drawn to consider the performative character of practices, that is, the ways in which they contribute to bringing about that which they seem to merely represent, but also the ways in which such performances dynamically negotiate their inventions with the entities they address, that is, the possibility of entities themselves challenging scientific practices (see Stengers 1997).

Among the diverse range of disciplines that one could attend to in order to construct an account of the role that practices play in the production of scientific knowledge, modern medicine arguably constitutes an especially fruitful field for studying these questions, for it entails a complex multiplicity of actors (human and non-human), settings and modes of knowledge. Thus, for example, Annemarie Mol (2002) has conducted an ethnographic study in which she shows that the different practices, techniques and apparatuses involved in the diagnosis of atherosclerosis (a condition involving the thickening of the arteries) within one and the same hospital actually cultivate a multiplicity of versions of the disease, each somewhat different, indeed, irreducible, to the other. Similarly, by drawing on the philosophy of Niels Bohr in researching the technological operations of ultrasonography, Karen Barad (2007, 201-202) argues that we should conceive of the sonogram not as a simple visualisation device which would " reflect," as it were, a fully pre-existent body of the foetus inside the womb, but rather, as an apparatus for the "making and remaking [of] boundaries" insofar as "the transducer does not allow us to peer innocently at the foetus, nor does it simply offer constraints on what we can see; rather, it helps produce and is part of the body it images."

As suggested above, it is important to keep in mind that to claim that scientific practices, technologies and techniques contribute to performing the entities and realities they seek to represent in no way assumes that the emerging object (e.g., the foetus; atherosclerosis) which is performed/represented has no bearing whatsoever upon the way in which such performance will present that object, or upon the very success of such a performance itself. On the contrary, "objects," as will become apparent in the next example, are very much recalcitrant to being acted upon, performed, in certain ways, and they are often able to force the researcher or scientist to change the questions that are posed to them (see Despret 2004; Stengers 2010).

What this means is that scientific knowledge is a constructive, that is, an inventive kind of practice in which a great diversity of actors, of natures and cultures, of times and spaces, dynamically negotiate the coming into being of new entities and values. If attention to practices not only obligates us to consider the relations, associations and entanglements among scientific actors but also the way in which what scientists and technicians do, the technologies and techniques that are put in play in order to generate knowledge about an object, become part of a heterogeneous negotiation that enacts a certain reality, can we think of time as being both productive of and also produced by practices and associations among entities? Let us explore this question by means of a detailed illustration.

In what follows, I will discuss a crucial event in the history of medicine that offers the opportunity to engage in a temporal reading of the way in which modern medical knowledge emerged by setting itself apart from what it defines as premodern healing practices. In an essay entitled "The Doctor and the Charlatan," Isabelle Stengers (1995) argues that, despite medicine being a practice with a very long history, it was in the eighteenth century that medicine began to constitute itself in the manner in which we conceive of it today, namely, as a scientific, rational practice. According to Stengers (see also Chertok and Stengers 1992 for a more detailed account of this episode), the event of the becoming-modern of medicine was directly related to the final assessments of the commission that was assigned to evaluate the practices of Mesmer and his "magnetic fluid."[2] After a series of studies, the commission concluded that, insofar as "[the fluid's] effects demonstrate its existence, it does not exist" (Stengers 1995, 132, own translation). In its stead, an alternative series of plausible and more "rational" explanations for the effects of Mesmerian practices was offered by the commission—influence, hope, the cessation of other medication, etc.

As Stengers (1995) argues, the effects of this commissioned study are inaugural in the becoming-modern of medicine as a science modelled on a theoretico-experimental framework, for they introduced a new value and with it a new form of disqualification: cure proves nothing. Thus, what defines modern medicine, indeed, what settled the longstanding debate

---

[2] This "fluid" acted as a material justification for the therapeutic practices which Mesmer himself carried out, oftentimes successfully, with patients. Again, I direct the reader to Chertok & Stengers 1992 for a more comprehensive study of Mesmerism and its fall.

around the competing figures of the "charlatan" and the "doctor," is a value whereby

> not all cures are equally valid. A cure as such proves nothing; a popular cure-all or a few magnetic passes can have an effect, but they do not qualify as cause. The charlatan is thereby disqualified as someone who takes this effect as proof. (Stengers 2000, 24)

In order to prove the existence of a therapeutic power, the cure needs not be only effective but "rational." It needs to conform to a series of requirements proper to every experimental procedure—replicability, stability, discreteness, etc.

If we perform a reading concerned with how temporality is managed in the aforementioned example, we can appreciate that one of the crucial aspects of this event is that what the creation of this new epistemic value—only the cures that conform to (theoretico-experimental) Reason are valid—entails is the production of a new type of temporality, one that, as suggested, ascribes a new present to scientific "rationality" and its alleged capacity to draw sharper distinctions, in this case, between "effect" and "proof," while at the same time disqualifying and relegating to an irrational past other healing practices for curing the living body. If in 1784, the year this study took place, the doctor and the charlatan were contemporaries, then the modern disqualification of the charlatan as someone whose healing practice is deemed irrational makes it today an archaism. The figure of the Charlatan embodies today the past against which the authority of the "contemporary" doctor is defined.

However, as argued above, it is not only authorities and institutions that have the power to invent the object of their study and how it responds. The body itself has a stake in how it is acted upon, indeed, how it is cured. Invention is always a complex negotiation among actors. In this sense, this recalcitrance of the living body, this propensity to be cured for "bad reasons" (Greco 2004, 694), puts medical knowledge and its modern temporality incessantly at risk. Stengers (1995) also points out that the controversy over the differences between cure and proof, far from being settled once and for all and relegated to a distant past which medicine could be said to have broken with, has been introduced into the heart of modern medical and pharmaceutical practices albeit in a transformed way, namely, through the so-called "placebo effect." As Monica Greco (2004, 694) argues in analysing this event:

> In the theoretico-experimental framework, the living body's ability to be healed through the imagination, through hope, through faith—in other

> words through 'bad reasons'—is clearly and crucially acknowledged, as the practice of clinical trials demonstrates. Simultaneously, however, and by the logic of the experiment itself, that very ability is explained away as irrational, as a 'parasitic and annoying effect' . . . that interferes with the pursuit of valid medicine. One might go as far as to speculate, on the basis of this argument, that scientific medicine could not acknowledge the placebo effect in a veritably positive way (that is, as a 'true' effect, due to 'good' reasons, worthy of investigation in themselves) without risking its claim to the ability to speak 'in the name of science.'

Thus, the placebo effect and its negative yet constitutive role in modern medical and pharmaceutical practices clearly illustrates the price that the living body's recalcitrance to conform to the values of the theoretico-experimental sciences has set for medicine in its striving to become modern, indeed, to be able to distinguish and disqualify, once and for all, the figure of the "charlatan," the one who is said to confuse effects with proof.

Furthermore, by returning to the question of temporality, the placebo effect also illustrates another case in point. Namely, the fact that the placebo effect clandestinely enacts, at the core of modern medical practice, the dark past from which modern medicine strived to break, a so-called *pre-modern* past in which bodies were cured for all sorts of reasons, "good" and "bad," suggests that, contrary to the progressive temporality upon which science asserts itself and which it invokes in order to disqualify and relegate others to its past, the past very much persists in the present. As Whitehead (1967, 18) claims, "there is no parting from our own shadows," the contemporary and the archaic share a co-presence which makes *our* present a coordinate at which a multiplicity of times crosscut and intersect.

What can we learn from this? Arguably, if we are to resist the parasitism to which modern temporality seems to force us, to claim that the past persists in the present is not something that can be mobilised in the same way the progressive time's arrow of modernity was asserted, that is, as a plain and simple matter of fact. On the contrary, I would suggest that such an analysis provides us not with fact but with a proposition that allows us to speculate, to gamble on the possibility of things becoming otherwise. Namely, what we can learn from this is precisely that time is no passive container of events, no matter of fact but a matter of concern (see Latour 2004; 2005), which is to say that it is dynamically entangled with entities and practices in a multiplicity of ways. Time is both productive of and produced by the practices and entities with which it is associated; it takes an active part in the constructive negotiations that give place to inventions. Moreover, what the previous discussion illustrates is that, despite its own claims to the contrary, modern knowledge is necessarily entangled with non-modern temporalities.

Thus, this brings us back (or forward?) to the question of revolution and change, for, if there really is no "parting from our own shadows," if the past is not definitely broken with and left behind, as the moderns would have it, but rather persists, becomes entangled with the practices and entities that are associated with it in such a way that we can move from a single conception of Time to a multiplicity of interweaved temporalities, how can we think of and celebrate epistemic and political revolutions?

Revolutions with a Memory: Change, Event and Betraying Modern Temporality

Up until this point, we have been engaging in a form of narration that aimed at destabilising modern temporality by making perceptible, that is, present, the already unstable and contingent nature of modern practices and also their equally unstable and contingent relations to time. Following Sousa Santos (2004), we could characterise this "making present" as a "sociology of absences" of sorts, one that is concerned with cultivating an ecology of times or temporalities by means of addressing the complexity of practices and hybrid temporalities that, while inhabiting the core of modern scientific knowledge, are still rendered residual and clandestine. Thus, such a sociology of absent temporalities would contribute to resisting the disqualifications and parasitic effects of modern time's arrow not by means of denunciation, as the post-moderns would have it, that is, through a critical operation that would attempt to challenge the hegemonic temporality of modernity, but through an immanent exercise of disclosing an element that both belongs to modern knowledge practices and connects them to an outside from which they seek protection; a strategy that, following Deleuze & Guattari (1987), we could associate with a form of betrayal.

Does this mean that such or other forms of betrayal should replace "revolutions"? Perhaps, or perhaps they do not need to. We have pointed out already that the notion of "revolution" seems to be too intimately associated with a break with what precedes it, too concerned with disqualifying its past and looking towards its irreversible future. The immediate move would thus be to do away with "revolution" altogether, to point the finger at it and accuse it of being not revolutionary but reactionary, of leading us back to where we started, to the time of the moderns. But this move is objectionable in two respects. First, because pointing the finger, judging, disqualifying, is exactly the operation of critical denunciation we have associated with modern temporality. Second, because there is something about the notion of a "revolution" that I, for the moment, would like to retain, for neither a scientific nor a political revolution implies *only* a break with the past. It also engages in forms of epistemic, social and political invention. It is this constructive, creative character of scientific and political revolutions that deserves, again, that the problem be "slowed down," in order to prevent us from throwing out the baby with the bathwater. We should reclaim that which testifies to revolutions' creativity, that is, what relates them to the becoming of an "event."

As Stengers (2000) argues, following Whitehead (1985), an event is a creator of a difference between a "before" and an "after;" it is about the creation of novelty in the world.[3] It should be noted, however, that unlike the notion of "revolution" associated with the moderns,

> the event is not for all that the bearer of signification . . . The event is not identified with the significations that those who follow will create for it, and it does not even designate a priori those for whom it will make a difference. It has neither a privileged representative nor legitimate scope. The scope of the event is part of its effects, of the problem posed in the future it creates. (Stengers 2000, 67)

The event cannot therefore become the carrier of progress, and thus, it cannot claim to have put its past behind. What it does bring about is difference. Moreover, the difference the event creates, and this is crucial, "belongs to a future *retaining some memory of its own novelty*" (Stengers 2005, 52, emphasis added).

Thus, through the notion of "event" as suggested by Stengers, we can reclaim the constructive, inventive character of "revolution" while escaping the modern temporality in which it places us. Because to retain a memory is to acknowledge that the creation, the difference that mattered, did not break with its past, that it only mattered *because* of its past, and thus, insofar as it might still matter, the past with respect to which it marks a difference must, in a way, persist (see Schlanger 1983). Moreover, insofar as time, or rather, a specific temporality, is not a passive container of events but something "that [is] to be explained by the contingent and

---

[3] Unfortunately, a more comprehensive exploration of the notion of event and its genealogy exceeds the scope of this text. However, one account is offered in Fraser 2010.

changing events from which [it is] abstracted," (Fraser 2010, 63) that is, insofar as it is produced by the event that creates a difference, it is the production of novelty, of new practices, new propositions and new entities that, in reconfiguring and transforming the associations among already existing practices, entities, and propositions from different, indeed, hybrid, times and spaces, may also reconfigure temporality itself.

Related to this is the fact that the becoming of an event and the temporality that thereof arises is never guaranteed; it has no privileged representative and does not belong to any right (Stengers 2000; see also Fraser 2010). Therefore, the event does not allow for explanations that would attempt to appropriate or reduce it away:

> No account can have the status of explanation, conferring a logically deducible character to the event, without falling into the classic trap of giving to the reasons that one discovers a posteriori the power of making it occur, when, in other circumstances they would have had no such power (Stengers 1997, 217).

What this implies, moreover, is that the coming into being of an event entails no normative stance as regards its future nor, for our purposes, a definite conception of temporality that would have the right to disqualify or reduce any other.

Thus, the practices of betrayal and revolution, in this event-related sense, must not necessarily be thought of as incompatible, nor need one replace the other. On the contrary, both of them can be regarded as ways to cultivate an ecology of temporalities whereby what is sought is not the denunciation and overcoming of a time that would, once and for all, be put aside and pushed back into a dark past, but rather an attempt at making perceptible the ways in which we inhabit worlds that are constituted by the co-presence of a multiplicity of beings and practices that belong to different times and that assert different temporalities. Indeed, insofar as the creation of an ecology involves, on the one hand, the suspension of certainty, and on the other, "the production of *new relations that are added* to a situation already produced by a multiplicity of relations," (Stengers 2010, 33, emphasis in original) an ecology of temporalities is an invitation to explore the possibility of inhabiting hybrid, multiple times.

Conclusion: Making Questions Proliferate

In this chapter we have explored the reciprocal relationships between western modern scientific knowledge and the specific temporality which underpins it and which in turn it comes to cultivate. By attending to medical practices I have attempted to destabilise and betray modern temporality by disclosing the hybrid temporalities and practices that, albeit clandestinely, inhabit its present. This has allowed us to reconsider the notion of "revolution" in a deterritorialised manner, that is, outside its modern understanding. In this non-modern, ecological account, revolution would not entail a radical break with the past but rather an affirmation of creativity and invention, a celebration of the "event" that, while potentially capable of reconfiguring the relations among entities, practices and temporalities into new and unforeseen arrangements, still retains a memory of its past.

Thus, I hope to have shown that even if no practice escapes the general parasitic effects of modern temporality, there are practices that remain vital tools of resistance (Stengers 2010, 10). Nevertheless, it should be crucially noted that, as such, the present discussion seeks no final solutions and provides no guarantees. Ecology is not a matter of settlements and consensus but rather a matter of care (Puig de la Bellacasa 2010), one of contingent and precarious yet sustainable relationships among others and ourselves.

Rather than providing solutions this chapter sought to multiply the questions, to make questions regarding scientific and other knowledges, regarding pasts and futures, proliferate. For it is by creating and making perceptible the multiple modes of existence of the entities and practices that inhabit our worlds, and not by opposing the power of modernity with some other power, that we may potentially envisage new forms of coexistence.

Works Cited

Barad, Karen. 2007. *Meeting the Universe Halfway: Quantum Physics and The Entanglement of Matter and Meaning*. Durham & London: Duke University Press.

Chertok, Leon and Isabelle Stengers. 1992. *A Critique of Psychoanalytic Reason: Hypnosis as a Problem from Lavoisier to Lacan*. Stanford: Stanford University Press.

Cohen, Bernard. 1985. *Revolution in Science*. Cambridge: Harvard University Press.

Deleuze, Gilles and Félix Guattari. 1987. *A Thousand Plateaus: Capitalism and Schizophrenia*. Minneapolis: University of Minnesota Press.

Despret, Vinciane. 2004. "The Body We Care For: Figures of Anthropo-zoo-genesis." *Body & Society* 10 (2-3): 111–134.

Fraser, Mariam. 2010. "Facts, Ethics, and Event." In *Deleuzian Intersections: Science, Technology, Anthropology*, edited by Casper Bruun Jensen & Kjetil Rödje, 57–82. New York & Oxford: Berghahn Books.

Greco, Monica. 2004. "The Ambivalence of Error: 'Scientific Ideology' in the History of the Life Sciences and Psychosomatic Medicine." *Social Science & Medicine* 58 (4): 687-696.

Hindess, Barry. 2008. "'Been there, done that . . .'." *Postcolonial Studies* 11:201-213.

Knorr-Cetina, Karin, Theodore R. Schatzi and Eike von Savigny, eds. 2000. *The Practice Turn in Contemporary Theory*. London: Routledge.

Latour, Bruno. 1988. *Science in Action: How to Follow Scientists and Engineers Through Society*. Cambridge: Harvard University Press.

—. 1993. *We Have Never Been Modern*. Cambridge: Harvard University Press.

—. 2004. *Politics of Nature: How to Bring the Sciences into Democracy*. Cambridge: Harvard University Press.

—. 2010. *On the Modern Cult of the Factish Gods*. Durham and London: Duke University Press.

Mol, Annemarie. 2002. *The Body Multiple: Ontology in Medical Practice*. Durham and London: Duke University Press.

Pickering, Andrew. 1995. *The Mangle of Practice: Time, Agency & Science*. Chicago: The University of Chicago Press.

Puig de la Bellacasa, Maria. 2010. "Matters of Care in Technoscience: Assembling Neglected Things." *Social Studies of Science* 41 (1): 85–106.

Schlanger, Judith. 1983. *Penser la bouche pleine*. Paris: Fayard.

Seth, Sanjay. 2004. "Reason or Reasoning? Clio or Siva?" *Social Text* 78 (1): 85-101.

Sousa Santos, Boaventura de. 2004. "A Critique of Lazy Reason: Against the Waste of Experience." In *The Modern World-System in the Longue Durée*, edited by Immanuel Wallerstein, 157–198. Boulder: Paradigm Publishers.

Stengers, Isabelle. 1995. "Le Médicin et le Charlatan." In *Médicins et Sorciers*, by Tobie Nathan and Isabelle Stengers, 127–177. Paris: Les Empêcheurs de Penser en Rond

—. 1997. *Power and Invention: Situating Science.* Minneapolis: University of Minnesota Press.

—. 2000. *The Invention of Modern Science*. Minneapolis: University of Minnesota Press.
—. 2005. "Whitehead's Account of the Sixth Day." *Configurations*, 13:35-55.
—. 2010. *Cosmopolitics I*. Minneapolis: University of Minnesota Press.
—. 2011. *Cosmopolitics II*. Minneapolis: University of Minnesota Press.
Whitehead, Alfred N. 1967. *Science and the Modern World*. New York: The Free Press.
—. 1985. *Process and Reality*. New York: The Free Press.

PART VI

REVOLUTIONARY TIME

A Time for Arab Revolutions: Walter Benjamin's Theses on History

Antoine Chollet

> Qui le croirait! On dit qu'irrités contre l'heure,
> De nouveaux Josués, au pied de chaque tour,
> Tiraient sur les cadrans pour arrêter le jour.
> —Walter Benjamin 1977, 262

It is a comforting fact to know that we can gain a better understanding of current events by reading past political thinkers. It shows that political thought is no mere esoteric discussion separated from any possible political reality; it can have some strength of its own, to initiate actions as well as replacing them with larger movements. Witnessing the Arab uprisings of 2011, I could not stop thinking about what Walter Benjamin wrote on revolutions. I would like to explain here how Benjamin can shed a precious light on these events and how they show, in turn, the extraordinary political relevance of Benjamin's historical and political interpretations. These uprisings soon became genuine revolutions, at least in Tunisia and Egypt: both countries experienced a tremendously accelerated course of events that overthrew their dictators in a matter of weeks, shattered existing centers of power (take the party-state RCD in Tunisia, for example) with incredible speed, changed political visions in the society and, finally, replaced the old fear of once all-mighty regimes with a new political imaginary (Castoriadis 1975). If these are not revolutions, then what is?

I will conduct this analysis through the concept of *time*. Time is indeed a central problem as far as revolution is concerned, and it is also central in Benjamin's thought. To be able to think about revolution, let alone to *do* it, one must first recognize and understand that revolutionary time is a *political* concept. There are always many times, and they are sometimes in conflict with one another. In these latter situations, we can properly speak of a politics of time, and in some even rarer instances, this politics becomes revolutionary—when people create, imagine, and hopefully institutionalize a new kind of time (Gurvitch 1964; Thompson 1967; Rifkin 1987).

Trying to conceptualize this new time, one has to confront the work of Walter Benjamin, and especially his very last text—the one he was probably carrying with him when he tried to escape the Nazi invasion of France in 1940, just before committing suicide. This text, "*Über den Begriff der Geschichte*" ("On the Concept of History"), contains one of the twentieth century's most radical accounts of time, both the one we are living in and the one we would need in order to make a revolution and change society (Benjamin 1977, 253–264). Benjamin famously describes the first of these times as "homogeneous and empty," without any quality attached to it and oriented only by a dreadful notion: progress, as stated in "Thesis XIII" (1977, 260–261). In "Thesis IX" he then quite as famously compares progress with a storm pushing toward the future the "angel of history," who can but see the ruins from the past piling up before his staring eyes (1977, 258–259).

The new time for revolution must be the exact opposite of the continuous and progressive time of the victors and oppressors; thus, it must be the time of the defeated and the victims. Here, Benjamin outlines some of the characteristics of a revolutionary time that could still be useful today. Since history, for him, is a sort of repetition in wait of a long-expected revolution, there is still some need for reading once more this famous text, unraveling a few of its mysteries, and trying to keep its explosive potency alive.

I want to emphasize three main characteristics of revolutionary time in Benjamin's theses. First, in that time, the past is exclusively dealt with through what Benjamin calls "remembrance" (*Eingedenken*). The past is not a continuous flow coming toward the present—this is academic history, according to Benjamin, and it always serves the oppressors—but a series of events with no apparent connection among them and usually much despised by established historians. Second, the center of temporality must be the present, what Benjamin names *Jetztzeit*, because a revolution can only happen there. Thirdly, the time for revolution has no future, or, to be more precise, no foreseeable and predictable future. Each of these characteristics can be recognized in the events in the Arab world that shattered long-established regimes in 2011. I do not know if the people there thought of Benjamin while doing the revolution—they most probably did not—but I am quite confident that I was not the only reader of Benjamin's theses "On the Concept of History" to link them to these events when they occurred.

This should not be interpreted as a sign of some sort of *bios theoretikos* (contemplative life) surrounded by books and incapable of action. On the contrary, I think it can be a truly emancipatory analysis of Arab revolutions:

not to give lessons or provide a definitive interpretation of them, but recognize that they continue a revolutionary thread (Arendt 1963) that goes all the way back to 1789 and travels through 1871 (Paris Commune), February 1917 (Russian Revolution), 1956 (Hungarian Revolution), and 1989 (movement of Tiananmen Square in China).

Critique of Progress

One of the main arguments in Benjamin's "Theses" is a radical critique of progress. This is stated most clearly in "Thesis VIII," in which Benjamin comments on the rise of Nazism in Germany:

> One reason why Fascism has a chance is that in the name of progress its opponents treat it as a historical norm. The current amazement that the things we are experiencing are 'still' possible in the twentieth century is *not* philosophical. This amazement is not the beginning of knowledge—unless it is the knowledge that the view of history which gives rise to it is untenable. (1977, 257)

If we agree with Benjamin that the belief in progress made the rise of fascism possible, a difficulty nevertheless remains: Replacing progress with another view of time and history, one "that avoids any complicity with the thinking to which these politicians [those who believe in progress] continue to adhere," demands a "high price" (Benjamin 1977, 258). Progress needs a certain sort of time in order to exist, one that is overly dominant in our societies, and which Benjamin famously describes as follows:

> The concept of the historical progress of mankind cannot be sundered from the concept of its progression through a homogeneous, empty time. A critique of the concept of such a progression must be the basis of any criticism of the concept of progress itself. (1977, 261)

This is the first critique of progress, the one that concentrates on the *ideology* of progress. There is a second critique in Benjamin's "Theses" that is a critique of the reality of a certain kind of progression. This progression or evolution, in fact, is the accumulation of victories in history: "Whoever has emerged victorious participates to this day in the triumphal procession in which the present rulers step over those who are lying prostrate" (1977, 256).

But the "triumphal procession" Benjamin writes about in "Thesis VII" is actually the pile of wreckage alluded to in "Thesis IX;" there is, of course, an inversion of values, "progress" being nothing but an ongoing catastrophe for Benjamin. This catastrophe is happening in a *continuous* time, the very one at which people in Paris shot in the revolutionary days of 1830 ("Thesis XV"). Progress, history of the victors, historicism, and continuous time all belong to the dominant temporal representation Benjamin wants to destroy, although he is well aware that it is a tremendously difficult task.

It is not difficult to see how *this* critique of progress might apply to Arab countries that experienced revolutions in 2011, most notably (but not only) Tunisia and Egypt. For militants in Tunis or Cairo, the idea that "progress" represents only the catastrophic continuation of the past was common thought since it was a kind of "progress" advocated by the regime—especially economic progress or growth—the one that was invariably referred to whenever one asked for some democratic reforms. This rhetoric had to be abandoned to imagine any sort of change in these societies. Revolutions do not lead to or accelerate progress, for only the generations of victors benefit from progress; revolutions halt what was previously *thought of* as "progress." The terror that the uprisings (especially the Egyptian one) provoked in Western governments originates in their Benjaminian quality: They stopped the continuation of the normal historical flow (i.e., "progress") that was to the advantage of oppressors here and there. In that sense, saying that the order of things as they go—and not some possible consequences of current changes—*is* the catastrophe (as Benjamin says in his text on Baudelaire) always has some revolutionary resonance. For societies in which the regime stood for modernization and economic advancement, Benjamin's critique of progress seems particularly relevant.

Yet, revolution cannot be limited to the overthrowing of a Tsar and the taking of the Winter Palace; it needs much more radically to *destroy time* as we know it. This may be undertaken in various ways, the first one being the transformation of the relation to the past.

The Past as Remembrance

For Benjamin, the past is not the accumulation of events that have happened up to the present (which is the historicist's position that reinforces past and present victors) but something one must *grasp from the present*. As he writes illuminatingly in "Thesis V":

> The true picture of the past flits by. The past can be seized only as an image which flashes up at the instant when it can be recognized and is

> never seen again. . . . For every image of the past that is not recognized by the present as one of its own concerns threatens to disappear irretrievably. (1977, 255)

This means that there is a past for every present, and not one long past that historians today should explore as a finite territory. Accordingly, the best historian—the character Benjamin calls a *historische Materialist* (historical materialist)—must be a "chronicler," someone who tries to conserve everything, for some part of it could, one day, prove to be decisive: "A chronicler who recites events without distinguishing between major and minor ones acts in accordance with the following truth: Nothing that has ever happened should be regarded as lost for history" (1977, 254).

Perhaps the tiniest event will be the one that has a revolutionary potential tomorrow, without any possibility to determine today which one it will be. For Benjamin, by recalling certain moments of the past, the historian can "blast open the continuum of history" (1977, 262), and the present might literally explode. This is why Benjamin always stresses the importance of the past in any political struggle, a past to which the future has often been substituted by the Left in its history, as "Thesis XII" explains:

> Social Democracy thought fit to assign to the working class the role of the redeemer of future generations, in this way cutting the sinews of its greatest strength. This training made the working class forget both its hatred and its spirit of sacrifice, for both are nourished by the image of enslaved ancestors rather than that of liberated grandchildren. (1977, 260)

"Enslaved ancestors" are the image of the past a historical materialist must remember, and, more importantly, make sure the oppressed remember. For it is the past, and not the hope in a glorious future, that can ignite a revolution. Very interestingly, remembrance seems to be the only way to revolutionize our world, an idea not far from the one Hannah Arendt advocates on education, when she writes: "Exactly for the sake of what is new and revolutionary in every child, education must be conservative" (1993, 192–193)—"conservative" being used here in its literal meaning: to keep as many things as possible from oblivion.

Was this reference to the past established during the Arab revolutions? Western societies tend to believe they are the unique bearers of a genuine democratic tradition. Therefore, when references were made by Tunisians and Egyptians to their own history, no one understood what they meant. And yet, Egypt as well as Tunisia has a long history of rebellion and popular political activity, one that many Western "democracies" certainly cannot claim. In 2011, these societies remembered they could rebel against a power oppressing them, and had already done so on numerous occasions in the past. The anticolonial struggles—against France in the 1950s for Tunisia, against Britain in the 1920s for Egypt—were obviously important moments, but by far not the only ones. Both countries have a long history of insurrections and revolutions. As Fisk reminds Western readers: "Egyptians in Cairo rose against Napoleon's army in 1798, fought the monarchy in 1881 and 1882, staged an insurrection against the British in 1919 and 1952, and rebelled against Sadat in the 1977 food riots and against Mubarak in 1986, when even the police deserted the government" (2011b). Samir Amin notices for instance that one of the symbols of *Wafd* (the Egyptian nationalist party that organized the fight against British imperialism in the 1920s), the crescent and the cross, representing Muslims and Copts, was seen in demonstrations on Tahrir Square in January and February 2011, making obvious the remembrance of past events and images during the revolution (2011, 15). If most Western commentators forgot Egypt had a revolutionary past, Egyptian protesters did not, and they perfectly understood the revolutionary potential of these references to a certain national past.

What is even more interesting, however, is the constant reference to a truly universal (that is, non-confined to Western societies) democratic and revolutionary tradition. When Alaa Al Aswani writes that "Tahrir Square became like the Paris Commune" (2011, ix), meaning the Paris Commune belongs to every revolutionary militant, anywhere and for ever, the universality of this tradition should be clear to everyone. Al Aswani's words convey a strongly Benjaminian interpretation of history and revolutions. Conversely Tunisians and Egyptians made revolutions for the entire world, activists everywhere contemplating with amazement what happened in these two countries. As one scholar notes about Tunisia: "It seems as if the whole world was staring at these Tunisians, delegating to them the defense of human dignity" (Béchir Ayari 2011, 58, my translation).

Revolutions Happen in the Present

As indicated earlier, the present is central in Benjamin's "Theses." It must not be understood, however, in the traditional sense as the mere passing between a massive past and an endless future but as *Jetztzeit*—"time filled by the presence of the now" as the translation says (Benjamin 1977, 261), or "now-time" (Osborne 2000). To really understand what Benjamin means by this *Jetztzeit* is one of the major interpretative difficulties of the "Theses." *Jetztzeit* is both a moment—the present (*jetzt*) —and a certain quality of time (*zeit*), precisely the time for revolution I try to examine here.

Apart from revolutionary moments, it is difficult to really understand this idea that major political shifts are grounded in the present, a present that is neither the continuation of a determined past (past remembered, as shown above, is *not* chronologically followed by the present) nor the preparation of the future. We usually see revolutions from a strategic—and I should say Leninist—point of view, because past revolutions have been reinterpreted that way afterwards. As Claude Lefort once said, "Specialists of retrospective rationality, marxo-seismologists and marxo-geologists will unfold their maps and calculations" (2008, 47, my translation) to show after the event why May '68 had to happen in France. But when we follow, even from a distance, a genuine revolution underway, let alone when we experience one as happened for so many people in Tunisia and Egypt in the first months of 2011, the strange idea that revolutions happen in a tremendously intensified present becomes evident. As Samir Amin, one of the 21st century's few remaining marxo-geologists, rightly writes, using a strongly Benjaminian image, albeit unknowingly: "The year 2011 opened with a series of *explosions* full of wrath from Arab peoples" (2011, 13, my translation and emphasis). What this shows very clearly is that Arab revolutions reactivated the idea that history is not written in advance, pre-determined by long tendencies and inertia. In his interesting analysis of the events, emphasizing this point, Bozarslan writes:

> Victor Hugo felt it in his time: a revolution is first and foremost an unpredictable storm that moves all present forces, but also initiates dynamics that did not exist before. This is why the heuristic meaning of a revolution can only be drawn *afterwards*, any discussion on its predictability thus being preposterous. (2011, 14)

Drawing primarily on the tradition of nineteenth-century revolutions in France (1830, 1848, 1871), Benjamin expresses the idea of an intensified present very powerfully, and, I should add, very convincingly. In an article written in *Le Monde* about the Arab revolutions, Alain Badiou, unknowingly following Walter Benjamin, asserted that: "the popular uprising we are talking about represents, in a very pure form, probably the purest since the Paris Commune, what we might call a communism in movement" (2011). There again we can see the parallel drawn with a revolutionary tradition that was suddenly rediscovered, as I have said before. With his idea of "communism in movement," Badiou also emphasizes the intensity of a political present that welcomes the "Event."

What happened in Arab countries in 2011 operated as a reminder of that truth. Following the events from a distance nevertheless allowed us to sense, albeit weakly, the intensity of what was happening. It also troubled journalists' conventional practice to grasp politics from the vantage point of the centers of power and not from the one of the people, or the oppressed. A revolution is a huge acceleration of activity by the people, along a different line of time, one that is nearly invisible from the usual position held by authorized commentators on powerful actors. Robert Fisk acknowledged that when he wrote, on January 29: "How does one describe a day that may prove to be so giant a page in Egypt's history? Maybe reporters should abandon their analyses and just tell the tale of what happened from morning to night in one of the world's most ancient cities" (2011a). And, following his own advice, Fisk describes what he sees among the people in Cairo, for this, and not the palaces of Mubarak or the army quarters, is the place where the revolution is happening.

Revolution Has No Future

The most immediate, but not the least intriguing, consequence of the centrality of the present in revolutionary time is the complete disappearance of the future. Reading Benjamin's "Theses," we can conclude—although it is not expressed so clearly in the text—that the future is a crucial dimension of temporality only in the "homogeneous and empty" time of the oppressors. Revolutionary time must abandon it. What this can possibly mean for a revolutionary process (such as the one we saw in action in Tunisia or Egypt) is that revolution must be done without fear or anticipation of its consequences in the linear time of "progress." Although this may appear rather foolish, for these consequences have great chances to be dreadful ones (let one think about the Paris Commune, for instance), we should see the problem from a different perspective: We can assume that, in most revolutions, if people had mainly thought of the consequences of the radical transformation of power they finally achieved to bring, they never would have begun any political action. You think of political consequences when you already hold power, or have high chances of holding it in the foreseeable future. For the weak, the defeated of history, the powerless, political action begins when they *stop* thinking about the consequences of their actions. And if they do not regret their revolutionary activity afterwards, in the event they were not killed in the course of it, it certainly is because of what the present looked and felt like during this very activity. It was *another* time, a *now-time*, and not the usual time, the empty and homogeneous one. To follow Benjamin, a revolution must be a pure means without an end other than its own perpetuation.

This Benjaminian principle applies very well to the Arab revolutions of 2011. If the militants assembled in Tahrir Square had continued to fear Islamists—the very people everyone in Egypt and in the entire Western world had learned to fear through the well-delivered lessons of Egyptian power and its proxies—can we be sure they would have demonstrated against Mubarak? A thorough analysis of the Egyptian (or Tunisian, for that matter) society in 2010 had an important opportunity to show that Islamists would have had a fair chance to win free elections. Would a democratic and responsibility-driven actor have chosen to act in these conditions, an imperfect dictatorship being better than an obscurantist theocracy? The decision would certainly have been harder to make with such concerns in mind. But once fear is overthrown in the mind of the people, all these calculations (to use Lefort's suggestive word again) become meaningless. Freedom and equality supersede complex anticipations and begin to actualize in the present. After all, this could be the simplest definition of "democracy."

In conclusion, a time for revolution in a Benjaminian sense should be understood as a kind of "vertical time," inexistent and meaningless according to dominant time but representing everything for revolutionists and people engaged in radical politics (Osborne 2000, 83). Revolutions are absent from history as much as history (in Benjamin's sense, that is as the victors' history) is absent from revolutions. They need only remembrance of revolutionary deeds from the past, the certainty that revolution is possible in the present, and no concern about its possible "future." Does that sound like an impossible task for today's political thought or action?

Works Cited

Al Aswani, Alaa. 2011. *On the State of Egypt*. New York: Vintage Books.

Amin, Samir. 2011. *Le Monde arabe dans la longue durée. Le "printemps" arabe?* Paris: Le Temps des Cerises.

Arendt, Hannah. 1963. *On Revolution*. New York: Viking Press.

—. 1993. *Between Past and Future*. Harmondsworth: Penguin.

Badiou, Alain. 2011. "Tunisie, Égypte: quand un vent d'est balaie l'arrogance de l'Occident. Les soulèvements des peuples arabes sont un modèle d'émancipation." *Le Monde*, Feb. 19, 21.

Béchir Ayari, Michaël. 2011. "Non, les revolutions tunisienne et égyptienne ne sont pas des 'révolutions 2.0'." *Mouvements* 66 (2): 56–61.

Benjamin, Walter. 1977. *Illuminations, Essays and Reflections*. Edited by

Hannah Arendt, translated by Harry Zohn. New York: Random House.
Bozarslan, Hamit. 2011. "Réflexions sur les configurations révolutionnaires tunisienne et égyptienne." *Mouvements* 66 (2): 11–21.
Castoriadis, Cornelius. 1975. *L'institution imaginaire de la société*. Paris: Le Seuil.
Fisk, Robert. 2011a. "A People Defies its Dictator, and a Nation's Future is in the Balance." *The Independent*, Jan. 29. Accessed Apr. 20, 2012. http://www.independent.co.uk/news/world/africa/robert-fisk-a-people-defies-its-dictator-and-a-nations-future-is-in-the-balance-2197769.html.
—. 2011b. "A Tyrant's Exit. A Nation's Joy." *The Independent*, Feb. 12. Accessed Apr. 19, 2012. http://www.independent.co.uk/news/world/africa/robert-fisk-a-tyrants-exit-a-nations-joy-2212487. html.
Gurvitch, Georges. 1964. *The Spectrum of Social Time*. Dordrecht: Kluwer Academic.
Lefort, Claude, Edgar Morin and Cornelius Castoriadis. 2008. *Mai 68, la Brèche*. Paris: Fayard.
Osborne, Peter. 2000. "Small-scale Victories, Large-scale Defeats, Walter Benjamin's Politics of Time." In *Destruction and Experience, Walter Benjamin's Philosophy*, edited by Andrew Benjamin and Peter Osborne, 57-107. Manchester: Clinamen Press.
Rifkin, Jeremy. 1987. *Time Wars, The Primary Conflict in Human History*. New York: Henry Holt.
Thompson, E. P. 1967. "Time, Work-Discipline, and Industrial Capitalism." *Past & Present* 38: 56–97.

REVOLUTIONARY TIME AS ENCOUNTER AND INTERSECTION OF DIFFERENCES

TANIA RISPOLI

> The awareness that they are about to make the continuum of history explode is characteristic of the revolutionary classes at the moment of their action. The great revolution introduced a new calendar. The initial day of a calendar serves as a historical time-lapse camera. And, basically, it is the same day that keeps recurring in the guise of holidays, which are days of remembrance. Thus the calendars do no measure time as clocks do; they are monuments of a historical consciousness of which not the slightest trace has been apparent in Europe in the past hundred years. In the July revolution an incident occurred which showed this consciousness still alive. On the first evening of fighting it turned out that the clocks in towers were being fired on simultaneously and independently from several places in Paris.
> —Walter Benjamin, 1940.

With these frequently quoted words, Walter Benjamin emphasized the difference between his conception of history and the traditional Hegelian philosophy of history that permeates the social democratic ideology but also the III International. Historical materialism and the working class expressed the revolutionary repudiation of the continuum of history imposed by those who had prevailed and were dominant. Introducing a "new calendar," which has its starting point in a new revolutionary day, the proletariat shatters the uniform flow of chronological time, thus reorganizing it in accordance to a program of commemoration and reactivation. Benjamin sees the anticipating signs of the future in the past. On the other hand, in the above-mentioned thesis, the idea that a revolutionary action is a condensation of time takes shape: it is an action that freezes time itself. If the first step defines the disconnection between the revolutionary spirit and reformist or conservative historicism, the second reveals Benjamin's characteristic messianism, in all its originality, though this does not necessarily represent a role model of revolutionary practice.

In general, following the current use of the concept in the framework provided by the philosophy of history, revolution is understood in accordance with the time of the Hegelian dialectic as future-oriented, or its opposite: the time of Benjamin, arrested, messianic, suspended, which emerges in the destituent insurrectional phase, contracted and entangled in a present that is immobilized as eternal, where we have the theological definition of eternity as *nunc stans*. How to avoid these two symmetrical perspectives on time in the Revolution—one which imagines a progressive temporal line and the other that shrinks temporality and concentrates it in a point?

The time of the current struggles (The "Arab Spring," Spanish Indignados, Occupy Movements, Italian students and researchers movements) has depth, it awakens the future in the past, making the future live in the present, without overwhelming it in the instant of an anarchic exemplary gesture. It builds on precariousness, but builds a temporality that is differential but organized in various moments of structured continuity, with no mythical trust in progress or messianic expectations of redemption. For instance, if we consider some of the best practices of the movements, we discover *tout de suite* some examples of what we are attempting to outline: the simultaneous resolution and execution of an occupation or the so called *mic check* used to give a speech in public where there is no amplification. In both situations, rhythm and time (shortened or extended) are shaped by a specific action and do not correspond to the imaginary and idealized time of progress or to that of a miraculous suspension. Time is, on the contrary, created in accordance to these specific events.

Moving critically away from the two dominant conceptions of revolutionary time—one connected with a ritual Hegelian and Marxist perspective and the other linked to a messianic view—is to outline a different concept of time of Revolution, which can better describe the present-day struggles. More analytically the issue can be approached in the form of three different questions that help explore the nebula of time. First of all: what it is meant by the word "Revolution"? And can this concept describe the nature of contemporary struggles? Secondly: is there a philosophical current that deals with the problem of time in a different way? What concepts can be transposed in our present? Thirdly: is it possible to imagine a link between time and the processes of subjectification?

In order to answer these questions it is useful to consider two main key-words: encounter and intersection. These terms, one used by Louis Althusser and the other borrowed from gender studies, can help to research an alternative idea of time that accounts for change and transformation.

Re-thinking the Concept of Revolution

First of all, it is important to discuss the theoretical construct of Revolution itself. The differentiation of this concept from the ideological one, where revolution is intended as transformation brought on by a final realization of class struggle, will be acquired only in a remote future. This conception of Revolution is closely linked to a theoretical approach that assumes subjectification in terms of a predefined identity, an often problematic view that can be found in both Lenin's theory of class struggles and party and in Schmitt's idea of a symmetrical reversal between dominated and dominators, or between friend and enemy. In Lenin there is a coincidence between the class before and after the event or process of struggle, guaranteed by the party that acts as an external vanguard of the class itself (Lenin 1964); in Schmitt politics as a whole is conceived as a "real possibility" of war (Schmitt 1964, 33), where the definition of enemy allows the identification of the self, or the group or nation, in foreign policy.

The revolutionary pattern—as it has been handed down from the history of philosophy or from an ideological tradition—is not able to adequately describe current conflicts that have recently taken place all over the world: from the Arab Spring to the riots in London, from the Italian struggles for universities and culture to the U.S. Occupy movements. The recent struggles are not the kind that believe and await a shining future and are carried forth by various and different subjects. Contemporary movements clearly identify their enemies (financial economy, the oppressors, governments), but their own identity has not been previously established and set out. As Hardt and Negri write:

> The multitude can develop the power to organize itself through the conflictual and cooperative interactions of singularities in common . . . the multitude must set the common in motion, opening up again its process of production. The political orientation also should be defined in the making of multitude, conceived not only as its political constitution but also as its economic production. (Hardt and Negri 2009, 175-177)

This means—and I will come back to this point further on—that class is not given before the conflict but is an *ensemble*—or better a "multitude"—that takes shape during the conflict itself.

A Machiavellian term, *tumulti* (tumults), allows one to re-interpret the revolutionary experiences of recent years. In the *Discourses*, Machiavelli explains that tumults are constitutive of all political bodies and, under certain conditions, produce positive and fruitful effects, as the history of the Roman Republic demonstrates. Furthermore, the tumults contribute to the imagination and transformation of a democratic process (Machiavelli 1998, 16).

It is not always appropriate to easily apply past theories to present occurrences; however some older concepts may illuminate some contemporary aspects or help conceptualize current events. The classical definition of Revolution is strictly connected to the period and to the kind of sovereignty that is being taken into consideration. In this way the Revolution fulfills the time given by the one and only God and by the Hobbesian or Hegelian state. It is the secularized figure of God as well as its reversal, the messianic interruption, "the exception" that suspends rights and power. This movement of disruption is symmetrical to the exception though which it becomes possible to recognize who is the sovereign. "Sovereign is he who decides on the exception" (Schmitt 2005, 5). The Revolution conceived in Benjamin's perspective is the opposite of Schmitt's exception, but it follows the same pattern of sovereignty. In fact "all significant concepts of the modern theory of the state are secularized theological concepts" and "the exception in jurisprudence is analogous to the miracle in theology" (Schmitt 2005, 36). The distinctive features of the monotheistic God—his omnipotence and his power to suspend the flow of time—characterize both sovereignty and the revolutionary class. The crisis of the theory of sovereignty and the decline of the nation state call for a parallel rethinking of the concepts and terms that describe conflicts.

Michel Foucault developed the idea of *gouvernamentalitè* (governmentality) that describes, in a genealogical perspective, the concrete techniques and technologies that produce and exert power. Furthermore according to Foucault there is, along with the theories of sovereignty, a whole plexus of local, effective and diffuse powers (Foucault 2009, 159-268). The current mechanisms of governance are expressions of these decentralized and multiple forms of control and exploitation of the force labor and of the entire *bios*.

Moving beyond this, it may be said that the financial capitalism of today exerts its power over life. Foucault explains that if the core of classical liberalism is the separation of the state from the market, the intent of neo-liberalism is to find the methods through which the state can support the market's mechanisms (Foucault 2010, 115-135). At present, the states not only support financial capitalism, but also have the duty to save the market, as shown by the problem of repaying debt: the states have saved the banks in economic crisis, aggravating the public debt. The concrete situation of Greece (compulsorily administrated from the EU, which lends money, but also established the rules of austerity), for example, overemphasizes the origin of the society of debt, in which from an individual and collective perspective the aim must be to sustain the financial economy.

Contemporary tumults are conflicts that are no longer directed against a sole sovereign: they are instead directed against the current forms of governance (multiple sovereigns). Tumults break out during a global crisis and mobilize a specific generational sector: a certain technical composition of labor becomes a political intersection, or in other terms, a political composition. A certain post-operaist literature points in this direction and helps re-interpret the concept of the working class. As Negri writes: "the making of class composition has changed profoundly" (2003). In Paolo Virno's words:

> The contemporary multitude carries in itself the history of capitalism. Moreover it is one and the same with the working class whose primary matter is constituted by knowledge, by language and by affects. . . . Working class is a theoretical concept, not a souvenir photo: it indicates the subjects that produce absolute and relative surplus value. (2003)

The multiple subjects of tumults bring into question the organization and the exploitation of labor, but also reclaim a different democracy. The extent of parliamentary and liberal democracy, in fact, during the economical crisis, becomes very low, because the main traditional institutions are devoid of real power. Through tumults, the multitude reclaim an alternative hypothesis of democracy, in which citizens and non-citizens may practice different forms of self-government and build institutions of commons (Illuminati and Rispoli 2011).

Tempore in uno...tempora multa latent

In the present context characterized by the decline of the traditional forms of sovereignty and of the symmetrical grammar of revolution, we do not find *one* exception by which power and its reversal can be defined. On the contrary, the forms of governance and the times of tumults are articulated in many different exceptions. In this situation Revolutions are less likely to be of an instantaneous kind and the same can be said for the impact these may have on a unified centre of power. Today's revolutions occupy material and symbolical places, not the centre of power, in the sense of an abstract statal power, a separated machine that can be applied for any use.

In his late works, Louis Althusser emphasized an underground current of the materialism of the encounter (2006). Recently Montag has edited a fragment that dates from 1966 in which Althusser already speaks about “the theory of encounter or theory of conjunction,” predating the development of these concepts. The ideological category of genesis is the same in ideology and in religion: in both we find the identification of “linear causality” resulting from a single cause. In this first fragment the French philosopher, investigating Marx’s concept of the mode of production, opposes a “structural causality” to a linear causality (Althusser 2012, 2-3), and later he introduces the idea that history and time are governed by the logic of encounter.

This linear concept of time where time is an arrow that moves from its origin to the effects it produces is a providential view of time, where necessity and teleology play an important role. Time rather responds to an aleatory mechanism—of deviation, encounter and capture. On the one hand Christian Monotheism and a certain historical materialism conceive time as being similar to an arrow, while the Newtonian philosophy imagines time as a void, a receptacle. Althusser examines past concepts of time and finds in Lucretius’s *De Rerum Natura*—present also in Machiavelli’s writings—a source of inspiration that enables him to re-think time. Nature is made up of atoms that move like rain drops in a void colliding and diverting their course and thus producing material compounds of all forms and shapes, through this *clinamen*, that is, an unpredictable swerve of the atoms. In this Epicurean and Lucretian perspective

> time exists not by itself . . . You may see clearly that all events from first to last do not exist, and are not by themselves like body, nor can they be spoken of in the same way as the being of the void, but rather you might justly call them the accidents of body and place, in which they are carried on, one and all. (Lucretius 1950, 42-43)

Therefore time adheres to events, indeed: “No man feels time by itself apart from the motion or quiet rest of things” (Lucretius 1950, 42). Time relates to this flexible contingency of things; it is neither neutral nor progressive. So we have a time for war and one for harmony, a time of poverty and another one of wealth: there are different kinds of time for each Occupy movement or experience of struggle. A time of revolutionary acceleration or, conversely, a time of ordinary economic struggle and of ordinary parliamentary debate are always articulate, not punctiform, but composed of different rhythms. In this sense also war and peace are associated with places, feelings and different practices (to kill and to love, to destroy or to build, to eliminate the resistance or to find a consensus: in these dissimilar actions, the moves of war are faster and more irreversible than those finalized to peace). In the movement experiences there are situations which are more linear and other more complex, homogeneous or differentiated subjects, diversified times of maturation and performing.

What is at stake in this genealogy is that the time of struggles is a time that refuses to wait, that does not believe the evolution and development of the modes of production have a necessary quality to them, but rather focuses on the encounter—successful, unsuccessful, reversible, revocable. The time of tumults also suggests the interweaving of different stories. As Lucretius wrote:

> Or will this be nearer truth? Because within a single time, which we perceive, that is, when a single word is uttered, many times lie unnoted, which reasoning discovers, therefore it comes to pass that in any time however small the several idols are there ready at hand in all the several spots. (Lucretius 1950, 170)

As with the process of feeling (the perception of *simulacra*), where each unit of time is a complex unit, with the revolutionary event different rhythms (of protagonists and antagonists, of single acts and collective organizations) are interwoven. The historical *continuum* is broken. There is a calendar of power, and a calendar for insurgencies.

Federation of Independent Times

To analyze the revolution of time means not only to assign a new status to this category, but also to imagine how conflicts and tumults bring into question the current time of the capital system, in terms of its organization and homogeneity. The revolution of time is theoretical, but practically acted out in the time of revolution (or tumults). The struggles are opposed to the articulation of the working time, which increasingly colonizes life-time, but also oppose the alignment of differences or the excluding production of differences.

In his 1967 work *The Society of the Spectacle*, Guy Debord defined "spectacular time" as a consumable "pseudocyclical" time—a mix between the traditional cyclical view and the Christian or Bourgeois idea of a linear time of accumulation of capital—in which the units become exchangeable and the qualitative dimension is suppressed. The spectacular time is the time of society transformed by industry and consumption. It is based on the ideology of progress that, assuming they are "free" producers and consumers, leads to the expropriation of the time of workers (Debord 2002, 43-44). In Debord's words, the opposite of this time is not a suspension or interruption of the ordinary flow, but the dissection and federate grouping of the times of living beings. In fact he writes:

> The natural basis of time, the concrete experience of its passage, becomes human and social by existing *for humanity*. The limitations of human practice imposed by the various stages of labor have humanized time and also dehumanized it, in the forms of cyclical time and of the separated irreversible time of economic production. The revolutionary project of a classless society, of an all-embracing historical life, implies the withering away of the social measurement of time in favor of a *federation of independent times*—a federation of playful individual and collective forms of irreversible time that are simultaneously present. This would be the temporal realization of authentic communism, which *abolishes everything that exists independently of individuals*." (Debord 2002, 45)

In this sense the historical time of change and transformation is not only dominated by a logic of encounter, but also results from an intersection of different and "independent" times. In order to give a broader account of the consequences this hypothesis leads to regarding the conception of subjectivities and struggles, it is necessary to explain the concept of the intersection of differences (and temporalities). In this respect it is useful to turn to gender studies and to postcolonial studies.

In Braidotti's works we find the idea that capitalism not only excludes what is different, but also produces and multiplies the differences, in the interest of commodification and profit. Referencing Crenshaw's theory, Braidotti writes about *intersectionality*:

> This means that all axes of differentiation—racialization, sexualization and naturalization—are internally differentiated in a complex manner. They also zigzag in and out of one another, triggering all kinds of combinations. The point is that the process, the movements and the trajectories of these lines of becoming deserve more attention than any of the specific identity formation they give rise to. The flows matter more than steady roots. (Braidotti 2006, 62)

Capitalism determines and sets borders to differences, and the identitarian subjectifications do not disappear spontaneously. On the contrary, these subjectifications re-emerge in new forms of fundamentalism (religious, ethnic, and also gender fundamentalism). In this context we find that figures of purity, martyrdom and *victimacy* are continuously being invented. All these cases show a time that is frozen in archetypes, stolen from a relational fluidification, that does not allow the emergence and the acknowledgement of new problems and plural subjects who carry these burdens. The aim of the intersection is to create innovative combinations of becomings, forming fresh alliances and translating subjectification into organizational forms.

The precarious and the migrant—not independently of their gender composition—are the two subjective figures who act out the temporal revolution and the actual tumults, challenging the temporality of the capitalist economy in crisis and blending together different narratives. The time as well as the space of migrants has been shattered, a great distance separates the land and culture of provenance from the one they have arrived in; furthermore, the continuity of experience and of a unitary lived time is made impossible by the kind of work they are forced to accept—that is, a paradigm for the working conditions of the national citizens. The fractioned temporality of tumults mirrors the biopolitical conditions of its protagonists and the devices of biopolitical exploitation.

The postcolonial theories, from Guha to Chakrabarty, opposed History 1, or rather the ethnocentric and necessarily "universal history" imposed by the logic of capitalism, instead presenting a History 2, or, to be more precise, the histories which are made up of the imaginations, feelings, beliefs and traditions of subalterns. Chakrabarty says that this opposition serves fundamentally to pluralize "the history of power in global modernity and separate it from any universalist narratives of capital" (Chakrabarty 2008, 15). History 2 splits the abstract Enlightenment of the historicism of Modernity—which makes a distinction between history and pre-history or between civilization and barbarism—and the homogeneous time of Capital. In *Provincializing Europe*, Chakrabarty, radically questioning the nature of historical time, writes: "Imaginations of socially just futures for humans usually take the idea of a single, homogeneous, and secular historical time for granted" (Chakrabarty 2008, 17). To criticize this perspective and to imagine a plural history of power is necessarily to develop other kinds of time, introducing discontinuities, ruptures, and shifts in the historical process, abandoning "the metanarrative of the nation state" (Chakrabarty 2008, 42).

Conclusion

In the course of this essay, a possible way to re-interpret the concept of revolution as a tumult has been shown. Overlapping tumults and revolution may account for the differences and diversities of contemporaneous struggles that are directed against governance mechanisms. These conflicts are not developed in a teleological and progressive course of history; instead they open a thousand plateaus of different histories, governed by a logic of encounter and open to various intersections.

In the case of the "Arab Spring," for example, the conflicts are over-determined by specific grounds, caused by geo-political reasons, by tribal differences (Libya, Yemen), by religious causes (Syria, Egypt), and by different forms of military forces (nonexistent in Tunisia, autonomous in Egypt, dependent on Baath in Syria). All these causes are related and together influenced the revolutionary situation, creating different results.

As far as the aspect of subjectification is concerned, the people who were involved in changing processes (as the case of Tunisia shows) not only tried to change their conditions in their own country, but also exported the desire of transformations into the lands of immigration. Sometimes it happened that the struggle of migrants became interweaved with the struggles of locals in precarious situations or women, and all these subjects in their multiple histories and segments of existence encountered each other, producing a revolutionary or tumultuous exchange.

The federation of independent times and singularities not only dissipates the irreversible time of production and consumption, that in the 21st century is the irreversible time of the crisis, of the inexorable tick tock of debt which gnaws away at life, but also inaugurates different forms of existence. In this subjectification process, the activities and the desires are rescued from an obligatory destiny of history, identity and exploitation. These existences are no longer useful to others, but are a free expression of life. Revolution is to change the relations of production and the times of life.

Works Cited

Althusser, Louis. 2006. *Philosophy of the Encounter: Later Writings, 1978-1987*. Translated and edited by Gary M. Goshgarian. New York: Verso.

—. (1966) 2012. "On Genesis." Translated by Jason E. Smith. *Décalages* 1 (2). Accessed Feb. 15, 2012. http://scholar.oxy.edu/ decalages/vol1/iss2/11.

Benjamin, Walter. (1940) 2003. "On the Concept of History." Translated by Harry Zohn. In *Walter Benjamin: Selected Writings Vol 4: 1938-40*, edited by Michael William Jennings, 389-400. Cambridge: Harvard University Press.

Braidotti, Rosi. 2006. *Transpositions: On Nomadic Ethics*. Cambridge: Polity Press.

Chakrabarty, Dipesh. 2008. *Provincializing Europe. Postcolonial Thought and Historical Difference*. Princeton: Princeton University Press.

Debord, Guy. (1967) 2002. *The Society of the Spectacle*. Translated by Ken Knabb. Canberra: Hobglobin Press.

Foucault, Michel. 2009. *Security, Territory, Population. Lectures at the Collège de France 1977-78.* Translated by Graham Burchell. Basingstoke: Palgrave McMillan.

—. 2010. *The Birth of Biopolitics. Lectures at the Collège de France 1978-79.* Translated by Graham Burchell. Basingstoke: Palgrave McMillan.

Hardt, Michael, and Antonio Negri. 2009. *Commonwealth*. Cambridge: Belknap Press of Harvard University Press.

Illuminati, Augusto and Tania Rispoli. 2011. *Tumulti. Scene dal nuovo disordine planetario*. Roma: Derive Approdi.

Lenin, Vladimir Ilic. (1917) 1964. "The State and the Revolution." In *Collected Works*, edited by Stepan Apresyan and James Riordan. Vol. XXV, 381-492. Moscow: Progress Publisher.

Lucretius, Titus Carus. (I. BC) 1950. *On the Nature of Things.* Translated by Cyril Bailey. London: Oxford University Press.

Machiavelli, Niccolò. (1531) 1996. *Discourses on Livy*. Translated by Harvey C. Mansfield and Nathan Tarcov. Chicago: The University of Chicago Press.

Negri, Antonio. 2003. "Multitude or Working Class?" Translated by Thomas Atzert. Generation-Online.org. Accessed Feb. 16, 2012. http://www.generation-online.org/t/negriESF.htm.

Schmitt, Carl. (1922) 2005. *Political Theology. Four Chapters on the Concept of Sovereignty*. Translated and edited by George Schwab. Chicago: The University of Chicago Press.

—. (1927) 2007. *The Concept of the Political.* Translated and edited by George Schwab. Chicago: The University of Chicago Press.

Virno, Paolo. 2003. "Multitude and Working Class." Translated by Arianna Bove. Generation-Online.org. Accessed Feb. 16, 2012. http://www.generation-online.org/t/multitudeworkingclass.htm.

REVOLUTION AND ITS AUTODESTRUCTION

THOMAS PIERAGASTINI

The Fog of Revolution; The Terror of Autodestruction

Everything that was solid has melted into air; everything seems to be wrapped in some sort of indeterminate fog. What had seemed like an inert politics, a lack of political conflict and rupture, only a few years in the past, seems like a distant memory now. Everything seems to be dying, but coming alive; from the streets of the Middle East to the streets of Oakland, from the so-called "Arab Spring" to the amplifications of conflict at the fringes of the "Occupy Movement," everything seems to be breaking out. This is a situation of total friction, an amplification of the crisis of the state almost to the point of total organizational rupture. Yet, at the same time it is a situation of almost total precarity; everything is collapsed yet everything still seems separated from our existences. Far from the "hope" spouted by the politicians, prison guards and mouthpieces of semiotic space, this is more a desperation that has hit a point of saturation.

Revolution is in the air, the magnitude of the insurgent is thick on everything like the residue of tear gas on the streets; but at the point of the utterance of the word "revolution" a black cloud descends, full of the stink of the rotting corpses of past "grand narratives," the fear, the possibility. Yet, in the moment where everything seems to be glimpsing into the possible, into the fog of this rupture in the state, we are coming into contact with the impasse of revolution. Stuck between the crushing repression of the police and the Terror, whether the Jacobin Terror or the Red Terror, revolution seems to be simultaneously possible, but also treacherous. With the utterance of the word "revolution" we immediately come into contact with its failure, the failure of the revolutionary project that appears when the jubilation on the streets quickly becomes the tribunals, guillotines and gulags of policing.

This is the enigma that we need to approach today. If we are to live in a time of revolution, if we are to be antagonists in this revolutionary rupture that has swept the globe, we need to be prepared for more than the gas and bullets; we need to approach the spectre of our own failure. It is not enough to come up with "better" theory, as if the failure of revolution is a failure of terminology scribbled on paper; we need to approach this problem directly. We need to come to understand what it is, within the assertions that plague the revolutionary project, that has come to operate as the very autodestruction of revolution; that point where riots become prisons, where the Russian Revolution becomes Leninism, where the French Revolution becomes Jacobin, where revolution becomes fascism. The revolutionary impasse and the attempt to "solve" it is the attempt to understand how a tactical amplification of crisis to the point of rupturing the operational capacity of policing quickly tends to be conflated with its opposite; or how it is that the amplification of crisis finds its end in the restructuring of the police, the gulag, the guillotine.

We find the roots of this problem in the conceptual attempt to conflate revolution with some result, or some conceptually teleological end point that somehow becomes a materially total political form; this is the conflation that occurs in the discussion of what will come about due to revolution. Unfortunately this question of "what comes next" has become a constant conversation, an annoyingly droning discourse of attempting to approach the question of what occurs after a series of events so catastrophic that the categories that were used to make sense of the world no longer make sense. It is not just the complete worthlessness of the question that is grating, it is not even the nauseating frequency with which this conversation occurs or the pretension that accompanies every "answer;" the problem with the question is the prophetic possibility, the elimination of the very possibility unleashed by insurrection through this conceptual essentialism. In this conflation revolution itself is removed from its material dynamics, the conditions of conflict, and inserted into some atemporal conceptual teleology, the discussion of what will result "from revolution." Conflict itself becomes obscured in the process of orienting revolution towards some materially impossible projection of events through time and space in a process which always is the attempt to define the very possibility unleashed in revolutionary dynamics.

The autodestruction of revolution is a specifically temporal issue. On the one hand revolution is a material amplification of conflict, one that occurs in particular material moments, and that has effects and resonances. On the other hand, in the very conceptualization of revolution, and in the insertion of revolution into a conceptual constancy of moments termed "revolution," we remove revolution from itself, conflict from itself, and generate a completely inert, atemporal concept of revolution. It is not in either of these manifestations that the problem of the autodestruction of revolution presents itself; it is in the conflation of these senses. To speak of revolution itself, without recognizing the disjunction between the particularity of the moment of conflict and the atemporality of the concept of revolution, is to remove revolution from revolution and to insert particular conflicts into a non-particular concept of "revolution." It is in this conflation, between particularly temporal conflict and the atemporal, non-particular, concept of "revolution" that we find the roots of this autodestruction.

The Dual Projection of the Dual Move

The autodestruction of revolution always already occurs as a dual move. Put simply, for the sake of concision, the first move is revolution itself; the inconceivable unleashing of possibility that occurs when there is a tactical amplification of crisis to the point of causing an operational rupture in policing. But it is not enough to just say that it is revolution itself, for we can never speak of what revolution "is." There are constant attempts to do so, but this is nothing but the attempt to speak the essence of moments, for revolution is a series of events which occur in a particular time and space that we speak of, artificially, through the atemporal term "revolution."

Through this attempt to speak of revolution as a conceptual "is," or through a supposed direct transmission between the conceptual and the material, we obscure the very particularity and temporality of events. Revolution is inconceivable due to the infinite distance that exists between the insurrectionary amplification of conflict in particular, temporal moments and the atemporal constancy of time and space posited in the concept of revolution itself.

Though this may seem simple we see this conflation play itself out in Leninism through the claim to understand revolution from the position of technician. In this construction of the technician we construct revolution as an object of analysis that we can stand outside of, and thus, we construct revolution as a static object, both as the basis of speaking of revolution conceptually as well as speaking of revolution in relation to a teleological strategy. This forms the core of the second move, the construction of revolution as conceptual object and material teleology separate from the particularly temporal events that we term revolution.

The Conceptual Projection

The conflation of revolution and teleology hinges on the attempt to conceptualize the possibility unleashed in revolution itself as separate from the particularity of this possibility in any moment. There is always this "process" posited, for example, in Leninism or anarcho-communism in which the possibility unleashed in conflict becomes this grand negation that lays the ground for some political form or structure; an absence is conflated with a presence through the conceptual teleology of the "revolutionary process," first conceptually then operationally. To understand this dual projection, this conceptual-operational projection, we need to first understand the construction of the inert concept of possibility that is the core of the attempt to conceptualize revolution. To glimpse into this infinite distance, this abyss of the concept itself, we should begin with Isaiah Berlin and the paradoxical attempt to speak of possibility in the *Two Concepts of Liberty*.

The core of this conflation, in Berlin, begins with this discussion of possibility, rather than actuality, as that which has not occurred, an absence, but also as a conceptual object termed "freedom," a presence; a conceptual object defined as an absence of "coercion" or force. This presents a dual implication. Firstly, all "coercion," or force, implies a sort of total presence or inert limit in which nothing can occur and through which all dynamics and possibility ends. Secondly, "freedom" is constructed as some field of total possibility, and thus a totally undefined void field of action; a notion of "freedom" as a total absence. A dynamic is posited between a total presence in which nothing can occur and a total absence in which nothing is occurring (Berlin 1969, 121).

"Coercion implies the deliberate interference of other human beings within the area in which I could otherwise act. You lack political liberty or freedom only if you are prevented from attaining a goal by other human beings" (Berlin 1969, 122). Here we must take note of a couple of aspects of this formulation. It is not a matter of the materiality of a specific total presence, or a specific effect of a specific action, it is only speaking of total presence that prevents one, not only from acting, but also from having this action manifest some necessary teleology or necessary trajectory of moments that are internal to the possibility to act. This absence of this presence becomes connected to a smooth teleology of action; that the absence of coercion necessarily implies the materiality of teleology. In this formulation the possibility of things occurring, but things that have not occurred, is fused to the teleology of the result of those actions if they were to occur; the dichotomy being constructed is one between the non-particular and teleological action, one with some necessary result before it occurs, and the impossibility of any action.

The conceptualization of this empty space of action is also the attempt to fuse the materiality of a particular time and space with some transcendental concept of absolute presence and absolute absence in which all moments are portrayed as the expression of some magnitude of this transcendental totality of absence and presence. In other words, to the degree that possibility is presented as a total absence, moments can be described only as an isolated expression of a magnitude of an atemporal and transcendental total absence. Only then can one be more or less "free" or differentiate between "freedom" and "coercion," and this qualitative characterization is necessarily joined to teleology. What is posited here is this space of action that oscillates between total presence, the total impossibility of action, and an action that emerges from a total absence that also serves as both the possibility of action and teleology of action.

This total absence becomes framed as a dual presence. Firstly, it becomes the presence of itself; to the degree that it is spoken of it exists as a conceptual object and, as such, it is always a constancy of absence, an atemporal absence. Secondly, this absence always exists as an essentially isolated and teleological absence; an absence that operates to the degree that some presence proceeds along a teleological path. Therefore, action becomes possible to the degree that it forms a feedback loop that implies its own trajectory, and as such, exists separate from history and effects. This dual presence not only removes absence from itself, in the positing of absence as a conceptual presence, but also removes it from the temporal dynamics of the time and space of action. Just as in the attempt to speak of revolution as a necessary trajectory moving from that which has not occurred, the pre-revolutionary, to some teleological end point, the only possibility of the event is in the manifestation of this teleology. There is no particularity to this concept of action, no particular action with particular effects; all moments are removed into a trajectory of moments which proceed through time in a closed loop, or they cannot occur at all (Berlin 1969, 122-130).

This is the core of the negationist tendency, what Berlin calls "negative freedom", in which, paradoxically, absence becomes conflated with a political presence; one can claim to be for negation. We see this play itself out in some concentrations in the anarchist "movement," but only in some concentrations, in which transgression becomes an injunction, a paradoxical declaration that one must transgress or that one can fight for "total negation." In this trajectory of thought, the possibility of transgression and the actualization of transgression become fused in a qualitative equivalence of transgression and "being revolutionary." Revolution becomes a qualitative term immanent in actions depending on the level of transgression actualized. The possibility of the action of transgression and the actualization of transgression must fuse in a direct teleological relationship or revolution does not occur.

While "negative" concepts of possibility operate to the degree that we speak of this absence as the absence of dynamics, even though this collapses into a positivist transcendental absence conceptualized as an atemporal object, "positive" concepts operate in a more directly paradoxical form. In the "positive" concept of "freedom" we are confronted with the attempt to construct this concept of absence as a presence directly, in the attempt to construct "freedom" as a total material terrain of action through the definition of "freedom," an absence, through a qualifying term, such as "equality." We also see this play itself out in capitalist "libertarianism" and "rational choice theory" in which the impossibility of being able to speak of "society" in a total sense comes to somehow imply the existence of a world of atomized, selfish individualism that finds expression in capitalist markets. To quote Margaret Thatcher, "there is no society, there are just individual men and women;" there is an absence, but only to the degree that it implies this presence. Here, just as in the attempts of specific tendencies of platformist anarchism, this absence comes to imply a total political form, an absolute material presence._In this attempt total absence and total presence are articulated through a singular term. What we are being presented with is a bifurcation that collapses at the very moment of its being stated. While "negative" concepts operate to the degree that all dynamics are eliminated, "positive" concepts operate to the degree that the moment is constructed as a total absence that attempts to operate as a total presence.

While it may seem that, in this paradoxical collapse, a conceptual difference is not posited, and this is the common critique of Berlin's text, it clearly is. This collapse of the "negative" into the "positive," and the paradox of the attempt to construct total absence as total presence occur as a methodological paradox. To the degree that "freedom," or possibility, is presented as a total absence then it cannot also be presented as a conceptual "is," an object for conceptualization, and thus a presence, a thing that is made sense of through the atemporal concept. This paradox forms the first pole of this dual projection, the dual projection that forms the core of the dual move of the autodestruction of revolution. Here possibility is always already presented as a totally present absence, an absence which occurs as "freedom," that is removed from the particularity of time and space and becomes an atemporal object projected conceptually across time and space as a presence (Berlin 1969, 131-134).

The Operational Projection

However, this is only part of the story. Up to this point we have only been speaking of the paradoxical conceptual projection of possibility as a total presence, the presence of total absence, the negationist tendency, or absence as a total presence, the positivist tendency, both of which collapse into a sort of positivism. But, here we are dealing with a very material issue, the question of how the material possibility unleashed in revolution, in tactical rupture, becomes its opposite, becomes the operation of policing, and this is a very material question. The dual move only finds its possibility in conceptual projection, not its actualization. In other words, this is not a question of "bad" concepts, it is not a question of "correct" theory or false consciousness, as if there is such a thing. Rather this is the question of the attempt to operate a transcendental, atemporal, conceptual universe in materially particular moments.

In Berlin this presents itself as the "social" question, the problematic that presents itself when this conceptual totality attempts to operate as a total "unity" of time and space materially (1969, 145-154). The conceptual projection was based in the elimination of possibility in this conflation of total absence with total presence; but this is merely a conceptual process. To the degree that the particularity of temporal moments exist at a fundamental distance from the atemporality of concepts then this conceptual process does not imply an actual material removal of revolution from itself, or the operational projection of the second move. Rather, this operational attempt always already exists as the attempt to span this infinite distance and operate this total presence materially. We see this in the very attempt to operate the Party Program, in the operational assertions of Kropotkin, in the Terror that was meant to "preserve freedom," in the Red Terror that was meant to "liberate" the working class.

In this attempt we come into contact with the very possibility of the organizational logistics of policing as a material operation, the attempt to operate an atemporal conceptual unity through particular material actions. In this operational projection two things occur. Firstly, there is an attempt to materially eliminate the particularity of temporal moments through the insertion of these moments into some transcendental presence or conceptual framework. We see this, again, in Leninism in the concept of the process of the transition to communism but also in constructions of revolution as expressive of some form of human essence or absolute "freedom;" this is also the condition of possibility for the state itself, the impossible attempt to construct a materially unified terrain. Secondly, this only occurs to the degree that a paradoxical material logistics is mobilized in particular moments. On one hand there is the attempt to construct a materialization of a non-particular action, but this can only be done, on the other hand, through particular actions. In other words, it is not that the Terror was one action, it was a collection of actions, all of which occurred in particular moments, that are conceived of as the Terror; but it is not that we can say that the Terror was the actual materialization of the ideals of "defending the revolution" in some unitary sense.

The dual move can never exist in just the conceptual removal of revolution from itself in the atemporal concept. Rather it exists in the impossible attempt to materialize this removal, to end revolution and the particularity of tactical conflict in some materially total, and materially impossible, non-particular and atemporal political form which always already exists to the degree that it operates in the temporal and particular. It is this paradox that constructs both the autodestruction of revolution in policing and the impossibility, and total crisis, of policing. We watched this play out live in Libya and Egypt where the formalized Western-backed state crushed street demonstrations and disarmed militias, usually by the force of "official" police operations, in order to attempt to operationalize some atemporal and total political form, but this attempt to construct a unified terrain failed dramatically. It is this material attempt to operate a material unity of time and space, which finds its possibility in the conceptual process of the removal of revolution from temporal particularity, that forms the materialization of this second move; this is the process of the Bastille becoming the Terror, the Russian Revolution becoming the attack on Kronstadt.

Whither Theory?

This disjunction, between the temporal moments that we term revolution and the atemporal concept of revolution, repositions the question. It is no longer a question of speaking about what revolution "is." The question is, rather, what we are doing when we think revolution or when we do theory. To the degree that we are positing an infinite distance between theory and particular moments we cannot also posit that the dual move of the autodestruction of revolution is a conceptual problem. It is not as simple as saying that theory is a plague, that the very attempt to conceptualize is cancerous; theory itself is the very possibility of saying "I," of making sense of the world, or the collection of phenomena that we term "the world." Rather, the task here is to try to reconceptualize the connection of theory to revolution, to rethink the role of thought in an oh-so-material, temporal, tactical collision of the dynamics of actions in a particular time and space. This is a question of the effects of theory as an action, as something that occurs; this is the question Sorel approaches in *Reflections on Violence*.

At the heart of the revolutionary project, spoken of as a conceptual whole, is conflict. The paradox, the tragedy, is the attempt to fuse the concept of conflict and conflict itself; the tragedy is always found in the attempt to remove conflict from itself, to remove the temporal particularity of a moment from itself. This occurs in abstract discussions of the "correct moment" for action or essentialized concepts of the "correct action," the fetishizing of property destruction as well as dogmatic pacifism, which removes action and its effects from the particular tactical dynamics in which it occurs. In this removal of conflict from itself, all that tactics serve to do is realize themselves outside of tactics, and this realization always removes tactics from the particular and temporal dynamics of conflict. We see this in pacifism where there are no dynamics of actions and tactics, there is only the isolated action that is "judged" within in some transcendental conceptual, atemporal, morality of action; actions become "right" and "wrong" but the very dynamic of its possibility and effects is completely obscured in this abstraction. However, this only operates materially to the degree that pacifists hand anarchists over to the police, to the degree that they become the police themselves, and we watched this play out in the anti-anarchist witch-hunts of the American antiwar and Occupy movements. In this form of movement policing, commonly called marshaling, the "plan of action," decided in some other moment by a particular group of people, comes to operate only to the degree that all other contingencies are eliminated, thus attempting to eliminate the particular dynamics of a particular moment of action and removing the possibilities of action from action itself (Sorel 2004, 65-66).

But, it is in Sorel's notion of the general strike that we can begin to glimpse another possibility for theory. "Every time that we attempt to obtain an exact conception of the ideas behind proletarian violence we are forced to go back to the notion of the general strike" (Sorel 2004, 119). At first glance this may seem like some attempt to combine particular conflict with some concept of conflict embodied in the general strike, but this would imply that "the conception of the ideas behind proletarian violence" and "proletarian violence" are fused. It is telling that the phrasing is so specific here; it is not the claim that proletarian violence and the general strike are the same phenomena, that there is some non-particular proletarian violence that is subsumed into the general strike, and that this is expressed in the "notion of the general strike:" Rather, all that is being claimed is that the concepts behind proletarian violence can be connected to the concept of the general strike, not that this in any way expresses conflict but is only a conceptual term for conflict (Sorel 2004, 120-123).

The term "general strike" exists as a form of speaking of conflict, of "proletarian violence" but not as equivalent to the particularity of actual proletarian violence. As removed from the actuality of proletarian violence the concept of the general strike becomes separate from actual proletarian violence and can, as such, exist as an expression of a notion, and thus an action of theory in itself. We see this in a certain thread in Sorel, the discussion of theory as an action, and thus an effectiveness separate from the attempt to exhaust the materiality of its conceptual objects. "The myth must be judged as a means of acting on the present" (Sorel 2004, 126).

It is not that conflict becomes infused with some conceptual core outside of conflict, as it does in pacifism, or embedded in some teleology, as it does in the conflation of revolution and political form. It is nothing more than the conceptual mythology of conflict that can never speak of conflict but does not attempt to speak of an "outside" or "end" of conflict. The conceptualization of revolution does not necessarily imply the attempt to operate this atemporal conceptual totality in materially particular moments. Rather, theory can be a means of "acting on the present," but only to the degree that there is a recognition of the infinite distance between the particular dynamics of "the present" and the concept of the present.

There is always this tendency in radical circles to speak of prefigurative politics, and in this we are coming to engage in this fusion of the dual move once again. In attempting to materialize atemporal concepts in material moments and inserting this fusion into a teleology, we are obscuring this gap that Sorel is so careful to posit between the "ideas behind proletarian violence" and proletarian violence. Though this attempt, to materialize the world we want to see, seems like a laudable goal, we are setting up the revolutionary project for one more failure. This removal of the moment from its temporal particularity, both into some direct relationship with theory and also into teleology, is nothing but the condition of possibility for this failure. To avoid this failure is nothing less than to abandon this idea of the prefigurative and correct form of politics, to separate revolution from teleology to be able to engage in revolution once again.

Once theory is removed from the attempt to speak of the world in direct ways, as well as the absurd attempt to speak of the future, then we can begin to relate to theory as something that occurs, and thus, as something that can be more or less effective. The removal of theory from the operational assertion and the assertion of its materiality allows us to approach theory as another tactic in the attempt to amplify conflict, in the attempt to generate this organizational rupture in policing through an amplification of conflict and crisis. In this we can realize something that theory has seemingly lost. Theory can be dangerous again, it can participate in the dynamic of amplification that we call revolution, but only to the degree that we abandon the idea of thinking revolution in a direct and teleological way. We, as theorists, must come to reject the role of the "scientists" of revolution, and come to participate in it once again.

Works Cited

Berlin, Isaiah. 1969. *Four Essays on Liberty*. Oxford: Oxford University Press.

Sorel, Georges. 2004. *Reflections on Violence*. Translated by T.E. Hulme. Mineola: Dover Publications, Inc.

Zamani to *Sasa*: Conceptualizing African Time and Revolution

Mickie Mwanzia Koster

Introduction

A clock beats for all revolutions. Although the beats vary, urgency is one common factor among revolutionists of social and political movements; they all express in some fashion or form that "it is time" or "the time is now." Time, especially as it relates to revolutionary moments, is a complex, dynamic, and spiritual phenomenon. The cultural analysis of time is even more challenging. In particular, scholars struggle with African notions of time. For example, Claudia Zaslavsky (1973, 62-63) treats this topic directly in her chapter, "The African Concept of Time," showing the African nuances of natural divisions of time, man-made divisions of time, age grades and oral traditions. African time differs strikingly from Western linear time, an ideology measured in relationship to the future. On the contrary, African time emphasizes the past and present, placing little value on the future, especially the intangible distant future (Mbiti 1969, 15-28). While African time emphasizes the past and present, it does so with the inherent goal of transforming the future. Likewise, revolutions envision a future different from the present. This work provides an insight into the complex relationship between time and revolution by providing a conceptual framework for analysis based on a specific revolutionary moment and time, the Mau Mau revolution from 1952-1960.

The Mau Mau revolution erupted in response to Britain's colonization of Kenya and represented an awakening in which Kenyans confronted the contradictions of British modernity and found a solution in the ways of their ancestors. The revolution resulted in the loss of thousands of African lives, including those questionably sentenced to hanging, and the detention of nearly a quarter of a million Africans (Elkins 2005). During Mau Mau, the oath was the power behind the revolution, providing members with the ability to traverse time while invoking ancestors and spirits, uniting fighters, and reminding participants of their civic and moral responsibilities (Hudson-Koster 2010). The ceremonies, when carefully examined, show that through symbolism and traditions, portals were opened, reconnecting members to the past and ways of the elders, while strengthening their collective commitment to the revolutionary struggle. This interpretation is important, especially as we consider the resurgence of revolutionary activity. In addition, the Mau Mau revolution was a particularly violent struggle that still occupies the attention of many scholars endeavoring to understand how Africans were able to shake the "happy valley" (Berman and Lonsdale 1992) of the British Empire (Anderson 2005; Elkins 2005). The Mau Mau revolution stood for hope and a time for change.

Both time and revolution are difficult terms to define because they are variable, conditional, and situational. Newell Booth Jr. questions whether "time moves or whether we move through time" (Booth 1975, 81). This study focuses on understanding the nature of time and revolution as they relate to each other under broad circumstances and during specific social and political moments, revealing that time and revolution are not fixed but respond and move. Using a combination of interviews from ex-Mau Mau fighters, archived data, and existing scholarship, this work argues that conceptual ritual interpretations are needed to represent the complex relationship and varied dimensions of time and revolution. Employing a framework enables us to examine the internally complex moment when fighters oathed to join the movement. The rebels collectively decided that it was time to stand and fight for freedom.

Exploring Time, Revolution, and the Oath

Modeling the relationship between oathing, time, and revolution reveals its dynamically complex nature. The relationship can be briefly examined from three perspectives. The first perspective, time and revolution as a continuum, represents time and revolution broadly and encompasses all of time and all revolutions past, present, and future. This perspective connects all revolutions over time, while also acknowledging their specific revolutionary circumstances such as the time, type of grievances, and the outcome of the revolution.

The second perspective, time and revolution Mau Mau, is specific to the Mau Mau revolution. Although there are many possible occurrences of time-revolution, this phase represents a specific relationship that connects the revolution, "Mau Mau," with the time, "1952-1960." Here, time is linked to the notion of *sasa*, which means "now" in Kiswahili. This is not a new term, but one that has been used by scholars to represent the present notion of time (Mbiti 1969, 21-22). Time, linked with *sasa*, is designed to convey the urgency of the revolution at the particular moment. The connection can be experienced in the context of a specific revolution, such as Mau Mau, but it also represents a connection to other possible revolutionary activities that took place during the same period of the 1950s and 1960s. This interpretation is important, as revolutions often occur in waves—one setting off or immediately following another.

The third perspective, time and revolution Mau Mau oath, is probably the most revealing yet least understood, partly due to the complex and dynamic system of relationships between time, place, oath type, role, statements, symbols, symbol type, curse, and justice. These elements are discussed below in detail. When the Mau Mau participants took or administered an oath, all of these elements were employed and operated together as a system. Time is instrumental in two dominant ways: as a marker to help define the oath and as a link to *zamani*, the Ancestors, and old traditions. According to Mbiti (1969, 22), *zamani* is a time element experienced by African people as a connection to the ancient past, the traditions of a long time ago, and the ways of African ancestors. In Kiswahili, *zamani* means a long time ago. This dimension of time is important because it shows that although the fighters took their oaths and pledges to respond to the urgency of the social, political, and economic issues in Kenya due to British domination, the ritual process connected them to the past. The relational nature of time is highlighted in this perspective.

The Mau Mau oath is specific to a time and undergirds a specific revolution. Its uniqueness is also characterized by the essential elements listed above. The date and time of each oath were significant, carefully considered, and brought attention to specific events that may have impacted oathing activities and needs of the movement. While the peak of the Mau Mau movement was between 1948 and 1954, the oath changed at times in response to the needs and conditions of the war. For example, participants took additional, more specialized, oaths in the later years as the war intensified (Kershaw 1997, 318).

Oathing ceremonies were also distinct based on their location. Oaths took place in very distinct urban or rural settings in specific districts and sub-locations. If a participant took more than one oath, each may have taken place in a different place. Symbolism and secrecy determined location. For example, oathing ceremonies in the forest possessed a certain level of intensity and mystery that was different from the oathing ceremonies in houses or in rooms. The forest location contains embedded symbolism as the place of the dead, unknown spirits, and obscurity. Thus, the forests were a sacred refuge space because it symbolized a place of isolation, a place of the unknown, danger, evil, and a place of mystical power. Oathing ceremonies in this setting were close to the natural and spiritual world.

The third element is the type of oath, which changed based on the period. Mau Mau oathing was distinct from oaths prior to 1950 because it often involved multiple oaths related to the movement. All participants underwent, at a minimum, the basic oath called the first Mau Mau oath. The Mau Mau oath was a tailored oath of resistance designed as an African response to colonial economic, social, and political injustices. The oath was a blend of the past with the urgency of the present that speaks to a relationship between time and revolution; it encompassed truth, secrecy, and unity. The first oath was vital as it was created "to guarantee the allegiance of its members . . . it demanded strict secrecy as well as total commitment" (Barnett and Njama 1966, 55). It was also referred to as the "muma wa uigoano" or the oath of the secret movement because of the danger and violence required to challenge British authority (Clough 1998, 97).

The overall structure and purpose of the Mau Mau oath was to establish a unified body that was in alignment with the goal of removing Europeans from Kenya. When questioned about the meaning of the Mau Mau oath and revolution, ex-Mau Mau respondents stated, "It meant unity in action" (Kakie 2009). The cohesiveness and energy gave a special power to the Mau Mau oath as a journey of participants over time and space. Secondary literature points to specific warrior oaths called batuni and muma wa ngero (the oath for killing) (Clough 1998, 109) that were noted in Kikuyu oathing practices as a second Mau Mau oath type. The second oath was described as oathing based on extreme conditions. These oaths were designed to prepare the select group of fighters to face death. The purpose of this warrior dimension of the oath was to strengthen group loyalty and allegiance while fighting. The additional oaths were usually conducted to promote the first or to strengthen a particular aspect of upcoming Mau Mau activities. The process was designed to show that these subsequent oaths were just as complicated and meaningful as the first Mau Mau oath.

Assuming specific oath roles was another critical component of the ritual. All Mau Mau participants played at least one role in oathing ceremonies that included the oath taker, the oath administrator, and the guard (Musuo 2009). The oath taker was the dominant character of the ceremony, responsible for repeating dictated words, statements, gestures, and acts. The oath taker followed the instructions of the oathing administrator who orchestrated all of the oathing activities and ceremony (Kenya National Archives 1953, 1954) and was typically an old or mature man (Malei 2009). The oath administrator was also the spiritual orchestrator. He helped move Mau Mau members in and out of different time ports spiritually to initiate them into the movement. The administrator was very keen on the meanings, purpose, and oathing process; he or she aimed to successfully transition the oath taker into being a full Mau Mau fighter who pledged to fight colonial injustice. This leader customized the oathing process based on the materials, artifacts, and resources available at the time. There was not one standard oath process because the administrator decided on the vows, the gestures and symbolic acts to follow, the duration of activities, and all of the other activities. One of the best ways to understand this role is to compare it to the current role of specialists in Kenya. The specialist is an individual knowledgeable about spiritual energy, traditional culture, and the invoking process (Munguti 2009). The different roles help reveal the interactions and responsibilities during the oathing process.

One of the most important aspects connecting the fighters to the *zamani* was oath statements. All oath ceremonies had oathing statements, vows, and utterances. The collection of oathing statements included short phrases, words, affirmations, and even songs that were used to move Mau Mau participants into full Mau Mau members. Oathing could consist of many different oathing statements. The vocal utterances of words, language, and actions were a key component of the oathing ceremony. The oath or oath spirits heard the words spoken by the oather.

Not all Mau Mau participants vowed the same oath statements. Generally, the statements were different based on all of the conditions that went into the Mau Mau oath experience. Some oathing statements were brief and others long; some included chants and singing, while others did not. However, all oaths did appear to be consistent in their meaning and purpose—the need for revolutionary change, liberation, and African unification crossing ethnicity, class, gender, religion, and age. Oathing administrators knew the core elements needed in the statements to ensure unity, secrecy, and allegiance to Mau Mau. Again, the oath administrator had flexibility in selecting the sequence, the accents, and the words as he was guided spiritually to know what would invoke the oath. One of the most important aspects of oathing statements was that they were being heard by the Supreme Being who was witness to the statements.

In several interviews, participants were questioned about the general meaning of oathing compared to ethnic specific meanings (Berman and Lonsdale 1992, 316-317). Generally, oathing was exercised to end colonial rule, to liberate people, to self-rule, to gain freedom, to unite, to retrieve land, and to seek independence (Kakie, Malei, Mulwa, Musuo, Wambua 2009). However, for some ethnic groups, the oath also meant it was time to retrieve their livestock and property, to receive equal wages, to expel the Europeans from Kenya, and to end European beatings so that they were able to return to a time of peace (Kakie, Malei, Mulwa, Musuo, Wambua 2009). The statements have multiple time dimensions; they signify a return to the traditions and beliefs of the past, while emphasizing the urgency of addressing present grievances.

Symbols helped shape the oath experience because they expressed the objectives and values of the ceremony (Draper 2000, 125). Victor Turner defines a symbol as "a thing regarded by general consent as naturally typifying or representing or recalling something by possession of analogous qualities or by association in fact or thought" (Turner 1967, 19). Turner discusses two types of symbols, dominant and instrumental. However, for the purpose of this time and revolution study, we will focus only on a few symbols that help show how fighters and oath administrators used symbols.

Many different symbols were used during the revolution. This section briefly highlights blood, banana leaves, and numbers as examples of how symbols were used to strengthen the movement. Blood was and is a powerful symbol type. The application of blood in the Mau Mau oath was a common symbolic gesture. The blood represented life and death (Zahan 1979, 36-52) in that oathing participants moved through time and space as they entered into a new life. Through the use of blood, the initiates underwent a rebirthing or initiation process. The ritual opened up new time and space dimensions, connecting to the long ago (*zamani*) past to foster present unity and allegiance. The relationships to death are seen on the individual and movement level. On the individual level, death was connected to the pre-oathed state of the person in that the old individual dies and will no longer exist. This concept is similar to the traditional initiation of young men and women. Death was also associated with the individual in the form of a curse for breaking oathing vows and revealing the secrets of Mau Mau. On the Mau Mau organization level, blood was symbolic of the eventual death of the Mau Mau movement resulting in the birth of a new Kenyan nation with the stolen lands being restored (Larson 1972).

Likewise, banana leaves were also an instrumental symbol in oathing. During the Mau Mau revolution, the leaves were used to form an arch symbolic of a gate through which participants went, representing a door to spiritual transformation. The physical movement through the banana arch represented the spiritual newness of the individual while also moving the individual through time in that the individual experienced a spiritual reconnection through time to traditions and ways when the society was at peace. The application of the banana leaves usually involved walking through the arch seven times (with seven as a symbolic number) giving it added meaning. In African culture, numbers like seven carried special spiritual power based on traditional knowledge. Thus, walking through the arch seven times was physical, but it was also spiritual in its ability to create a spiritual opening allowing participants to journey across time and space. The act allowed participants to remember and connect to the ways of the ancestors and the values of the people as a source of strength and renewal. All revolutions have embedded within their structure different types of symbols that are used to bond members and connect them to varied dimensions of time, present or past.

One of the most significant and powerful facets of the revolution was that membership was bound by a deadly curse. The oath curse is prevalent in most East African cultures. The curse, like the oath, always existed; it was known, feared, and used as a means for ensuring societal order and justice (Ueda 1977). As a result of this tradition, the ritual acts were sacred, and the curse was revered. A failure to abide by the oath statements meant severe punishment. For example, consider the following oathing statement from one oather, J.M Wambua: "I will not reveal any secrets to any government officials. If I do, I should die. I will never report Mau Mau to the government. If I do, I should die" (Wambua 2009). This spiritual curse was anchored in the ways of the ancestors as a means for establishing justice (Dundas 1913, 510-514). The oath worked because the oathers truly believed in the power of the statements and vows. The curse was designed to persecute violators: a traditional form of justice summoned from the past to support the Mau Mau revolution. The cursing nature of the oath was timeless; it served as a power binding force in the revolution. Oath curses are still feared in Kenya today.

Conclusion

The time, revolution, and oath model represents a complex system that was fluid, dynamic, circumstantial, and highly customized. The revolutionary moment captured during the oath was based on the time, place, oather, oath administrator, symbols or artifacts available, the location of the ceremony, statements, and deadly curses. It was a very elaborate system full of meaning, structure, memory, dimension, and circumstance. As a result, this study argues that conceptual ritual interpretations are needed to represent and capture the relationship between time and revolution. The framework presented in this study, although rooted in the case of the Mau Mau revolution, represents a few of the possible interpretations of time and revolution.

During revolutions, African rebels protested, remembering a glorious or desirable past and time of peace, harmony, stability, and righteousness. In this study of the Mau Mau, fighters joining the revolution found strength in embracing African culture. Mau Mau candidates experienced an initiation oath ritual that took them on a journey through time to reconnect them with the traditions and ways of the ancestors and to remind them of the ancient *zamani* ways of justice through spiritual curses as outlined in this study. These oath practices were invoked as a hidden source of power for fighters to embrace the *sasa* moment of the struggle. The relationship between time and revolution connects to a larger liberation narrative that is still being shaped, which is evident based on the resistance fervor in Africa since 2011 (Koster 2011). The revolts should not be limited to the activities of Tunisia and Egypt heavily covered by Western media, but should include the protests and actions pervasive in the continent during this revolutionary time (Ekine 2012, 2). Thus, the treatment of time and revolution as depicted in this study offers a subaltern and relevant African interpretation of power used as a system for social, economic, and political change. There is still much more to uncover in this time and revolution story; after all, the struggle for freedom is never really over.

Works Cited

Anderson, David. 2005. *Histories of the Hanged: Britain's Dirty War in Kenya and the End of Empire.* London: Weidenfeld and Nelson.

Barnett, Donald L. and Karari Njama. 1966. *Mau Mau from Within: Autobiography and Analysis of Kenya's Peasant Revolt.* Dublin: MacGibbon and Kee.

Berman, Bruce and John Londsdale. 1992. *Unhappy Valley: Conflict in Kenya and Africa*, Books One and Two. Oxford: James Currey.

Booth, Newell. 1975. "Time and Change in African Traditional Thought." *Journal of Religion in Africa* 7:81-91.

Clough, Marshall S. 1998. *Mau Mau Memoirs: History, Memory and Politics*. Boulder: Lynne Rienner Publishers.

Draper, Jonathan A. 2000. "Ritual Process and Ritual Symbol in 'Didache' 7-10." *Vigiliae Christianea* 54 (2): 121-158.

Dundas, Charles. July-December 1913. "History of Kitui." *Journal of the Royal Anthropological Institute of Great Britain and Ireland* 43: 480-549.

Ekine, Sokari. 2012. *African Awakening: The Emerging Revolutions*. Cape Town: Pambazuka Press.

Elkins, Caroline. 2005. *Imperial Reckoning: The Untold Story of Britain's Gulag in Kenya.* New York: Henry Holt and Company, LLC.

Hudson-Koster, Mickie. 2010. "The Making of Mau Mau: The Power of the Oath." PhD Dissertation, Rice University.

Kakie, S. [ex-Mau Mau]. 2009. Interview by M. Musyimi. Tape recording and notes. February 2009. Machakos District.

Kenya National Archives. 1953, 1954. Ministry of Legal Affairs 1/791.

Kershaw, Greet. 1997. *Mau Mau from Below.* Ohio: Ohio University Press.

Malei, J. M. [ex-Mau Mau]. 2009. Interview by M. Musyimi. Tape recording and notes. February. Machakos District.

Mbiti, John. 1969. *African Religions and Philosophy.* London: Heinemann.

Mulwa, K. [ex-Mau Mau]. 2009. Interview by M. Musyimi. Tape recording and notes. February. Machakos District.

Munguti, M. N. S. [ex-Mau Mau]. 2009. Interview by M. Musyimi. Tape recording and notes. February. Machakos District.

Musuo, P. [ex-Mau Mau]. 2009. Interview by M. Musyimi. Tape recording and notes. February. Machakos District.

Turner, Victor. 1967. *The Forest of Symbols: Aspects of Ndembu Turner.* London: Cornell University Press.

Ueda, Hitoshi. 1977. "Kithitu among the Kamba of Kenya – The Case Study of Kilonzo's Kithitu." *Journal of African Studies*, Japanese Association for African Studies, 16 (March): 23-55.

Wambua, J.M. [ex-Mau Mau]. 2009. Interview by M. Musyimi. Tape recording and notes. February. Machakos District.

Zahan, Dominique. 1979. *The Religion, Spirituality, and Thought of Traditional Africa.* Chicago: The University of Chicago Press.

Zaslavsky, Claudia. 1973. *Africa Counts.* Boston: Pindle, Weber and Schmidt, Inc.

MODERNITY, REVOLUTION, SILENCE: ON SCHMITT, ADORNO, AND DERRIDA

LEWIS LEVENBERG

When confronted with exceptional ruptures in the flows of modernity, our conceptions of regularity, temporality, and reason come under fire. Carl Schmitt (1888-1985), Theodor Adorno (1903-1969), and Jacques Derrida (1930-2004) all argued that such exceptions transform into the very rules of rationality and politics. A genealogy of alterity, aesthetics, and futurity in these three theorists' work demonstrates the emergence, articulation, and continuity of a totalizing philosophical discourse of modernity. Radical shifts in historical perspective can only be thought possible at the limits of this discourse. In the recognition of an existential enemy, or of the dissolution of rationality into technicality, or of the inscription of history on oneself, revolutions exceed metaphorical, punctual occasions for contemplation. Struggle and bloodshed, especially over ideas, demand our honest attention to persistent silences.

Eminent philosophers have dealt with certain aspects of the relationships between Schmitt, Adorno, and Derrida. But none have placed all three of them at once into conversation, or suggested that the non-relation between Schmitt and Adorno even indirectly influenced Derrida's decisions to study Schmitt in depth, or to largely avoid discussing Adorno until late in his life (Agamben 1998; Mouffe 1999; Banisch 2002; Boersma 2005; Agamben 2005; Moazzam-Doulat 2008; Huysmans 2008).

History, too, shows more division than potential interaction between Schmitt, Adorno, and Derrida. The rise of Hitler and National Socialism was Schmitt's ticket into professional academic success. His time in Berlin lasted until the fall of the third Reich in 1945, at which point he was sent to an internment camp, and from there returned to his hometown of Plettenberg, until his death in 1985. Adorno, meanwhile, had his academic credentials revoked by the Nazi regime, and left the country until after the end of the third Reich. He then wrote and taught in Frankfurt until his death in 1969. During the twilight of these two careers, Derrida's was on the rise in France. He published three masterworks in 1967. While teaching in Paris during the 1990s, he wrote two books that deal directly with Schmitt, and when he won the Adorno Prize in 2001, he finally publicly remarked on Adorno's work (Derrida 2005). The intersection of their thought, though a challenge to describe, affords contemporary theoreticians a serious opportunity to address political and temporal problems.

The motivating contradiction of the conjunction between capitalism and liberalism makes the question of modernity continue to resonate. Capitalism's rise depends upon the machination of a market unfettered by law or national boundaries, while a modern liberal state must depend upon its independence from concerns of capital. So the two regimes, in principle, decide not to interfere with one another, to remain parallel. However, they remain deeply embedded in one another. The obscured or disavowed connection between the economic and the political realms forms its contradictory basis for development and progress. On the same basis, radical disruptions of that linear march—revolutions—have no place in the scheme, either historically or conceptually.

By the same token, argues Adorno, philosophy has developed, since the Enlightenment's gathering together of capitalism and liberalism, along contradictory lines (Adorno and Horkheimer 1947). The formal, rational systems, upon which modern thought has depended, have also combined epistemological concerns with technical matters. In its continual and accelerating technicality, rationality itself becomes irrational. Its dialectical revelation of itself as its absolute other is marked, for Adorno, by the rise of the culture industry, the domination of late capitalism, and the emergence of totalitarianism, especially in the form of German National Socialism. These "limits" to Enlightenment, what he calls the "pure contradiction between the forces and the relations of production," express a potential radical breaking point for modernity. But events in the world cannot lead to transcendence of the world; no event can serve as a determinate negation to the forces and relations that produced it (Adorno [1966] 1990). Even contradictions must give rise to further developments in the material world. Unforeseen consequences of worldly negations certainly occur, but none of them afford such a radical break in either philosophy or world history. Revolution remains speculative while negation avoids determination.

Adorno never adopts Schmitt's works or themes as central tutor-texts, but he demonstrates an indirect conceptual engagement. The clearest indication that he is aware of Schmitt, as both a political philosopher and in his historical context, comes in *Minima Moralia*, concluding an essay that critiques "impoverishment of the relation to others" by way of extreme visual scrutiny of all others:

> The New Testament words, 'He who is not for me is against me', lay bare the heart of Anti-Semitism down the centuries. It is a basic feature of domination that everyone who does not identify with it is consigned for mere difference to the enemy camp: it is no accident that catholicism is the Greek word for the Latin totality, which the National Socialists have realized. It means the equation of the dissimilar, whether it be the 'deviationist' or the members of a different race, with the opponent. In this respect National Socialism has attained to historical consciousness of itself: Carl Schmitt defined the very essence of politics by the categories of friend and enemy. Progress to such consciousness makes its own regression to the behaviour patterns of the child, which either likes things or fears them. The *a priori* reduction to the friend-enemy relationship is one of the primal phenomena of the new anthropology. Freedom would be not to choose between black and white but to abjure such prescribed choices. (Adorno [1951] 1974, 131-2)

Adorno assumes that his readers know that Schmitt was an excommunicated Catholic, as he links Schmitt's work to the Nazi regime with which it was so deeply intertwined. Ironically, Schmitt's reliance on Nazi support for his academic career meant that he had to endure investigations and harassment by those benefactors, precisely on account of his previous Catholicism, repudiated after his divorce from his first wife. More importantly, Adorno here addresses an earlier claim by Schmitt that "all significant concepts of the theory of the modern state are secularized theological concepts" (Schmitt [1922] 2005, 49). Schmitt places the burden of politics, its existential ground, on a collective agency. He gives political credence to the entity (usually but not exclusively a State) that can decide who is a friend and who is an enemy (Schmitt [1932] 2007). In this way, the thought of politics always becomes the thought of alterity; there can be no world-state in Schmitt's view, since that would eliminate enmity, hence politics, hence the state itself.

What Adorno calls "the equation of the dissimilar . . . with the opponent" elides an important distinction in Schmitt's argument, between existential "enemies," whose recognition as such is the genesis of politics, and ideological "foes," with whom one merely disagrees or quarrels, but who do not threaten one's (group's) life. Indeed, Adorno concedes the "historical consciousness" of Nazism. However, that consciousness is carefully qualified: Adorno notes that it is only a self-recognition, a consciousness "of itself." This indicates that the all-important alterity demanded by Schmitt's decision-making capacity has embraced not only all others, but Nazism itself: Nazism, in other words, recognizes itself in history because it has become unrecognizable to itself.

The Hegelian strands that link Schmitt and Adorno now give way to Kantian terms. Their use indicates Adorno's mistrust of using material (in this case, visual) criteria to draw metaphysical conclusions. Having already noted the highly visual process of distinguishing "others" in Nazi self-consciousness, he also highlights a central contradiction of that process: its "progress," he writes, "makes its own regression."[1] The childish, "primal" grounds of Schmitt's politics become not only metaphysical and regressive, but tautological, because by relying on the recognition of enmity, State and Law justify their own existence as the only entities capable of such a distinction. Finally, Adorno's invocation of "freedom" in response to Schmitt's authoritarianism implies that political formations, like any other relation of production, cannot be considered independently of their actions.

Other indications that Adorno reads and responds indirectly to Schmitt occur in two other works from around the same time. Schmitt argues in his earlier works that sovereignty is not a quality invested in a leader, but the function of a "state of exception" from constitutional bureaucracy, because in the state of exception, the State gains the ability to make existential decisions.[2] For Adorno, the justification of and for a State is not its capacity to decide who its enemies are, but how it acts towards those enemies. In *The Authoritarian Personality* (1951), Adorno and his co-authors bring to bear both the psychoanalytic and the materialist critiques sketched here. When social conditions allow one leader to take a position as a sole sovereign or tyrant, that person must on the one hand come to terms with their radical fungibility, their ability to be replaced by any other single person with compatible personality traits. On the other hand, they must allow and encourage the identification of each person in the polis with their singular subjectivity.

Finally, in "Freudian Theory and the Pattern of Fascist Propaganda," also composed in 1951, Adorno addresses the mass-psychological conditions of possibility for a totalitarian state. He reiterates the regressive, explicitly oral character of the call for each member of the state to identify with the leader. He also compares the production of fascist propaganda to the mechanisms by which the culture industry, which he and Horkheimer

---

[1] Cf. Adorno's arguments, with Horkheimer, that pleasure becomes regressive, the culture industry provides only an illusion of choice, and that technical rationality leads on to totalitarianism and genocide, in *Dialectic of Enlightenment*, 1947.

[2] Cf. Foucault's (1976) conception of biopower as the existential stakes of political decisions, as against Agamben's rather more popular adoption of Schmitt's term.

described in *Dialectic of Enlightenment* (1947), claims to provide choices among cultural products while actually limiting them. Throughout the essay, Adorno refers to fascism's eradication of individual psychological freedom, that is, the individual's autonomy and ability to call upon their own superego, in favor of a collective "social hypnosis," mobilized precisely through the repetitive stirring of those personal psychological agencies on a massive scale. Adorno's critique of fascist propaganda as inverted social psychology carries with it his argument that Schmitt's intellectual production constitutes an inverted, metaphysical, and regressive Hegelian dialectic.

Other secondary analyses of the theoretical relationship between Adorno and Schmitt also dwell on the significance of a largely obscured or implied response. William Winstead (2010) argues that in the work of the Frankfurt School, and especially of Adorno, Schmitt's concept of the enemy as the existential grounds of politics takes on a utopian function. The "friend" that Schmitt posits by negation becomes the very condition of possibility for community. A.G. Duttman (2006) takes a similar stance, placing Adorno and Schmitt into a negative relation that, he argues, characterizes the problems of political theory from both the Left and the Right. Enough is preserved to directly refute Rolf Tiedemann's claim, "that Adorno was not interested in Schmitt and knew very little about him" (Adorno 2006, 325-6). In fact, it appears that Adorno not only understood in great detail how Schmitt thought, but took multiple opportunities to respond to him.

The very sovereignty that Schmitt theorizes as the function of the state of exception occurs, for Adorno, alongside the emergence of a totalizing and totalitarian culture industry. That industry turns potentially political subjects, with the autonomy and rationality to recognize alterity and even enmity, into objectified and commodified parcels. These reconfigure themselves to construct a more efficient consumer base for that very industry, precisely because the conjunction of sovereign and industry ameliorate their immediate material conditions. This blinds the population to the reality of an irrational totalitarianism, and convinces them that such conditions make their lives better, even while it leads them to choose the systemic domination that impoverishes their ability to recognize that domination as an existential threat, that is, as their own enemy.

In short, Adorno corrects Schmitt's contention that a political entity arises at the moment of recognition of its enemy, by *including* the concept of differences between individual and collective political agency. He shows how that multiplicity of alterities narrows into the apparently insuperable choice between totalitarianism or death.

Adorno's refusal to engage with Schmitt's literary and dramatic theories of aesthetics indicates nothing so much as a point of similarity. The inversion to remember, however, takes place not at the level of interpretation, but at the level of agency. While Schmitt, particularly in *Hamlet oder Hekuba* ([1956] 2009), considers a work of art on the merits of its internal temporal coherence and its reflection of the intention of its producer, Adorno approaches works of art based on their affective power (Adorno [1955] 1967, [1970] 1988). He traces not how the work appears, but how it operates in relation to its mode of production and to its effect on the beholder. In this way, the aesthetic differences between Adorno and Schmitt mirror their conceptions of politics, because while Schmitt justifies art tautologically, Adorno approaches art materially.

Comparing their concepts of futurity yields further significance. Schmitt imagines the permanence of the state of exception, and the subsequent normalizing of sovereignty and authority, as the marker of entrance into futurity. Thus, he justifies the Third Reich's suspension of constitutional rule as one transcendent moment in the punctuated fabric of history (Schmitt 1934). Conversely, Adorno, while retaining a certain mystical messianism in his scattered calls for a communist, classless society, and for the revolution towards that end, remains materially grounded in his conception of the negations of traditional social formations. In other words, he counters Schmitt's notion of a state of exception that justifies authoritarian sovereignty with the argument that—even during revolution—history continues apace, not punctuated but continuously (Adorno [1966] 1990). In this way, even the limits of Enlightenment or modernity do not indicate an end to history.

The lack of historically verifiable interaction between the persons of Schmitt and Adorno forces this analysis to dwell on the textual and conceptual interactions between them. The silence of history makes possible nothing more and nothing less than a possibility. That possibility, of engagement, provides the point of transition to think through Derrida's position in this genealogy.

Derrida's interest in Schmitt earned him some antipathy in the 1990s. However, a closer look shows that Derrida's readings of Schmitt end neither in nihilism nor in a justification for authoritarianism, but raise a more interesting political question than either. Unlike Adorno, Derrida had the luxury of treating Schmitt's work after the latter's death. He does so with two books, both from the 1990s. *Politics of Friendship* (1992) takes Schmitt's enemy thesis as a direct provocation to think through, in positive and differential terms, what the figure of "friendship" can mean or do, rather than allowing it to exist only as a negative concept, the absence of a threat. Composed a few years earlier, *Force of Law* (1992) indirectly but in great depth examines Schmitt's influence on Benjamin's political theory, and cites Schmitt's work in a discursive footnote.

These two points of departure allow Derrida to approach Schmitt as a figure in intellectual history much in the same way as does Adorno, albeit taking the concepts of sovereignty and social justice to a more radical point of deconstruction. Derrida recognizes in Schmitt (and, I would suggest, in Adorno's reading of Schmitt) the revolutionary potential of locating political agency in the capacity to decide on existential threats. He makes this clear by extending the possible distinctions that can be made from a simple and dualistic enemy-friend division, to one that incorporates Adorno's individual-society layer, and even goes so far as to explain the stakes of recognizing oneself as the existential enemy (Derrida 1992). The recognition of an absolute alterity within the absolute singularity of one's own being marks the auto-immolation of a dialectical rationality in response to the increasing irrationality of Enlightenment modernity. This is the logic of a revolution: suspension of historical consciousness in decisive, momentous violence against the conditions of one's own subjectivity.

As Mark Bahnisch (2002) notes, Derrida remains concerned throughout his body of work with determining an essence of political thought and behavior. Drucilla Cornell (1992) argues that Derrida's contribution is to explicitly engage with the limits of Schmitt's project, and to think where political consciousness turns in upon itself. Malek Moazzem-Doulat (2008) traces the residues of Schmitt's messianism in Derrida, residues that also permeate Adorno's political thought. The continuation of the secular theology that Schmitt positioned at the root of political activity accentuates this genealogy, as Michael Naas (2008) has also pointed out.

These attempts to articulate the direct engagement between Derrida and Schmitt, in clearer and more accessible terms than those allowed by Derrida's own writing, all point to the conversion from a thought of politics as an historical inevitability, in Schmitt, to an historical possibility, in Derrida. Why, then, did Derrida write about Schmitt in such direct terms, and build on the responses made possible by Adorno, but leave any comment on Adorno aside for so long?

Derrida's pattern of avoiding detailed discussions of Adorno carried through his career. Only in 2001, upon his receipt of the Adorno Prize awarded by the Frankfurt Institute, did he situate Adorno's work as an explicit tutor-text for his own thought. "Fichus" (2005) engages with and cites three of Adorno's seminal books: the *Dialectic of Enlightenment* (1947), *Minima Moralia* (1951), and *Prisms* (1955). Incidentally, these three works hold the majority of Adorno's conceptual engagement with Schmitt. "Fichus" also deals in depth with Walter Benjamin, as did *Force of Law* (1992). Even when he discusses Adorno's work explicitly here, Derrida takes care to note that he remains at a critical distance from his "adopted father." The necessity of this distance results from Derrida's acknowledgment that post-structuralist thought maintains a tenuous relationship to critical theory. The combination of elements of literary, psychoanalytic, sociological, and otherwise "phallogocentric" disciplines inevitably diverges into myriad strains. In this context, argues Derrida, to try to eliminate differences among the strains and to produce a unified concept of either critical theory or post-structuralism would be to miss their most productive contributions to social thought.

Before this, Derrida had only made fleeting reference to Adorno. For example, he mentions Adorno's name as part of an Ashkenazi Jewish philosophical tradition, in an extended footnote to *Monolingualism of the Other* (1998). Adorno's name also appears in the midst of a long footnote to *Force of Law.* However, despite the dearth of explicit citations, Derrida's literary and philosophical techniques often mirror Adorno's, as in *Politics of Friendship,* which takes up implicitly and expands upon many of the themes of Adorno's dense dismissal of Schmitt in *Minima Moralia*. When Derrida addresses the latter text during his "Fichus" speech, he fails to connect their three trajectories—even while his conceptual approach merges and extends them.

"Fichus" allowed other scholars to comment upon the problematic relationship between Adorno and Derrida. Shane Weller (2006) argues that the speech demonstrates the debilitating effects on Derrida's thought of the admission of an "affinity," between his interpellation of philosophy with literature, and Adorno's of language with materiality. Jean-Phillipe Deranty (2006) takes the speech as an occasion to unpack the complex genealogy that post-structuralism, as exemplified by Derrida, inherits from Adorno's critical theory. Christoph Menke (1998) also weighs in on the text, taking it as explicit evidence for a continuation of a critique of aesthetic negativity that binds Derrida to Adorno.

However, Derrida does take deconstruction to a point beyond the recognition of contradiction, where Adorno leaves off; in addition, his conception of a radical futurity, the thought of what comes next in an increasingly irrational society, moves beyond Adorno's exhortation that to think is all that one can or must do, opening the possibility of embodied practice and becoming more personally involved with radical political struggles. Both Derrida and Adorno recognize a dialectical relationship between an individual's perception of subjectivity, and their concomitant technical objectification in and as labor. Derrida, however, remains willing to reclaim, at the limits of his own thought, the contradictory mechanism of a transcendental apperception, precisely in order to achieve an immanent critique of literature, politics, and philosophy as such. On this account, his long silence about Adorno illustrates the degree to which the situations that Adorno had described as the very limits of Enlightenment philosophy led not to Enlightenment's end, but to its continuity and even intensification. His longer silence—on whether he was aware of, or consciously avoiding, the tension between Schmitt and Adorno—speaks to the stakes of such a disagreement: the life or death of a revolution.

Between citation and silence there develops a specifically dialectical relationship. This dialectic makes explicit three lines of inquiry: aesthetics, alterity, and futurity. These demarcate the conditions under which it is possible to think revolution in its historical context. First, our minds must be quieted, and attend to difficult problems of contradiction and of absolute rupture. Second, we must examine revolutionary phenomena beyond their ephemeral noises, to their political roots. And finally, our success or our failure to think through our revolutions will depend on our ability to forsake speculation, when blood flows.

Works Cited

Adorno, Theodor, and Max Horkheimer. (1947) 2002. *Dialectic of Enlightenment: Philosophical Fragments.* Translated by E.N.F. Jephcott, edited by Gunzelin S. Noerr. Stanford: Stanford University Press.

Adorno, Theodor, Else Frenkel-Brunswik, Daniel J. Levinson, and R. Nevitt Sanford. 1951. *The Authoritarian Personality.* New York: Harper & Row.

Adorno, Theodor. (1955) 1967. *Prisms*. Translated by Samuel and Shierry Weber. Cambridge: MIT Press.

—. (1951) 1974. *Minima Moralia: Reflections from Damaged Life.* Translated by E.N.F. Jephcott. London: Verso.

—. (1951) 1985. "Freudian Theory and the Pattern of Fascist Propaganda." In *The Essential Frankfurt School Reader*, edited by Andrew Arato and Eike Gebhardt, 118-137. New York: Continuum.

—. (1969) 1989. "The Culture Industry Reconsidered." Translated by Anson Rabinach. In *Critical Theory and Society*, edited by Stephen Bronner and Douglas M. Keller, 128-135. New York: Routledge.

—. 1989. "Society." Translated by Frederic Jameson. In *Critical Theory and Society*, edited by Stephen Bronner and Douglas M. Keller, 267-287. New York: Routledge.

—. (1966) 1990. *Negative Dialectics.* Translated by E.B. Ashton. New York: Routledge.

—. (1970) 1998. *Aesthetic Theory.* Translated by Rolf Tiedemann. Edited by Robert Hullot-Kantor. Minneapolis: University of Minnesota Press.

—. 2006. *Adorno, History and Freedom: Lectures 1964-1965.* Edited by Rolf Tiedemann, translated by Rodney Livingstone. Note 8 to Lecture 25. Cambridge: Polity.

—. (1956) 2010. Dialogue with Max Horkheimer. "Towards a New Manifesto?" Translated by Rodney Livingstone. *New Left Review* 65: 33-61.

Agamben, Giorgio. 1998. *Homo Sacer: Sovereign Power and Bare Life.* London: Verso.

—. 2005. *State of Exception.* London: Verso.

Bahnisch, Mark. 2002. "Derrida, Schmitt, and the Essence of the Political." Refereed paper presented as part of the Political Theory Stream Jubilee Conference of the Australasian Political Studies Association Australian National University, Canberra, October 2002.

Boersma, Jess. 2005. "What About Schmitt? Translating Carl Schmitt's Theory of Sovereignty as Literary Concept." *Discourse* 27(2-3): 215-227.

Buck-Morss, Susan. 1977. *The Origin of Negative Dialectics.* New York: The Free Press.

Cornell, Drucilla. 1992. *The Philosophy of the Limit.* New York: Routledge.

Deranty, Jean-Phillipe. 2006. "Adorno's Other Son: Derrida and the Future of Critical Theory." *Social Semiotics* 16(3): 421-433.

Jacques, Derrida. (1967) 1973. *Speech and Phenomena: and Other Essays on Husserl's Theory of Signs.* Evanston: Northwestern University Press.

—. (1967) 1978. *Writing and Difference.* Translated by Alan Bass. Chicago: University of Chicago Press.

—. (1972) 1982. *Margins of Philosophy.* Chicago: University of Chicago Press.

—. 1988. *Limited Inc.* Translated by Jeffrey Melman and Samuel Weber. Evanston: Northwestern University Press.

—. 1992. "Force of Law: 'The Mystical Foundation' of Authority." In *Deconstruction and the Possibility of Justice*, edited by D. Cornell, M. Rosenfeld and D. Carlson, 38-44. New York: Routledge.

—. (1992) 1997. *Politics of Friendship.* Translated by George Collins. New York: Verso.

—. (1967) 1997. *Of Grammatology.* Corrected Edition. Translated by Gayatri C. Spivak. Baltimore: Johns Hopkins University Press.

—. 1999. "Marx and Sons." In *Ghostly Demarcations: A Symposium on Jacques Derrida's 'Specters of Marx,* edited by M. Sprinker, 213-263. New York: Verso.

—. (1998) 2000. *Monolingualism of the Other.* Translated by Patrick Mensah. Stanford: Stanford University Press.

—. 2005. "Fichus." Speech delivered 22 September 2001, on receiving Adorno prize. In *Paper Machine*. Translated by Rachel Bowlby, 163-181. Stanford: Stanford University Press.

Duttmann, AG. 2006. "Adorno's Rabbits; or, Against Being in the Right." *New German Critique* 33:179-190.

Foucault, Michel. (1976) 1980. *Power/Knowledge: Selected Interviews and Other Writings, 1972-1977.* Edited by Colin Gordon. New York: Vintage.

Gerhardt, Christian. 2006. "The Ethics of Animals in Adorno and Kafka." *New German Critique* 33:159-178.

Habermas, Jurgen. (1985) 1998. *The Philosophical Discourse of Modernity: Twelve Lectures.* Translated by Frederick Lawrence. Cambridge: MIT Press, 1998.

Huysmans, J. 2008. "The Jargon of Exception—On Schmitt, Agamben and the Absence of Political Society." *International Political Sociology* 2: 165–183.

Jay, Martin. 1973. *The Dialectical Imagination: A History of the Frankfurt School and the Institute of Social Research, 1923-1950.* Los Angeles: University of California Press.

—. 1993. *Force Fields: Between Intellectual History and Cultural Critique.* New York: Routledge.

Kellner, Douglas. 1989. *Critical Theory, Marxism, Modernity.* Baltimore: Johns Hopkins University Press.

Lash, Scott. 1993. "Reflexive Modernization: The Aesthetic Dimension." *Theory, Culture, and Society* 10 (1): 1-23.

Menke, Christoph. 1998. "Aesthetic Sovereignty." In *The Sovereignty of Art: Aesthetic Negativity in Adorno and Derrida,* translated by N. Solomon, 161-180. Cambridge: MIT Press.

Moazzam-Doulat, Malek. 2008. "Future Impossible: Carl Schmitt, Jacques Derrida, and the Problem of Political Messianism." *Philosophy Today*, 1 Apr.

Mouffe, Chantal, ed. 1999. *The Challenge of Carl Schmitt.* London: Verso.

Naas, Michael. 2008. *Derrida from Now On.* New York: Fordham University Press.

Schmitt, Carl. 1934. "Der Führer schützt das Recht." *Deutsche Juristen-Zeitung*, Aug. 1. http://www.flechsig.biz/DJZ34_CS.pdf .

—. (1922) 2005. *Political Theology: Four Chapters on the Concept of Sovereignty.* Translated by George Schwab. Chicago: University of Chicago Press.

—. (1950) 2006. *The Nomos of the Earth in the International Law of Jus Publicum Europaeum.* Translated by G.L. Ulmen. New York: Telos Press.

—. (1932) 2007. *The Concept of the Political.* Translated by George Schwab. Expanded Edition. Chicago: University of Chicago Press.

—. (1963) 2007. *Theory of the Partisan: Intermediate Commentary on the Concept of the Political.* Translated by G.L. Ulmen. New York: Telos Press.

—. (1928) 2008. *Constitutional Theory.* Translated by Jeffery Sietzer. Durham, NC: Duke University Press.

—. (1956) 2009. *Hamlet oder Hekuba: The Intrusion of Time into the Play.* Translated by David Pan and Jennifer Rust. New York: Telos Press.

Weller, Shane. 2006. "When the Other Comes Too Close: Derrida and the Threat of Affinity." *Kritikos* 3 (June). http://intertheory.org/weller.htm.

Winstead, William. 2010. "The Utopian Function of the Enemy in the Thought of the Frankfurt School." Western Political Science Association, Annual Meeting. Accessed Dec. 31, 2011. http://ssrn.com/abstract=1580687.

Wolin, Richard. 1992. *The Terms of Cultural Criticism: The Frankfurt School, Existentialism, Postmodernism.* New York: Columbia University Press.

Because I Have Nothing Better to Do: Occupy Wall Street and the Dance of Time

Cecile Lawrence

> **Occupy Wall Street** is a leaderless resistance movement with people of many colors, genders and political persuasions. The one thing we all have in common is that We Are The 99% that will no longer tolerate the greed and corruption of the 1%. We are using the revolutionary Arab Spring tactic to achieve our ends and encourage the use of nonviolence to maximize the safety of all participants.
>
> The #ows movement empowers real people to create real change from the bottom up. We want to see a general assembly in every backyard, on every street corner because we don't need Wall Street and we don't need politicians to build a better society.
>
> —http://occupywallst.org

For the first time, on the 16th of April, 2012, in the face of Capital as represented by the New York Stock Exchange, Occupy Wall Street (OWS) occupied Wall Street itself. George Washington, the first President of the U.S., "had their back," so to speak. His statue at the top of the steps leading into Federal Hall, as well as the building itself, for a while gave the Occupiers some modicum of protection from arrest by the New York City Police because of the status the building has as occupying federal land which, of course, is land stolen from its original occupants. In contrast, the New York Stock Exchange, directly opposite, the source of decades of repeated cycles of national economic crashes by then gone global, was owned by New York City, on land also stolen from its original occupants. One could say, therefore, that theft is deeply rooted in U.S. culture, theft of land, theft of money, theft of culture, theft of time above all. And because Capital is so deeply rooted in U.S. culture, theft is deeply rooted. And because theft is deeply rooted in the culture, deep economic and socio-cultural crises rear up cyclically.

In this cramped environment in the man-made echoing canyons of lower Manhattan, significantly a few blocks away from the site of the September 2001 attacks on the Twin Towers of the World Trade Center, people identifying as Occupy Wall Street highlighted and manifested the histories, juxtapositions, confluences, connections and conflicts over physical and figurative space that the movement quickly came to embody since this new time of revolution began.

The move to Wall Street itself came after people first gathered on September 17, 2011 in Zuccotti Park, to live 24/7 a hard life on a sloping multi-level gravestone-like slab of granite, a few city blocks away from Wall Street itself, with tents, mattresses, a kitchen, library, medical support, all the basic necessities of life as the Occupiers saw it. The originators of the idea for OWS—Micah White, a senior editor of Adbusters Magazine living in the U.S. and Kalle Lasn, in Canada, the magazine's editor in chief—chose that date as it marks Constitution Day in the U.S. (Greene 2011). They wanted to simulate an American version of the Tahrir Square "January 25 Revolution" protests in Egypt earlier in 2011 (Said). The mainstream U.S. media, in the customary reductionist habit of dominant U.S. culture, had labeled the various uprisings in Tunisia, Yemen, Morocco, etc. and especially Egypt, as a group as "The Arab Spring" instead of the labels the protesting people of those countries used for their own revolutions such as "January 25 Revolution" (Egypt), "February 20 Movement" (Morocco). So powerful is the domination of the West and the right to name, granted or taken, that even media sympathetic to the uprisings in these countries as well as media sourced in those countries but written in English used the term "Arab Spring." One example is the title of an article about the January 25 Revolution and the February 20 Movement (Bhatia 2012). And even a powerful governmentally connected U.S. NGO like the Wilson Center for International Scholars used the terms "Arab Spring" and even "American Winter" unsparingly in its already long list of presentations on the topic in 2012 (Woodrow Wilson 2012a,b,c). The people in the Arabic speaking countries who are engaging in uprisings use months and days, while the West tends to use seasons of the year. This tendency of the West may signify an orientation towards blurring, sweeping up into an amorphous grouping, lack of precision, unwillingness to deal with details, especially or including when it comes to naming time.

White and Lasn also created some of the images and themes, such as the ballerina dancing on the muscular oversized Wall Street bull, that have come to be associated with OWS. An idea, even if it is a call to action, does not become a revolution without people, the same people as the idea creators or others motivated by them, spending time in planning and organizing, forging alliances, resolving differences, even if temporarily, and so on. Various individuals would have to come together, as they did, with or without prior knowledge of the Adbusters Magazine call to action, in order to give voice to their resistance to the austerity measures being implemented by those in power in New York City, across the country and the world in response to yet another of the increasingly frequent catastrophic economic crashes. Some of the memes and chants taken up by the movement, such as "We Are The 99%," had their origin in this time of planning.

This essay seeks to analyze what it was that constituted the match that lit the uprising that came to be called Occupy Wall Street. Progressive academics and activists in the U.S. had been questioning over the years, more frequently and urgently as 2000 came and went, why people in the U.S. had not already taken to the streets. Not even the call to action of the simultaneously beloved (by the Left) and reviled (by the Right) Frances Fox Piven in December 2010 was enough to light the start of revolution in response to millions of people being officially unemployed (Piven 2010) along with the millions more unofficially so. What was it that allowed people to tolerate or even succumb to the increasing amounts and depths of economic assaults upon the U.S. population, in combination with the wars on Iraq and Afghanistan, the rapidly growing economic divide between the wealthy and the rest of the population, the disappearing of the middle class, the pro-corporate stance of the holdings of the U.S. Supreme Court to the detriment of the health and welfare of not just human beings but also of the land, air and water, the evidence of global climate change, with the U.S. being categorized as a leading contributor to the greenhouse gasses being blamed as the chief causative agent, just to name some of the concerns? One theory has it that the U.S. cultural psyche, its race memory, had been so thoroughly traumatized by the carnage of the U.S. Civil War that even just thinking about revolt was enough to freeze citizens into inaction.

A persistent undercurrent in U.S. culture is a terrifying fear of race war. As an eminent U.S. Civil War historian put it: "race [is] the central problem in how Americans make choices to remember or forget their Civil War" (Blight 2001, 2). Howard Zinn, true to form, expressed it more directly: "As long as we have a politics of race in America, we will have a politics of Civil War memory" (Zinn 2003, 4). White supremacists have played the race war card from time to time so as to get their interests to prevail. Furthermore, moments in the past that brought the country close to rebellion, such as the Civil Rights movement, did generate token measures of reform by the government that acted as a pressure relief valve without there being any real intention of engaging in deep structural change (Zinn 2003, 443, 601 passim). Equally important was the fact that to confront the injustices in the country would call into question the deeply held belief in U.S. exceptionalism. Winfried Fluck holds that this myth is the essential glue that holds the country together through a long history of conflict (Fluck 2012). Therefore, many reasons or excuses for inaction prevailed. The people had internalized these fears of rebellion. Amongst the comments on progressive blogs, the term "sheeple" appeared with depressing regularity to describe the general public. This quiet appeared in stark contrast to people in other countries who, with seemingly fewer but definitely deeper compelling reasons, had already taken to the streets. What single event or web of events made this time in September 2011, the right time?

As an important feature at the start of the movement, as a means of exerting power that occurs in any major movement for change, the Occupiers restored part of their first occupied space's original name, Liberty Plaza Park, using the label "Liberty Park," and not Zuccotti Park, in all their communications when referring to their new home, work, and play space. Zuccotti Park presented the strange phenomenon of a privately owned public space, emblematic of the ongoing privatization of the commons. And it is this focus on exerting the power of naming and taking control, as much as possible, of modes of communication about themselves amongst themselves as well as globally that has defined just a portion of this new time of revolution. The power to name is, after all, the ultimate power (Bible, John 1:1).

When the movement sprang to life on that day in September, naysayers pontificated loudly, spurred on by mainstream media, impatiently demanding instant responses to questions abut the goals and "demands" of this new group/movement, while at the same time labeling the Occupiers as lazy, dirty and shiftless, or even mentally deficient, all the names for othering and diminishing that the dominant culture had developed over decades to concretize the Hobbesian or Ayn Randian pecking order of mass control and oppression. According to Mészáros, when a "radically new dimension of time enters the picture" the "question of *meaning* appears" which "extricates individuals from the brutish constraints of their earlier 'hand-to-mouth' existence" and with meaning comes choice (2008, 36). With choice comes freedom. Thus began the new struggle for whose framing and definition would prevail, who would win the fight to define the meaning of OWS. One instantly iconic image appeared soon, in the photographs and videos of workers, possibly traders, in business dress, on a balcony overlooking the Wall Street area, sipping from champagne glasses while laughing, taking camera/phone photos of and gesticulating towards the Occupiers on the street below (Yakas 2011). It was easy to label these images as "The 1% sneer at the 99%."

The reality of the revolution manifested much differently from the frames presented in the dominant media and their corporate underwriters, as well as by those well indoctrinated into fear of any resistance against the corporate-governmental hegemony, even as they themselves suffered under the policies implemented by this 1%, the term invented to complement OWS's "We Are The 99%." This reality produced, from the New York City General Assembly (NYCGA), within 12 days of the start of the occupation, a "Declaration of the Occupation of New York City" (2012), a comprehensive document listing 23 separate grievances, covering multiple points of existence as a human being on the planet, as a start. The document is not addressed to the power structure, but to the people, the 99%, throughout the world, exhorting all to join OWS in solidarity. Nor does the document come authored from a so-called leader, or small committee set up to be the representatives of the masses, as is the norm in existing capitalistic structures, but instead, in order to emphasize that the OWS revolution had launched a new model for relationships and products, the document is sourced in a new structure of cooperative, collaborative decision-making based on consensus. Therefore, those within the frame of the dominant structure had difficulty recognizing this comprehensive but not all-inclusive non-supplicating document as a statement describing the purpose of OWS. Neither in this document nor in their chants does OWS beg or ask, refusing to go along with the status of prisoners, supplicants, or enslaved with which the 1% would want to continue to categorize them. As if this radically different way of communicating the nature of their existence was not enough, the NYCGA also provided a visual representation of the document's words, created by the Arts and Culture Group of OWS, as a companion, not an appendix, to the document, thereby recognizing and validating the reality of different modalities of communication, denying, to some extent, the hegemony of the written word.

As the movement developed, a dance with time as well as with place came into play. Everything would be, had to be, new. Because the entire structure of the existing society had been so thoroughly corrupted, Occupiers felt their only recourse was to start from scratch. No wonder the people who felt so deeply that the time had definitely come for a new world were motivated to abandon their former places of residence, if they had such, as a fair number of people already without shelter found in

Occupies not only community, but a place to sleep. Occupiers were moving to a new land, but in contrast to the history of previous times where newcomers would invade and occupy territory already inhabited by people not much different from them in many respects except perhaps for skin color, language and culture. Beginning in 2011, a besieged people were taking back the physical and mental space that the 1%, the corporate-governmental complex, had wrested from them. Furthermore, the Occupiers were intent on sharing their time and space with all who would come and join them, again in contrast to the dominant paradigm of greed, selfishness and self-centeredness. As István Mészáros reminds us, "the concept of *free time* is totally meaningless to capital" (Mészáros 2008, 49). Corporations had spent decades and lots of money in public relations and marketing, determined to convince the general public, starting in the U.S., that they should no longer be in community, that the commons must no longer exist, that a corporation is a person, that they should no longer trust each other, even love each other, but that where true love and community resided was in the purchase of products and in heeding the beck and call of the corporate siren.

The uprisings in Tunisia, Morocco, and Egypt came from years of thinking and planning (Daragahi 2011). But the question was, as with OWS, when would the definitive actions, the tipping point actions, take place? What would be the spark? The structures against which the Occupiers protested themselves came from years of careful planning. According to István Mészáros, the problem with the system and its structure is not so much with capitalism but with Capital (Mészáros 2010). The corporate siren, through the influence of money and the influence of influence, had almost completely taken into its grasp entire governmental structures. The rules of engagement had been changed over time to increasingly favor private corporations over the needs and dreams of actual human beings. The justices of the Supreme Court of the U.S. (SCOTUS), in their holding in the 2010 Citizens United vs. Federal Election Commission case, granting to corporations the free speech rights of human beings set out in the First Amendment of the U.S. Constitution was just the latest egregious expansion of corporate control over the entire U.S. if not the world. This case allowed unlimited corporate funding of political campaigns, thereby facilitating almost limitless corporate control and even foreign governmental control over the U.S. electoral process, in effect allowing if not encouraging the corporate purchase of a U.S. government most amenable to its agenda. In the activities in early 2012 leading up to the November 2012 U.S. Presidential elections, as well as for other seats at the national and state levels, the term "Super-PAC" came into the language, to describe the political action committees funding political campaigns with mind-numbing amounts of money (Hickman 2012). The fallout from this court decision also buttressed the ever-increasing expansion of U.S. based corporations globally. The term "transnational corporation" had become passé culturally and economically, giving way to "multinational corporation," an entity with no single definite geographical home office, such that it became a rabid amoral amorphous "Vampire Squid" (Taibbi 2010) for which neither conceptual structures much less language had yet been developed. Along with the statement of the occupation of New York City, Occupiers also later provided in one sentence who they were and what was their purpose.

> Occupy Wall Street is a people-powered direct action movement that began on September 17, 2011 in Liberty Square in Manhattan's Financial District. OWS is part of a growing international movement fighting against neoliberal economic practices, the crimes of Wall Street, government controlled by monied interests, and the resulting income inequality, unemployment, environmental destruction, and oppression of people at the front lines of the economic crisis. (Occupy Together)

People who came to call themselves Occupiers sensed or knew that not only was the situation dire in 2011 but its complexity and its structure of layers of webs one on top of another and even threaded through each other faced them with a daunting task. The term "another world is possible" was already being used by the World Social Forum in 2001 and its offshoot in the U.S. Social Forum first in Atlanta in 2007 and then in Detroit in 2010. Excerpts from The Charter of Principles of the World Social Forum:

> 1. The World Social Forum is an open meeting place for reflective thinking, democratic debate of ideas, formulation of proposals, free exchange of experiences and interlinking for effective action, by groups and movements of civil society that are opposed to neoliberalism and to domination of the world by capital and any form of imperialism, and are committed to building a planetary society directed towards fruitful relationships among Humankind and between it and the Earth.
>
> . . .
>
> 10. The World Social Forum is opposed to all totalitarian and reductionist views of economy, development and history and to the use of violence as a means of social control by the State. It upholds respect for Human Rights, the practices of real democracy, participatory democracy, peaceful relations, in equality and solidarity, among people, ethnicities, genders and peoples, and condemns all forms of domination and all subjection of one person by another.
>
> . . .

> 14. The World Social Forum is a process that encourages its participant organizations and movements to situate their actions, from the local level to the national level and seeking active participation in international contexts, as issues of planetary citizenship, and to introduce onto the global agenda the change-inducing practices that they are experimenting in building a new world in solidarity. (World Social Forum)

and the description of the U.S. Social Forum:

> The US Social Forum (USSF) is a movement building process. It is not a conference but it is a space to come up with the peoples' solutions to the economic and ecological crisis. The USSF is the next most important step in our struggle to build a powerful multi-racial, multi-sectoral, inter-generational, diverse, inclusive, internationalist movement that transforms this country and changes history.
>
> We must declare what we want our world to look like and we must start planning the path to get there. The USSF provides spaces to learn from each others (sic) experiences and struggles, share our analysis of the problems our communities face, build relationships, and align with our international brothers and sisters to strategize how to reclaim our world. (U.S. Social Forum)

resonate with and presage the themes in the OWS Declaration of the Occupation of New York City. Since here we have extensive, clearly carefully deliberated statements of purpose both expressing the need for urgent action, as early as 2001, why was it that neither the World Social Forum nor the U.S. Social Forum provided the impetus for people to take to the streets? At the U.S. Social Forum in 2010 in Detroit, amongst the overwhelming number of workshops, street marches around the city blocks surrounding the main gathering site at the city's convention center, tablings, etc. was one large area inside for the Beehive Collective, which in hindsight seems a clear progenitor of the type of detailed creative collaborative work demonstrated in the visual representation of the Declaration of the Occupation of New York City (Beehive Design Collective). So very many of the features of the U.S. Social Forum, such as the use of the consensus method of decision making, the movement assemblies, the choice of the devastated city of Detroit as a gathering place, seem to have generated the seeds of what later developed into OWS. Once OWS came into being, activists were well aware of these connections between the World Social Forum, the U.S. Social Forum and OWS but the discussion started out being relatively low key (Steisslinger 2012). There were many differences that made U. S. Social Forum more like a conference than the "space to come up with the peoples' solutions to the economic and ecological crisis" that the organizers used as their event description, such as there even being organizers in advance of the actual event, the use of a city convention center as the main site, the use of Wayne State University classroom spaces several miles away from the main site, all of which undoubtedly involved the payment of fees, thereby rendering the U.S. Social Forum as operating within the structures of Capital.

In 2011, communications technology had advanced so rapidly that dependency on government or corporate controlled communication methods had given way, albeit not completely, to the likes of Twitter and Facebook, along with wireless and satellite connection. Therefore, people could communicate with each other in larger numbers and more quickly than ever before, and not just in their own country or part of the world but globally. Obviously, and it has been noted before, including in this book, these technologies played a very important role in these revolutions that started in late 2010, picking up speed and frequency in 2011 and on. And for those who critique this use of communications technology to protest against the very structures that created that technology, there is always the appropriate response that it is right and historically based for the enslaved to use the master's tools against the master.

A young man burning himself to death in Tunisia on the 3rd of January 2011 concretized and focused the desperation of the situation, followed by the sight of women in leadership roles in front of massive crowds in Egypt and Morocco, countries perceived by the West to have so thoroughly silenced all their women. This need for concretization embodies how deeply embedded in the cultural mind-set of the U.S. population is the capitalist orientation towards materiality, how closely connected to each other are the ideologies of materialism and U.S. culture. The stark reality of the deaths of protestors in other countries either through self-immolation, bullets, being trampled by horses and camels, led no doubt to a deeply experienced and viscerally felt realization in the minds of young white people in the U.S. possibly for the first time, of how fiercely the powers that be hold onto their power, how courageous protestors would have to be and for an indeterminately long time. African-Americans, First Nations/indigenous people, and others trampled in the dust of the drive of Capital, already knew this well but people of the dominant white culture in the U.S. needed to feel this for the time to be right. A connection of solidarity finally was made. From the formation of the country, the white male landed wealthy power structure in a capitalist system rooted in slavery had separated blacks and poor whites from joining in rebellion by placing the latter as overseers of the labor of the former. With the increasing frequency and enormity of the economic crashes, whites in the U.S. began to really see themselves as not much different from blacks. With the murder, for example, of the young black Trayvon Martin in February 2012, it became easier for people of all colors, including Occupiers, to march shouting "We are all Trayvon Martin."

The psychological need to deal with the enormity of the structural revolution envisioned by the Occupiers manifested in play and dance and not just with the ballerina on the bull. Creative works, scenes and actions of spectacle and carnival popped up wherever Occupiers had erupted. At one point, a group of Occupiers or perhaps members of the serious prankster YesLab (Bilchbaum 2011), dressed as clowns, distracted the NYPD guarding the Wall Street bull by cavorting around it gradually drawing the officers away from the bull, while a person garbed in bullfighter tights approached, red cape in hand, mounted first a police car then the bull, all successfully videotaped and uploaded to YouTube before the police arrested them (del Signore 2011). Often, impromptu dancing in the streets took place, complete with musical accompaniment of horns and drums. To signify the simultaneity of individualism with community, a concept unthinkable in a culture of Capital, many OWS signs were handmade and personal. Signs also "lovingly" addressed Capital, with wording such as "Dear 1%, we fell asleep for a while. Just woke up. Sincerely, the 99%" (Fleming 2011). "Only the creative use of free time by the social individuals, in pursuit of the objectives freely chosen by them, can bring about the much-needed beneficial result" (Mészáros 2008, 59).

For the first seminal occupation to start, instructions came definitively to people of all colors across the U.S. from two disaffected white men, one old, one young, specifying a time and date complete with a name "Occupy Wall Street." As the now iconic chant reply to "Mic check" goes "I occupy because I have nothing better to do [with my time]."

Works Cited

Beehive Design Collective. 2012. Accessed Apr. 28. http://www.beehivecollective.org/english/front.htm.

Bhatia, Juhie. 2012. "Young Moroccans Keep Arab-Spring Spirit Alive." *WeNews,* April 25. Accessed Apr. 28, 2012. http://womensenews.org/story/leadership/120424/young-moroccans-keep-arab-spring-spirit-alive.

Bible, New International Version. 1984. Accessed Apr. 28, 2012. http://niv.scripturetext.com/john/1.htm.

Bilchbaum, Andy. 2011. "Bullfight on Wall Street." *Yes Lab*, Nov. 9. Accessed Jan. 9, 2012. http://yeslab.org/bull

Blight, David W. 2001. *Race and Reunion: The Civil War in American Memory*. Harvard University Press: Cambridge, Mass.

Daragahi, Borzou. 2011. "Tunisia's Uprising was Three Years in the Making." *Los Angeles Times,* Jan. 27. Accessed Apr. 28, 2012. http://articles.latimes.com/2011/jan/27/world/la-fg-tunisia-uprising-20110127.

Day, Anna Therese. 2012. "Women and the Egyptian Revolution: An Interview with Mona Eltanhawy." *Ms. Magazine Blog*. Apr. 27. Accessed Apr. 28, 2012. http://msmagazine.com/blog/blog/2012/04/27/women-and-the-egyptian-revolution-an-interview-with-mona-eltahawy/.

Declaration of the Occupation of New York City. 2012. Accessed April 28. http://www.nycga.net/resources/declaration/.

Del Signore, John. 2011. *Gothamist*, November 9. Accessed Dec. 9, 2011. http://gothamist.com/2011/11/09/matador_jumps_on_cop_car_by_wall_st.php#photo-1

Federal Hall National Memorial. 2012. Accessed Apr. 28. http://nyharborparks. org /visit/feha.html.

Fleming, Meghan. 2011. "The 2nd American Revolution." Pinterest. Accessed May 1, 2012. http://pinterest.com/meghanlfleming/the-2nd-american-revolution/

Fluck, Winfried. 2012. The Essential Glue is American Exceptionalism." *The European*, April 16. Accessed May 1, 2012. http://theeuropean-magazine.com/627-fluck-winfried/628-americas-self-imag

Greene, Brian. 2011. "How Occupy Wall Street Started and Spread." *U.S. News,* October 17. Accessed Apr. 28, 2012. http://www.usnews.com/news/washington-whispers/articles/2011/10/17/how-occupy-wall-street-started-and-spread.

Hickman, Blair. 2012. "The Best Watchdog Journalism on Campaign Finance." *Pro-Publica.* Apr. 27. Accessed Apr. 28, 2012. http://www.propublica.org/article/the-best-watchdog-journalism-on-campaign-finance.

Mészáros, István. 2008. *The Challenge and Burden of Historical Time: Socialism in the Twenty-First Century*. New York: Monthly Review Press.

—. 2011. *The Structural Crisis of Capital*. New York: Monthly Review Press.

Occupy Together. 2012. Accessed Apr. 28. http://www.occupytogether.org/occupy-wall-st/.

Occupy Wall Street. 2012. Accessed Apr. 28. http://occupywallst.org/.

Piven, Frances Fox. 2010. "Mobilizing the Jobless." *The Nation.* January 10. Accessed Apr. 27, 2012. http://www.thenation.com/article/157292/mobilizing-jobless

Said, Carolyn. 2011. "Micah White - An Architect of Occupy Movement." *San Francisco Gate*, Dec. 5. Accessed Apr. 28, 2012. http://www.sfgate.com/cgibin/article.cgi?f=/c/a/2011/12/04/BUOG1M75GR.DTL.

Steisslinger, M.B. 2012. "Occupy, the World Social Forum and the Commons: Social Movements Learning From Each Other." *The Future of Occupy.* March 13. Accessed Apr. 28, 2012. http://thefutureofoccupy.org/2012/03/13/occupy-the-world-social-forum-and-the-commons-social-movements-learning-from-each-other/.

Taibbi, Matt. 2010. "The Great American Bubble Machine." *Rolling Stone*, April 5. Accessed Apr. 28, 2012. http://www.rollingstone.com/politics/news/the-great-american-bubble-machine-20100405.

U.S. Social Forum. 2012. Accessed Apr. 28. http://www.ussf2010.org/.

Verveer, Melanne. 2012. "Why Women Are a Foreign Policy Issue." *Foreign Policy May/June.* Accessed Apr. 28, 2012. http://www.foreignpolicy.com/articles/2012/04/23/why_women_are_a_foreign_policy_issue.

Woodrow Wilson International Center for Scholars. 2012a. "The Arab Spring, a Year On: How's America Faring?" Accessed Apr. 28, 2012. http://www.wilsoncenter.org/event/the-arab-spring-year-how%E2%80%99s-america-faring.

—2012b. "Is the Arab Awakening Marginalizing Women"? Accessed Apr. 28, 2012. http://www.wilsoncenter.org/event/the-arab-awakening-marginalizing-women.

—2012c. Search of site using term "The Arab Spring." Accessed Apr. 28, 2012. http://www.wilsoncenter.org/search/site/The%20 Arab%20Spring.

World Social Forum. 2012. Accessed Apr. 28. http://www.forumsocialmundial.org.br/main.php?id_menu=4&cd_language=2.

Yakas, Ben. 2011. "Anti-Occupy Wall Street Champagne Toast Later Today?" *Gothamist.* September 23. Accessed Apr. 28, 2012. http://gothamist.com/2011/09/23/antioccupy_wall_street_champagne_t.php.

Zinn, Howard. 2003. *A People's History of the United States: 1492-Present.* New York: Harper Collins.

CONTRIBUTORS

Anupa Batra is a lecturer at the California College of Arts. She received her Ph.D. in philosophy, and her dissertation was on the relationship between Kant and Deleuze. Her research interests include the question of subjectivity, time, theories of materialism, and aesthetic theory, among other things.

Alexander Brown is a Ph.D. student currently conducting research on contemporary Japanese social movements at Hitotsubashi University in Tokyo. He has a strong interest in autonomist theory and in 2010 he co-organised the *From Empire to Commonwealth: Communist Theory and Contemporary Praxis* conference at the University of Wollongong.

Antoine Chollet is a research and teaching fellow in political theory at the University of Lausanne, Switzerland. He earned a Ph.D. from Sciences Po in Paris in 2009, with a dissertation on time in democratic theories. He has written on direct democracy, Castoriadis and Lefort, and nationalism. He is currently working on presentism, and also on lotteries and random selection of officials.

Natalie Churn is currently completing her Ph.D. at the University of Freiburg. Her research focuses on marginal concepts of time in Australian popular culture, and she has presented on aspects of this topic at many conferences. She has taught in Australian Cultural Studies and British News Media, and has received prizes and several scholarships for past degrees.

Amanda Conroy is a Ph.D. student at the Gender Institute and the Centre for the Study of Human Rights, London School of Economics and Political Science. A recipient of the Frederick Bonnart-Braunthal Scholarship (2011-2014), Amanda's research interests include feminist and post-colonial political theory; right-wing and conservative social movements; the intersections between gender, race, and nationalism; and the gendered dimensions of citizenship.

Rhiannon Firth completed her Ph.D. at the University of Nottingham with the Department of Politics and International Relations. She is now Postdoctoral Fellow at the institution's Centre for Advanced Studies. Her research interests include transformative theory, utopianism and radical subjectivity. Her thesis was recently published as a monograph by Routledge.

Christian Garland is a writer and theorist; he is also a Ph.D. candidate at the University of Warwick and Fellow of the Institut fur Kritische Theorie, Freie Universitat Berlin. He has research interests that include the original Frankfurt School and Critical Media Theory.

M.F.N. Giglioli obtained his Ph.D. in Political Theory from Princeton University in 2010, focusing on issues of power, legitimacy, and revolution in fin-de-siècle Continental political thought. His current project traces the relationship between the social and political condition of women and the practice of war in the long 19th century.

Rochelle Green is an Assistant Professor of philosophy at the University of Arkansas, Little Rock. She specializes in the areas of social and political philosophy, feminist philosophy, and German Idealism. Much of her work focuses on the relationship between hope, justice, and social change.

Alison Hulme's empirical work so far involves working with wholesale buyers and waste peddlers in China. She has written and spoken on various aspects of China's development including entrepreneurialism and thrift, and recently developed a course on China's socio-economic landscape for the University of Iceland. She also consults on public and private-sector research projects.

Brad D. Hume is a historian of science who has taught at numerous universities and has been a research fellow at various venues. He focuses on temporality and human heredity and has published numerous articles and has a book manuscript under review, *"We the People": Heredity, Human Science, and Identity in America, 1776-1861*.

Łukasz Kamieński is an Associate Professor at the Faculty of International and Political Studies, Jagiellonian University, Krakow, Poland. He holds a Ph.D. from the Jagiellonian University; M.Sc. in International Relations, London School of Economics and Political

Science; and M.A. in Political Science, Jagiellonian University. His work concentrates on military technology, armed forces, and strategy.

Bradley Kaye is an independent scholar working as an editor at Edwin Mellen Press. He has a Doctorate from Binghamton University's Philosophy, Interpretation, and Culture Program, and has taught philosophy and sociology at Binghamton University and Broome Community College.

Mickie Mwanzia Koster is an Assistant Professor of History at the University of Texas at Tyler. She received her M.A. and Ph.D. in History from Rice University. She teaches pre-colonial, colonial and Modern Africa, East African History, and African Historiography. Her manuscript is entitled, *The Power of the Oath: The Making of Mau Mau in Kenya, 1952-1960.*

Cecile Lawrence, originally from Jamaica and currently residing in the U.S., has a Ph.D. from Binghamton University and a J.D. from the University at Buffalo. Published in anthologies and journals, she has worked as a college and non-profit administrator and has presented at numerous conferences. She has been a social change activist for over 15 years, speaking at rallies and giving testimony at numerous hearings in multiple cities.

Lewis Levenberg holds a Presidential research scholarship at George Mason University in the department of Cultural Studies. His research interests include the social conditions of technological architecture, alongside interactions between poetics, historiography, and critical theory.

Linda Levitt is an Assistant Professor in communication studies at Stephen F. Austin State University. Much of her research investigates the intersection of media and cultural memory, and she has published work on television studies, celebrity culture, and public acts of commemoration.

Allison M. Merrick completed her Ph.D. at the University of Southampton, England, with a thesis entitled "On Nietzsche's Genealogical Mode of Inquiry." She is currently an Assistant Professor of Philosophy at the University of Arkansas at Little Rock.

Bryan Nelson is a doctoral candidate in the Social and Political Thought programme at York University, Toronto. Under the supervision of Dr. Brian Singer and Dr. Martin Breaugh, his doctoral research concerns problems of radical democracy in contemporary French political philosophy. His most recent publication addresses Karl Marx and subjectivity (*Subjectivity*, 2011, vol. 4 (4): 395-412).

Thomas Pieragastini is a recent graduate of the Philosophy, Interpretation and Culture Program at Binghamton University, completing a Ph.D. in Philosophy with concentrations in economics, tactical theory/security studies and radical "political theory." He currently lives and works in Pittsburgh, Pennsylvania.

Kamilla Pietrzyk is a Ph.D. candidate in Political Science at York University. She obtained a B.A. and M.A. from the Faculty of Information and Media Studies at the University of Western Ontario. Her interdisciplinary doctoral research project applies a temporal lens to the study of collective action. Her work has appeared in *Fast Capitalism and Convergence.*

Maria Prodromou holds a Ph.D. in Continental Philosophy from the University of Essex. Her current research explores the relationship between ethics, aesthetics and the political, the relationship between art and philosophy, and the role of the imagination in articulating radical politics of resistance.

Christos Retoulas (DPhil, Oxford) is the founder of the notion of the Judeo-Christiano-Islamic "Byzantine-Ottoman Continuum" of Ecumenical Romanity based on the reality of the theantropic Perfect Human Being. He is a member of the scientific board of the Dimitri Kitsikis Foundation.

Tania Rispoli is a Ph.D. student in Philosophy at the University of Rome, Tor Vergata. In 2011 she wrote a book with Prof. Augusto Illuminati (*Tumulti. Scene dal nuovo disordine planetario/Tumults. Scenes from the New Planetary Disorder.*) She is also a political and social activist who took part in the movements of the last years in Italy and Europe.

Andrew Robinson is a freelance researcher and activist based in the UK. He is co-author of *Power, Conflict and Resistance in the Contemporary World* with Athina Karatogianni, and over a dozen papers on radical theory and social movements.

Martin Savransky is a Ph.D. student at the Department of Sociology, Goldsmiths, University of London and holds an International Fellowship at the Institut für Soziologie, Albert-Ludwigs-Universität Freiburg. His research interests involve the Production of Knowledge in the Social Sciences; Continental Philosophy and Social Theory; New Empiricisms; Constructivist and Speculative Thought.

Dalbir S. Sehmby is an English instructor at Campus Saint-Jean at the University of Alberta, in Edmonton, Alberta, Canada. Dr. Sehmby has taught in the fields of film/media studies, comparative literature, and English; he pursues an academic interest in media representation, popular culture studies, and Sikhism.

John Streamas, an immigrant Japanese American, is the first in his family to complete high school. Now an associate professor, he has published critical work on race during wartime, Terrence Malick's films, and the racial politics of time and space. He also writes poems, stories, and informal essays.

Enqi Weng completed her Masters of Global Media Communications at the University of Melbourne in 2012. Her thesis was on the Australian media's representation of Christianity and her research interests are in media/religion and new media/Singapore politics. She has nine years of experience in marketing communications from commercial and not-for-profit sectors.

INDEX